LUBRICATION
and
MAINTENANCE
of INDUSTRIAL
MACHINERY

Best Practices and Reliability

LUBRICATION *and* MAINTENANCE *of* INDUSTRIAL MACHINERY

Best Practices and Reliability

Edited by
Robert M. Gresham
George E. Totten

Society of Tribologists
and Lubrication Engineers

CRC Press
Taylor & Francis Group
Boca Raton   London   New York

CRC Press is an imprint of the
Taylor & Francis Group, an **informa** business

The material was previously published in *Handbook of Lubrication and Tribology: Volume I Application and Maintenance, Second Edition* © Taylor and Francis 2006.

CRC Press
Taylor & Francis Group
6000 Broken Sound Parkway NW, Suite 300
Boca Raton, FL 33487-2742

© 2009 by Taylor & Francis Group, LLC
CRC Press is an imprint of Taylor & Francis Group, an Informa business

No claim to original U.S. Government works
Printed in the United States of America on acid-free paper
10 9 8 7 6 5 4 3 2 1

International Standard Book Number-13: 978-1-4200-8935-6 (Hardcover)

Library of Congress Cataloging-in-Publication Data

Lubrication and maintenance of industrial machinery : best practices and reliability / Robert M. Gresham, George E. Totten.
    p. cm.
Includes bibliographical references and index.
    ISBN 978-1-4200-8935-6 (alk. paper)
    1. Machinery--Maintenance and repair. 2. Industrial equipment--Maintenance and repair. 3. Lubrication and lubricants. I. Gresham, Robert M. II. Totten, George E. III. Title.

TJ153.L845 2008
621.8'16--dc22
    2008033439

Visit the Taylor & Francis Web site at
http://www.taylorandfrancis.com

and the CRC Press Web site at
http://www.crcpress.com

Preface

In years past, most industrial operations had a lubrication engineer on staff who, although somewhat of a jack-of-all-trades, was responsible for the lubrication maintenance of industrial equipment. His or her skills extended well beyond changing the oil and greasing the equipment. Rather, he performed, at a rudimentary level, many of the practices that have now become the basis for today's proactive maintenance programs. Modern manufacturing operations must have reliable equipment to maintain stable delivery schedules and operate with the greatest overall efficiency. This reliability is a key element of overall global competitiveness. To get maximum benefit of the advanced maintenance reliability-based operational strategies, an excellent understanding of equipment lubrication is a prerequisite. The goal of this book is to demonstrate the key role of effective equipment lubrication practices in a proactive reliability-based maintenance program and the best practices for achieving the cost reduction and the inherent resultant increase in operational reliability.

The book begins with a chapter written by Mark Castle, a certified maintenance reliability professional, on "Full Circle Reliability," which sets the stage for the rest of the book by demonstrating the critical role of effective lubrication in competitive operations. Subsequent chapters explore how lubricants degrade in service and the methods for detecting and measuring the extent of this degradation. There are chapters on lubricant cleanliness (contamination control), environmental implications of lubricants, centralized lubrication systems—theory and practice, conservation of lubricants and energy, storage and handling, and used oil recycling. The book also covers critical elements of the reliability puzzle, lubrication program development and scheduling. Thus, this book covers from A to Z the key role of effective equipment lubrication practices in a proactive reliability-based maintenance program and the best practices for achieving maximum cost reduction and the inherent increase in reliability.

This volume was written by a peer-recognized team of expert contributors from a wide variety of industry segments. Each chapter was written by an expert both knowledgeable and active in the subject area. Thanks go to these individuals; without their expertise and hard work this book could not be possible. Thanks must also go to their employers for their support of this effort and their contribution to industry.

Because of its emphasis on the practice of lubrication engineering, this book is an excellent reference for those preparing for STLE's Certified Lubrication Specialist® certification examination. As such, it has been recommended for the body of knowledge for STLE's Certified Lubrication Specialist Certification. This volume belongs in the reference library of all maintenance reliability professionals and other practitioners in the field.

The Editors

Robert M. Gresham, PhD, is the director of professional development of the Society of Tribologists and Lubrication Engineers. His technical concentrations include molecular photochemistry, emulsion polymerization, size reduction, and solids classification as well as the field of lubrication. Dr. Gresham gained 12 years of practical experience with the Dupont Company in a broad range of functions including manufacturing, customer service, and polymer and dye research. He has 17 years experience in the field of lubrication as vice president of technology with E/M Corporation, a manufacturer and applicator of solid film lubricants. He was responsible for new product development, quality control, pilot plant production, and grease and oil manufacturing. Dr. Gresham has been a member of STLE for more than 20 years, serving as chairman of the Solid Lubricants Technical Committee, chairman of the Aerospace Industry Council, Industry Council coordinator, the Handbook Committee, the board of directors, treasurer, and secretary of the society. He has also served on several ASTM and SAE committees concerned with standards and specification development. Dr. Gresham is currently responsible for STLE's education and certification programs. He received his PhD degree in organic chemistry in 1969 from Emory University in Atlanta.

George E. Totten, PhD, is the president of G.E. Totten & Associates, LLC in Seattle, Washington, and a visiting professor of materials science at Portland State University. Dr. Totten is co-editor of a number of books including *Steel Heat Treatment Handbook, Handbook of Aluminum, Handbook of Hydraulic Fluid Technology, Mechanical Tribology,* and *Surface Modification and Mechanisms* (all titles of CRC Press), as well as the author or co-author of over 400 technical papers, patents, and books on lubrication, hydraulics, and thermal processing. Dr. Totten is a Fellow of ASM International, SAE International, and the International Federation for Heat Treatment and Surface Engineering (IFHTSE), and a member of other professional organizations including ACS, ASME, and ASTM. Dr. Totten formerly served as president of IFHTSE. He received Bachelor's and Master's degrees from Fairleigh Dickinson University in Teaneck, New Jersey, and a PhD degree from New York University.

Contributors

Mark Barnes
Noria Reliability Solutions
Noria Corporation
Tulsa, OK

Dennis W. Brinkman
Indiana Wesleyan University
Marion, IN

Mark Castle
Chrysler Corporation
Kokomo, Indiana

Paul Conley
Lincoln Industrial
St. Louis, MO

James C. Fitch
Noria Corporation
Tulsa, OK

Malcolm F. Fox
IETSI
University of Leeds
Leeds, UK

Ayzik Grach
Lincoln Industrial
St. Louis, MO

Mike Johnson
Advanced Machine Reliability
  Resources (AMRRI)
Franklin, TN

Robert L. Johnson
Noria Corporation
Tulsa, OK

Barbara J. Parry
Newalta Corporation
North Vancouver, Canada

Jacek Stecki
Subsea Engineering Research
  Group
Department of Mechanical
  Engineering
Monash University
Melbourne, Australia

Allison M. Toms
GasTOPS Inc.
Pensacola, FL

Larry A. Toms
Consultant
Pensacola, FL

Table of Contents

1

Full Circle Reliability

Mark Castle, CMRP
Chrysler Corporation

The plant's equipment is to manufacturing what an engine is to an automobile; it is the key factor in getting to your destination. The main enemy of mechanical failures is friction. With proper lubrication, friction is reduced to a minimal impact in moving parts. Having an active total lubrication process significantly reduces the risk of encountering a friction-related mechanical failure of your plant's equipment.

It is imperative for organizations wishing to achieve financial stability and growth to manufacture products for sale or use in the global marketplace. World economies move through individual peaks and valleys at different times, and providing products to the world allows you to have a prosperous market somewhere around the globe at any time. Manufacturing is normally most efficient working to a level and balanced delivery schedule in order to fulfill global distribution requirements. A plant must have reliable equipment to maintain stable delivery schedules and operate with the greatest overall efficiency. Companies who begin the path to be globally competitive have the prerequisite of finding the optimum balance in both quality and cost of manufacture in order to be the most competitive producer. Equipment reliability is a key component in overall competitiveness. The stakes necessary to become a competitive global producer are high and have led manufacturing management to seek out advanced maintenance strategies to positively affect their current quality and overall cost structures.

The final objective for the maintenance group to be successful requires personalizing a mix of advanced maintenance strategies to fit their individual corporate requirements to generate reliability into the plant's equipment. Maintenance organizations have a tremendous impact in achieving high reliability of the plant's equipment to improve quality and lower operational costs. The cost of equipment downtime is normally higher than the cost of a well-designed and maintained piece of equipment. Management's search for a magic potion or cure-all for a defective maintenance system is common but the search can lead to enlightening results. The key enabler for an advanced maintenance system to function efficiently is a core foundation rooted in the basic fundamental maintenance practices specifically to reduce the equipment's total life-cycle costs. All advanced maintenance strategies are wasted without a firm foundation in fundamental maintenance practices. A good lubrication process is the fundamental way to reduce the effects of friction. Friction deteriorates the ability of the equipment to deliver high quality and low total life-cycle costs. To get maximum benefit of the advanced maintenance operational strategies, an excellent understanding of equipment lubrication is a prerequisite. We now explore some of the most common proactive maintenance strategies.

Lean manufacturing is a common strategy in the current manufacturing environment. Lean manufacturing and lean maintenance share a common goal of doing more proactive maintenance work with fewer overall resources. The elimination of waste is at the heart of all lean strategies. When waste is eliminated from the traditional maintenance systems, there is still a need for enough personnel to complete the necessary tasks at the appropriate time. Any activity that is more than necessary is also wasted resources. When equipment is serviced or repaired at more frequent intervals than optimal, this is also wasted resources. Competitive maintenance organizations need a proactive organizational strategy to reduce waste.

It is common for organizational leaders at the highest levels to hear a new strategic buzzword for maintaining equipment and adopt it for their organizations, hoping it is the magic potion to solve their complex equipment reliability problems. We explore several of the most common advanced maintenance strategies used in industry today.

The first strategy to be explored is Preventive Maintenance (PM). This is a system that has been around for over 100 years. It involves following the manufacturer's recommendations written in the equipment manuals and performing the recommended maintenance tasks listed at the recommended intervals. Following the advice of the engineer who designed the equipment is a great starting point for PM. The equipment's design engineer knows the most about the design weak points, wear points, and lubrication requirements necessary to prolong the equipment life cycle. It is the most basic system to listen to and follow the manufacturer's suggestions which are engineered into the equipment. The PM strategy has matured since the early days of its use. Although following the recommendations of the designer gets you started, you must then use your own judgment, equipment data, and experience to design PM checks that can detect, reduce, or eliminate commonly found equipment failures your organization has experienced.

The modern PM strategy optimizes all equipment experience for early detection of equipment abnormalities. If, while replacing a filter, checking a bolt for tightness, or checking the equipment lubrication levels, an adverse condition can be detected early before a breakdown occurs, then an opportunity exists to resolve or repair a known condition before the equipment will fail to operate. Early detection usually allows the corrective action to take place when the equipment is not in use. This allows a maintenance planner to kit (gather together) the parts necessary for the corrective action, plan the necessary repairs, and schedule the work to take place at the next available interval, possibly at lunch-time, off-shift, or a weekend. There are many maintenance organizations today that use a team concept to review, evaluate, and upgrade the PM checks performed on an individual piece of equipment. When the different craft or skilled trades personnel are brought together with the specific task of increasing uptime for a piece of equipment, the teams draw on their trained theoretical knowledge, intimate knowledge of machine construction, and experience with the common corrective actions and repairs completed that have led to a specific machine downtime in the past. Specialized teams of skilled trades and engineers who are engaged in specific equipment problem solving can drastically improve a PM system in a plant.

Reliability Centered Maintenance (RCM) is another strategic approach to increasing reliability by disciplined analysis of each machine component in a ranking system to determine risk of failure and propose proactive solutions. The RCM process begins on a broadscale approach in a plant or process. Questions are asked to determine which processes or machines are the primary process bottlenecks in reducing product output or are a primary cost driver to the plant. The equipment used in the process is ranked to establish the specific equipment's risk profile for catastrophic breakdown. The ranking system typically contains three ranking classifications: low, moderate, and critical. In an effort to focus resources, all equipment with a critical ranking would be prioritized to allow the improvement to the most critical equipment first. This is to improve the weakest link of the chain within the production process and provide the largest benefit first. After the equipment is chosen with the highest value associated with improving its current condition, a detailed root cause analysis can be performed on each component level part in order to develop the appropriate proactive response to detect and prevent the failure from occurring. This analysis could lead to an enhanced preventive measure for early detection and correction of the abnormality or could warrant a rebuild of a component before a failure occurs.

A detailed analysis on a specific piece of equipment would be investigating specific machine conditions that might lead to early detection of a future problem or deteriorations in the operational performance of the equipment. If there are air pressure gauges, large particles embedded in filters, high grease or oil usage, or excessive amperage draw, indicating a motor working on the high end of its operating parameters, then these indicators would be conditions that can be adapted into a PM check that would be useful in finding potential machine problems early in the equipment failure process. The corrective measure can then have any necessary parts kitted and planned for the next appropriate machine downtime period. With proper analysis, and if you can measure the critical inputs during the PM process, you can predict the output's effect on the equipment.

The analysis of specific equipment may also benefit from conditional inputs for teardown and rebuild maintenance. As an example, a matured oil analysis program can sample and analyze suspended particles in lubrication oil and then reveal specific individual components, such as bearing or brass spacer wear, early in the deterioration phase. Early detection would allow the necessary parts to be obtained and a restoration plan developed to rebuild a component before catastrophic failure occurred. This saves both downtime and wasted resources to restore the equipment. The primary goal is to seek ways to detect potential equipment failures while the equipment is still in use to allow a proactive restoration plan to be developed in the early stages of failure.

Total Productive Maintenance (TPM) is another maintenance strategy of continuous improvement used to engage the whole team who operate, maintain, and support the equipment. Operators, skilled trades, engineering, and management work as a team to identify and root-cause equipment problems, brainstorm and determine the best solution, and implement the best course of action to eliminate the problem from reoccurring. Ownership in the equipment generates a strong goal alignment to the health and welfare of the equipment. Pride develops and fosters a new era of increased cooperation in the overall maintenance of the equipment. The difference in perspective is astounding. Treat the equipment as if you own it! The days are behind us when a company can afford to let their employees treat the equipment as a rental car. The high-tech equipment of today is expensive and delicate and the owners must treat it with great care.

All TPM initiatives are based upon three primary principles: concept of zero waste (safety, scrap, downtime), employee involvement, and continuous improvement. The primary Key Process Indicator (KPI) used to measure the effectiveness of TPM is Original Equipment Effectiveness (OEE). OEE is comprised by multiplying the Performance Availability (PA) by the Performance Efficiency (PE) by Quality Rate (QR).

$$OEE = PA \times PE \times QR$$

The whole equipment process is broken down in the KPI. Performance availability is the ability of the maintenance organization to have the machine available to run when needed. The performance efficiency is the ability of production to utilize the machine at capacity during the available time. Quality rate is the percentage of scrap or waste parts produced by the equipment. Each of the owners has a stake in improving the process for maximum OEE to be obtained. World Class OEE is 85%. To obtain 85% OEE for the equipment, PA, PE, and QR must each average 95% individually.

$$PA = 95\%, PE = 95\%, QR = 95\%$$
$$OEE = PA \times PE \times QR$$
$$OEE = .95 \times .95 \times .95 = .0857 \text{ or } 85.7\%$$

All owners of the process must have their individual goals aligned to average above 95% in each area of the process. Any lower than expected sections can have root-cause analysis input from the TPM team in order to continuously improve the process. No one knows more about the individual nuances of the equipment than those closest to it every day. Secondary KPI measures for the equipment are Mean Time to Repair (MTTR) and Mean Time Between Failures (MTBF). MTTR is an indicator of maintenance's efficiency in repairing the machine over time. The sooner that early detection methods are used, maintenance planning can plan the job and kit the parts to reduce repair time for the equipment, thus reducing MTTR. The MTBF key indicator is data to show whether the equipment is experiencing downtimes closer together or farther apart over time. A total lubrication process is imperative to have in place to reduce mechanical failures to the equipment. Mechanical failures have a large impact on both MTTR and MTBF on the equipment. The goal is for MTTR to trend down and MTBF to trend up over time.

The final common maintenance strategy is Predictive Maintenance (PdM). PdM is a proactive strategy focusing on four primary maintenance specialties. The first is vibration analysis. Vibration analysis captures a baseline vibration signature on a rotating component when new, and continually compares the current data to the baseline when the equipment was new. Each vibration signature is a composite sampling of the frequency of each rotating part used in the equipment. An experienced vibration analyst can narrow down any abnormal reading to the most likely component in the actual process that is beginning to fail which causes a spike in the specific frequency of the component as compared with the baseline data. Each frequency change tells a specific story. The need for an experienced vibration analyst cannot be understated. The equipment is revalidated with a new baseline after any rebuilds are completed. The validation process also enhances the analysis skills of the vibration analyst. Confirmation of correct issue detection and corresponding recommended repairs are confirmed with a vibration signature returning back to baseline readings. The experience gained in this process helps in detecting common problems more quickly with a more definitive solution as experience improves on each piece of equipment. A lubrication process is imperative for eliminating internal wear on rotating or sliding components of a machine, which results in a negative change in the vibration signature.

Thermography is the process of scanning equipment using infrared (heat) technology to determine abnormal hot spots in components of operational equipment. Using infrared technology detects temperature differentials between components to detect abnormal expectations. As an example, if a thermographer found that a three-phase electrical connection showed one incoming fuse block lug at a significantly higher temperature than its other two lugs, then an infrared and regular picture of the abnormality would be taken and attached to a maintenance job order for immediate repair. It would be likely for a loose connection or faulty component such as a stripped thread of the hold-down lug to be found in this example. In both vibration analysis and infrared technology, the equipment is still in production while being tested and analyzed. It is one of the least intrusive checks that can be done to maximize data collection and problem identification while minimizing the impact of collecting data on production.

The next portion of the PdM strategy is ultrasound. Ultrasonic testing detects pressure differentials on equipment by listening for its high-pitched sound waves that occur as it is trying to equalize pressures. Ultrasonic testers are inexpensive and can detect gas leaks, air leaks, and almost any turbulent flow constraints in a system. A primary use in a manufacturing plant can be to identify air leaks. Compressed air is one of the highest cost utilities used in manufacturing and can show substantial cost savings to the companies who use it.

The last portion of the PdM strategy is tribology, which is the analysis of lubrication properties. Lubrication monitoring is fundamental on large or specialized systems to extend the life of the equipment. The lubrication process is engineered for each piece of equipment to reduce friction and prevent mechanical wear. With any deterioration in the designated lubrication process for any piece of equipment, the life cycle of the equipment will deteriorate. Lubricant condition monitoring for viscosity, contamination, oxidization, and wear particle count is imperative for the operating envelope of the equipment. Proper analysis can indicate rubbing, cutting, rolling, sliding, and severe sliding wear of the equipment. Being proactive in determining lubrication problems and concerns is a much less expensive alternative than allowing excessive friction to ruin a piece of equipment.

A common thread for all maintenance strategies is a proactive approach to prevent equipment problems from occurring, early detection of equipment problems by performing PM checks on the equipment, and continuous improvement to the maintenance planning and restoration process to increase equipment reliability. The primary key to reliable equipment is preventing mechanical equipment problems from occurring by utilizing an active lubrication process on all assets of the plant. The facility-required support equipment, such as air handlers and mist collectors, are especially critical to ensure the plant is mission capable in our environmentally friendly world. Without a comprehensive lubrication program in operation, all the maintenance strategies discussed above will never reach their full potential within an organization. The best magic potion to resolve equipment downtime is to focus on the basic maintenance fundamentals. The correct lubricant properties for the application, contamination control and analysis, and lubrication handling process are imperative to achieve the best life-cycle costs for the equipment. Full circle reliability requires the perfection of basic maintenance practices to enable advanced maintenance strategies to reach their full potential in preventing mechanical equipment failures. The following chapters will enable you to perfect your lubrication process for full circle reliability to occur in your organization.

Suggested Reading

1. Association for Facility Engineering, Certified Plant Maintenance Manager Review Pak, Association for Facility Engineering, Reston, VA, 2004.
2. Liker, J., *The Toyota Way*, 1st edition, McGraw-Hill, New York, 2003.
3. Moubray, J., *Reliability-Centered Maintenance*, 2nd edition, Industrial Press, New York, 1997.
4. Nakajima, S., *Introduction to TPM: Total Productive Maintenance*, Productivity Press, Philadelphia, 1998.
5. Smith, R. and Hawkins, B., *Lean Maintenance: Reduce Costs, Improve Quality, and Increase Market Share*, 1st edition, Butterworth-Heinemann, Oxford, UK, 2004.
6. Williamson, R., *Lean Maintenance: Doing More with Less*, Strategic Work Systems, Columbus, NC, 2005.

2

The Degradation of Lubricants in Service Use

Malcolm F. Fox
IETSI
University of Leeds

2.1 Introduction

The very nature of lubricant service means that lubricants deteriorate during their service use. It is normal for lubricants to degrade by partial evaporation, oxidation, and contamination. The purpose of lubricant formulation for a defined application is to control the deterioration of that lubricant in a planned manner over an established period of time, work, distance, or operation. The deterioration of a lubricant can either be planned and controlled by various means or be uncontrolled. Modern practice is strongly directed to the former.

2.1.1   Controlled Deterioration of Lubricants

The way that a lubricant is changed in service use addresses the two extremes of one of the following:

- A time- or distance-defined period of lubricant replacement, such as 500 h operation, annually, or 10,000 km, without regard to the actual state of the lubricant. But custom and practice show that the service interval set is sufficient to ensure that excessive wear does not occur — a precautionary principle. This approach does not require sampling and analyses or "on-board" sensors and therefore is low-cost. The issue is that the lubricant is replaced with a substantial amount of remaining "life" in it, therefore tending to be wasteful of resources.
- At the other extreme, a quantitative appreciation of the state of the lubricant is done by sampling at regular intervals and monitoring various parameters to give a collective assessment of the condition of the lubricant, that is, "condition monitoring."

  The time interval of sampling should be, at most, half of the anticipated service interval. The database built up over time has value for long term and is concerned with long-term trends in lubricant parameters such as wear metal concentrations, viscosity, and particulate levels. For a full condition monitoring program, the lubricant is replaced when its condition reaches a lower bound of aggregated parameters and it is judged to be, or close to being, unsuitable for its purpose of lubricating and protecting the mechanical system.
- An interim position is to sum the overall performance of the system, be it engine or machine, from its last service interval by integrating power levels used in time intervals/distances traveled/time elapsed. The underlying assumption is that the level of performance and its time of operation are related to the degradation of the lubricant. Thus, 100 km of unrestricted daytime high-speed driving on an autobahn in summer is assumed to degrade a lubricant more than 100 km of urban driving in autumn or spring. Thus, the aggregates of high-power level operation over time are weighted more than the same period of low-power operation. Integration of the high- and low-power level operation is already used in some vehicles to indicate to the operator when the system's service is due and the lubricant must be replaced.

The objective at the end of the service period must be that the lubricant still be "in grade," therefore specification, and that the engine or machine not to have suffered "excessive wear" or component damage. This "state of grace" is readily achieved by the vast majority of lubricants in operational service through the development and testing of formulations. The major current development is for service intervals to increase in terms of hours operated or distance traveled. Thus, for light vehicles, service intervals are progressively increasing to 20,000, 30,000, and 50,000 km for light vehicles. A target of 400,000 km is envisaged for heavy duty diesel engines or their "off-road" equivalent.

2.1.2   The Effects of Deterioration

Lubricants are formulated from a base oil mixture and an additive pack, as described elsewhere in this volume. The base oil is usually a mixture of base oil types and viscosities chosen for their physical and chemical properties and their costs. The additives form part of an additive pack to protect oxidation, wear, acidity, and corrosion, to remove and disperse deposits, maintain a specified operating viscosity range, and minimize foaming. A filter in the lubricant circulation system should remove suspended particulates above a certain diameter.

Lubricant degradation occurs throughout its service life and the baseline for change is reached when its further deterioration would lead to a level such that it cannot protect the system from further excessive wear. This occurs because the lubricant has become physically unsuitable for further service use for several separate or joint causes:

- It has become too laden with particulate dirt.
- Its viscosity has increased/decreased beyond its specification limit.

- Its additive pack has become depleted in one or multiple components. Often the additive component actions are interdependent, thus oxidants may protect other additive actions.
- Abrasive and corrosive materials can cause bearing damage, or bore polish by removing the cross-hatched honing marks, which maintain the lubricant film, or in extreme cases, "scuffing" of piston and bore.

These effects are often interdependent and will cause further changes either directly related or through catalytic effects. When these lubricant deterioration effects occur in such complex systems as lubricant formulations, then a structured approach is needed to understand and solve the problem.

2.1.3   Physical Causes of Deterioration

A lubricant formulation becomes physically unsuitable for further continued service use through a range of the following causes:

1. Internal sources: Internal contributing sources are those which are either introduced into a system by the *production* or *repair* process, as:
   (a) Textile materials such as (production line) cleaning cloths, contributing "lint," which compacts into obstructions of oilways.
   (b) Metallic materials such as metallurgical cutting residues and welding repair particulates or production grinding processes, or by the operational process, of either fuel or oxidative use, as follows:
   - Harder/softer particulates from the partial oxidation of lubricants as harder particulates from longer, C_{30} hydrocarbons, as in lubricant hydrocarbons, and softer particulates from shorter, C_{15} hydrocarbons, as in diesel fuels hydrocarbons.
   - Through defective sealing systems, which allow ingress of silicaceous abrasive sources.
   - Fuel condensing into the lubricant and reducing its viscosity, or together with condensed water, forming an emulsion of low lubricity value. Cooling water ingress into the lubricant system through defective seals is another source of water contamination.
2. External sources: External contributing sources, predominantly grit and dust, are those either introduced into a mechanical system by:
   (a) Infiltration through exhausted and inefficient oil filters
   (b) Filling through unclean filler pipes/tubes
   (c) Lubricant reservoirs open to the (unclean) atmosphere
   (d) Through overwhelmed air filters, as in desert area operations

The debris of system wear, abrasive wear products from combustion processes, and defective sealing materials are physical causes of lubricant deterioration. Another obvious physical cause of degradation is to add an incompatible lubricant to an existing formulation in an existing system — while the base fluids may be miscible, their additive packs may be incompatible and precipitate ("drop out"), leaving the circulating fluid as a simple base oil system with little mechanical/tribological protection.

In most cases, the physical causes of lubricant deterioration are simply related to good maintenance, or the lack of its meaningful application, simply put as "good housekeeping."

2.1.4   The Effects of Lubricant Chemical Deterioration

Of all the chemical causes of lubricant deterioration, oxidation is the most important. It has extensive onward connections to the formation of organic acids, usually carboxylic acids, sludges that lead to resins/varnishes, which in turn bond carbonaceous deposits onto system components. Oxidation forms hard carbon from heavy hydrocarbons such as lubricant base oils, engines become very dirty and if the oxidation is sufficiently severe, then essential small orifices such as filters, minor oilways, and crucial orifices such as undercrown cooling jets become blocked and rapidly cause severe wear problems.

Oxidation is temperature dependent and, as a chemical reaction, is subject to the Arrhenius effect of reaction rates doubling/trebling for every 10°C increase in temperature. Thus, a reaction rate of unity at 300°C will increase to between 2 and 3 at 310°C, to between 4 and 9 at 320°C, and between 8 and 27 at 330°C, and so on — a compound increase. This has important implications for trends in increasing engine power densities, smaller lubricant volumes, and reduced cooling effects due to vehicle aerodynamics, which lead to increased engine operating temperatures, including its lubricant system. Future lubricants must withstand higher operating temperatures using smaller volumes for longer service intervals. Advanced lubricant formulations must be developed, which can operate at consistently higher temperatures to prevent their deterioration below levels that protect power train systems for extended, longer, service changes.

The reserve concentration of unused, effective antioxidant in the lubricant during its service life is a crucial factor. Exhaustion of the antioxidant in the continuous use of a lubricant rapidly leads to the mechanical deterioration of the system.

It is not sufficiently appreciated that heavier hydrocarbons, as used in lubricant base oils, have up to 10% of air dissolved or entrapped within it, the difference is semantic. The mechanical movement of the lubricant, as flow, agitation, or foaming, will maintain the air/oxygen concentration in the oil and increase the rate of oxidation.

High temperatures will also affect the base oil molecules and additives directly. Thermal degradation is selectively used in refineries to reform hydrocarbons at temperatures similar to those by lubricants experienced within engines. Under the relatively uncontrolled thermal breakdown conditions within an engine, base oil molecules can break down into smaller molecules, "cracking," or become functionalized with carbonyl groups, particularly, and undergo polycondensation to form varnishes and gums, which trap and sequester carbonaceous particles. The thermal stability of base oils is an important parameter in their selection.

Additives are destabilized by high engine operating temperatures, dependent upon the extent and duration of their exposure to these high temperatures within the engine system, such as the ring zone and valve guides. The term "additives" covers a wide range of compounds, which can contain sulfur, phosphorus, and chlorine. Complex additives can break down to form a range of smaller compounds; thus, Zinc Di-alkyl Di-thio Phosphates (ZDDPs), antioxidant and antiwear agents break down in the ring zone of diesel engines to form organic sulfides and phosphate esters [1]. But reaction between additives — additive interaction — caused by exposure to high temperatures, not only depletes those additives but can also generate sludge deposits. The intermediates may also be corrosive to the system.

There are several overall tests for the antioxidant reserve/antioxidancy of an oil, new or used, as the ASTM 943, 2272, and 4310 tests, and also the IP 280 tests. Of these are the following:

- The *Rotating Bomb Oxidation Test (RBOT)*, ASTM 2272 where a rotating bomb is loaded with a lubricant oil charge, pressurized with oxygen in the presence of a copper catalyst and water within a glass vial. The time recorded for the oxygen to deplete, by reaction, and its pressure to fall by a specified increment of 25 psi (1.74 bar). This method is operator-intensive and has a range of random errors greater than the other.

- *Pressurized differential scanning calorimetry (PDSC)* method, CEC-L85T-99-5 is a relatively low-cost test with much improved reproducibility, where a small (8 mg) sample within a very small cup is held under 35 atmospheres pressure of air in a differential scanning calorimeter at 190°C. The time for the overall additive function to be exhausted by the combination of high temperature and the diffusion controlled oxidizing atmosphere and the residual hydrocarbon combustion to give an exotherm, as in Figure 2.1, is the "induction time." New lubricant formulations will have longer induction times, which will gradually reduce for used samples of that formulation as its service life proceeds. A "zero" value for an antioxidant "induction time" indicates that the lubricant sample is substantially degraded and unprotected against further, and substantial, oxidative attack.

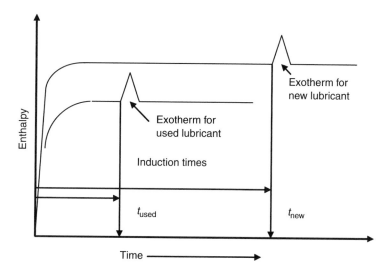

FIGURE 2.1 PDSC induction time plots for new and used lubricant samples.

2.1.5 The "Bath-Tub Curve"

All systems wear but at different rates in their serviceable life. The pattern of wear is well described by the "bath-tub" curve, which is a plot of "wear" against time (Figure 2.2). It can also be regarded as a plot of system failure against time.

A "bath-tub" curve does not describe "wear" (or "failure") for individual systems but is a statistical description of the relative wear/failure rates of a product population with time. Individual units can fail relatively early but with modern production methods, these should be minimal; others might last until wear-out, and some will fail during a relatively long period, typically called normal life. Failures during the initial period are always caused by material defects, design errors, or assembly problems. Normal life failures are normally considered to be random cases where "stress exceeds strength." Terminal "wear-out" is a fact of life due to either fatigue or material depletion by wear. From this it is self-evident that the useful operating life of a product is limited by the component with the shortest life. The "bath-tub" curve

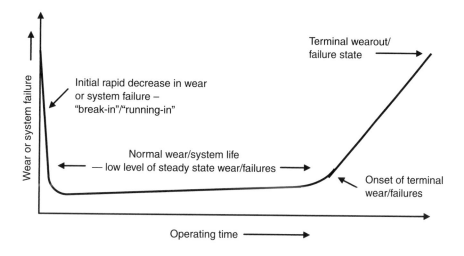

FIGURE 2.2 The "bath-tub" curve for wear/system failure.

is used as an illustration of the three main periods of system wear/failure, and only occasionally is wear and failure information brought together into a database and the initial, normal, and terminal phases of system wear failure measured and calibrated. The timescales for these phases usually vary between one system and another. However, when condition monitoring is used to monitor the wear of a system, then a gradually increasing level of iron in each sample taken at service interval lubricant changes indicates that an engine has entered the final phase of its service life and its replacement and overhaul is becoming due. The necessary replacement arrangements can be made without failure or unexpected interruption of service. This saves costs because the engine is worn, but not damaged, readily and economically overhauled, the operation is planned and service interruption is minimized. Informed replacement of worn systems or components is usually estimated to have a direct benefit/cost ratio of 10:1, rising to 20:1 when indirect costs of unexpected interruptions of service are included.

2.2 Field Tests for Lubricant Deterioration

Laboratory analyses of lubricants are necessarily done in laboratories; they are accurate but delayed unless, unusually, an operating site has its own laboratory. There is a good case for simple field tests, which may be less accurate but gives an immediate indication. Often the operation is physically separated from a laboratory, as in a merchant or naval ship, and needs reliable, simple tests.

2.2.1   Direct Observation of Lubricant Condition

An experienced observer of lubricant condition will give considerable attention to the color of a lubricant sample — it is helpful to compare with an unused sample. Oxidative and thermal breakdown of a lubricant, often beyond exhausting its antioxidant reserve, gives a darker, more brown, color. The deepening in color is also associated with a very characteristic "burnt" odor, which is recognizable when experienced. The viscosity of the sample will also increase.

2.2.2   Field Kits for Lubricant Condition

Various "field kits" are available to measure the essentials of lubricant condition, such as viscosity, water content, particulates, and degree of oxidation. These were called "spot tests" in the past but have improved in reliability to be acceptable for continuing analyses where access to laboratory tests is limited, such as on ships or isolated sites.

Viscosity is readily measured by using a simple "falling ball" tube viscometer in the field on site. Comparison with an identical apparatus, often in a "twin arrangement" containing a new sample of lubricant gives a direct comparison of whether the used lubricant viscosity has relatively increased or decreased by the respective times taken for the balls to descend in their tubes.

The simplest method to determine particulate levels in a sample of a degraded lubricant is the blotter test, where a small volume of oil sample is pipetted onto a filter paper or some other absorbent material. This test can take various forms, either using a standard filter paper or a thin layer chromatographic (TLC) plate. The measurement concerned is the optical density (OD) of the central black spot. The higher the level of particulate, the denser (darker) the spot. The assumption is that the spread of the lubricant sample disperses carbon particulate within an expanding circle and that the OD of the carbonaceous deposit is a direct measurement of the mass of particulate present in that sample. The system can be quantified by use of a simple photometer, for field-based simple systems, or a spectroreflectometer for laboratory measurements. Methods of automating these types of systems have included the following:

- Automated, accurate, constant volume pipetting of the oil samples
- Video measurement of the oil sample blot on the filter paper, thus its OD
- Data recording of these results

Despite many attempts and applications, these advanced methods have not achieved universal acceptance, possibly because of the increased complications built onto an initially simple test. Another, and major, problem is the heterogeneous nature of the samples presented for analysis, which give different responses, arising from:

- Different base stocks, such as the differences introduced by the mineral, semisynthetic, and synthetic base stocks used in modern lubricant and hydraulic formulations.
- Different formulations, such as the differences between hydraulic, automotive, aerospace, and marine fluid formulations, a high dispersancy oil spreading its carbonaceous matter over a greater area than a low dispersancy oil.

Marine lubricant formulations are an interesting case to consider. The lubricant volumes used per engine/vessel are very large, of the order of 10^3 l. The fuel used is high in sulfur, not being controlled to the same extent as land-based automotive diesel fuel, causing extensive additive and base oil degradation. The general case is for vessels to pick up the available top-up lubricants whenever they dock in various ports, leading to heterogeneity of base stock and additive formulation. These factors lead to scatter in the particulate signal/concentration plot.

A further development of the blotter test is to use TLC plates, which are more uniform than paper. The intensity of the black spot from a 50 μl aliquot can be measured and, if its image is captured electronically, may be integrated across its area. But the black carbonaceous spot will also have a base oil ring extending beyond it, seen either as a change in white shade or a fluorescent area under UV illumination (Figure 2.3).

The diameter of the white oil ring measures the movement of the lubricant and the black soot ring measures the movement of the soot particulate. This can be developed into a measurement of dispersancy for the oil sample. Dispersancy is a difficult property to measure; analysis of the dispersant concentration may indicate the amount of free dispersant in the sample together with a variable amount of dispersant desorbed from the particulate, an unsatisfactory measurement. The most effective way to measure dispersancy is to measure the dispersancy ability of a sample, not the concentration of dispersant.

The dispersancy of a sample can be measured by the ratio of the black soot ring to the white oil ring. While this is not absolute, the change in dispersancy over the course of an engine test or the service life of a lubricant can be followed by the change in the spot/ring ratio, as the CEC97-EL07 development method. The method is very reproducible, provided that all of the following are considered:

- Multiple samples are taken, which is much easier than the previous methods.
- The sample images are captured using high resolution optical electronic methods and the area of each spot integrated, as the edges of the spots are often uneven in detail.
- Each micropipetted sample is accurately and reproducibly dispensed.

The ratio of the "spot" diameters for the white oil ring and the particulate measures the ability of the lubricant sample to disperse carbon particulates, a high ratio indicating a high level of dispersancy remaining in the lubricant. Equally, a low ratio of carbonaceous black spot to the radius of the oil blot indicates a low level of dispersancy.

Dilution of a used lubricant sample with a light hydrocarbon such as Petroleum Ether 60/80 and subsequent filtration through a standard filter paper will indicate the nature of the larger particulate

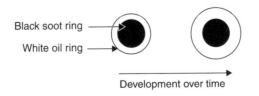

FIGURE 2.3 TLC plate soot spot/oil ring dispersancy test.

debris, emphasizing metal particulate debris. Microscopic examination of the metal debris can show the nature of the larger metallic debris, which indicates the pattern of wear.

Water content can be measured "in the field" by mixing a lubricant sample with a carbide tablet in a sealed stainless steel bomb. The measurement of water content is through the reaction of the carbide tablet (or calcium hydride in an alternative model) to generate gas pressure within the bomb. The pressure level generated is a measure of the water content of the sample. An alternative quick test for water content is the "crackle test," where a small lubricant sample is suddenly heated. This can either be done by suddenly inserting a hot soldering iron bit into the sample — if water is present, a "crackling" noise is heard, which is absent for dry samples (the noise comes from steam generation in the sample) — or a small drop of sample can be dropped from a syringe onto a "hot" laboratory hot plate, when again a crackle will be heard if the sample is wet. From experience, the limit of detection is taken to be 0.1% water.

The degree of oxidation can be measured by a simple colorimeter using a standard sample to measure color, ASTM D1500. The trend compared to previous values is the important observation. If the change occurs early in the service of the lubricant, then the antioxidancy reserve of the lubricant is being rapidly depleted or the lubricant is being contaminated. It is important to consider the change in color in combination with values and changes determined for acid number and viscosity for the same samples.

Other simple tests are available in addition to those described above, as a suite packaged into a portable package for measurement of lubricant degradation in isolated situations such as remote mines and on-board ships.

2.3   Laboratory Tests for Lubricant Deterioration

Some introductory general remarks are useful:

- Results from laboratory tests for lubricant deterioration are of much greater value if the original, virgin, unused lubricant is used as a benchmark.
- Similar tests apply to most forms of lubricants as the deterioration challenges they face are chemically and physically similar.
- However, the results from similar tests for different forms of lubricants must be considered in the context of each lubricant's application.

The advantage of laboratory tests is that they should have a background of both quality assurance and control. From this, they have serious weight in solving problems, assessing oil change intervals, what preventive maintenance is required from condition monitoring to conserve the system, as well as lubricant resources. The primary objective of a laboratory analysis program for lubricant samples is to ensure that they are fit for further service. If the lubricant is unfit, or becoming unfit, for further service, then it must be replaced.

The benchmark for a laboratory program of sample analyses to assess a lubricant's deterioration is to offer a rapid turnaround for analytical results, their assessment against limit values, and reporting back to the client. Isolated heavy plant mining operations can have lubricant analytical sample reporting times of weeks due to transport and communication issues; intensive transport systems in developed countries can expect less than 24 h reporting, such that a sample taken one day will be analyzed and reported upon before the next day's operation commences and the appropriate action taken.

An equally important benchmark is for a laboratory to meet the various national or international standards, such as the ISO 9000 series. The use of certified analytical standards and accredited solutions is part of a complete package, which best involves a collaborative program of regular analyses of samples sent from and collated by a central standards body. All apparatus and substances used should have an audit trail for standards and calibrations that are maintained. This is not only a good practice but necessary to respond to any implied liabilities, which may arise later.

Of the many tests available, the major issues of lubricant deterioration are addressed by analyses of viscosity and viscosity index (VI), trace metals, particulates, ash, acidity/base reserve, and water contents.

Other minor issues are color, demulsibility, foaming, rust testing, infrared spectroscopy, and to a certain extent, x-ray fluorescence (XRF). Gas and liquid chromatographies, x-ray diffraction, interfacial tension, and density are peripheral techniques that might be used to investigate unusual occurrences.

2.3.1 Viscosity and Viscosity Index

Viscosity is the foremost quality of a lubricant to be measured. A lubricant must maintain its viscosity to effectively protect a system against seizure. Variations in viscosity are usually associated with effects, which show up in other analyses; therefore, a multidimensional approach is needed to consider the root cause of the change.

Viscosities of lubricant samples are now measured by automated systems, taking samples from multiple sample trays, either circular or linear, and injecting them into either kinematic or absolute viscometers thermostated at either 40 or 100°C. If separate viscometers at these temperatures are used, then the VI of the sample can be calculated. Standards are inserted into the flow of samples through the system for quality control. Individual measurements using manual stopwatches and U-tube suspended viscometers are now rarely used in laboratories.

A small increase in lubricant viscosity may be due to evaporation of the lighter ends of the base oil after prolonged high level operation. Beyond that, significant increases in viscosity, up to 10 to 20% being regarded as severe, result from the inadvertent replenishment with a higher viscosity lubricant, extensive particulate contamination, and extensive base oil oxidation. The particulate contamination as well as extensive oil oxidation will be readily seen, the latter on its own as an increasing dark brown coloration. The black particulate contamination will obscure the brown oxidation color. Oxidation effects will also appear in the Fourier Transform Infrared (FTIR) spectra and decrease in the PDSC antioxidant reserve time.

A decrease in the viscosity of operating engines is usually due to fuel dilution, a characteristic occurrence when an engine idles for a prolonged period. A locomotive used for weekend track maintenance train duties will run its diesel engine at idle for periods of several hours and its lubricant will show a significantly decreased viscosity afterward due to fuel dilution. If subsequently used for normal, higher power duties, the increased lubricant temperature will evaporate the condensed fuel and the viscosity returns to its previous value. A more serious occurrence is when fuel and water are extensively condensed in the crankcase of a very cold engine at start-up. During short journeys, when the engine lubricant rarely becomes warm enough to evaporate the condensed fuel and water, the two contaminants can combine to cause the additive package to precipitate out from the lubricant formulation. The engine may then have "oil" but is left with considerably reduced protection wear. Measuring fuel dilution in diesel lubricants is difficult and is discussed later in Section 2.4.2, Flash Point of Degraded Lubricant.

Fuel condensed into a lubricant has the role of a solvent and the same effect of decreased viscosity is found when a solvent becomes entrained, such as a refrigerant fluid. Chlorofluorocarbons (CFCs) are well on their way to removal and nonreplacement from refrigeration systems but their replacements, the hydrochlorofluorocarbons (HCFCs) and hydrofluorocarbons (HFCs) have the same effect of reducing viscosity if allowed to leak through seals or rings and dissolve in a lubricant.

Viscosity index improvers (VIIs) are long chain polymers of various basic units. Their different structures resist high rates of mechanical shear, as in bearings or in the ring pack/bore wall interface, to different extents. While there is a separate effect of temporary viscosity shear loss, lubricants with VIIs can suffer permanent viscosity shear loss due to breaking of the polymer chains. The initial lubricant selection process should have considered how robust the formulation was to permanent shear thinning. Tests for this include high temperature and high shear procedures such as ASTM D4683 and D4741.

If the viscosity of a lubricant changes during its service use, then its VI will change necessarily. The major cause of a reduced VI is breakage of some of the polymeric VII polymer molecules to give smaller chains of less effect. There are two effects — reduction in the molecular weight of the VII additive will reduce the viscosity of the lubricant formulation at both 40 and 100°C and also reduce the temperature related VII effect. The latter effect normally has the greater weight so that the permanent shear breakdown of polymeric VII additives reduces the lubricant's VI. It is not unknown but rare for the VI of a used lubricant to increase in service use, often associated with extensive oxidation.

2.3.2   Trace Metals

The term "trace metals" in a lubricant sample not only covers metals generated by wear in the system but also the elements from the additive pack. While the determination of trace metals for a "one-off" sample gives some insight into the condition of a lubricant, the major value of trace metal determination lies with long-term condition monitoring. The "bath tub" curve of Figure 2.2 is recalled here — following the level of iron fine particulate in a series of regularly sampled lubricant from a system is an essential part of condition monitoring. The "break-in" or "running-in" phase, normal wear, and the gradual increase in wear element determination can be followed running over many hours and lubricant service changes. The onset of terminal wear can be detected and followed, with arrangements put in place to remove and replace the engine system. Levels of wear elements measured are usually iron (from bores and crankshafts), lead and copper from bearings, aluminum from pistons, and chromium from plating on piston rings. Others may be added to follow specific effects; for example, sodium levels indicate the ingress of cooling water and its additives, silicon levels indicate the ingress of sand and rock dust.

It is important to recognize that the level of wear elements in a system's lubricant is individual both to system design and to individual systems. Thus, levels of iron in the normal wear phase of engines will be different from one design to another; in addition, there will be some variation between the normal wear phase iron levels of engines of the same design. The quality of the lubricant used will also affect the level of wear metal; the higher the quality of lubricant, the lower the level of wear elements. The emphasis for assessing the condition of lubricated systems is placed upon the trend in wear element levels. While the iron level in the lubricant of one engine may be higher than another, it is the trend for successive samples over time in the measured levels that is important.

Wear processes in lubricated systems rarely occur for one metal. Increases in the levels of several wear metals can indicate the occurrence of a particular wear process or contamination. Table 2.1 describes wear elements found in lubricants in service life.

Wear metal analyses have additional effectiveness when combinations of enhanced element rates are considered, such as for a diesel engine. Combinations of enhanced wear elements are unique to each operating system design and its pattern of use. "Expert systems" applied to an extensive data system can be used to develop "rules," which indicate which main assemblies or subassemblies are developing enhanced rates of wear and require attention for certain engine designs. The examples given in Table 2.2 are typical for certain applications — other systems may have different combinations for wear patterns, it is for the expert system to recognize them. More extensive combinations of elements indicating particular wear patterns by system components can be developed, such as using the "principal indicator" and associated "secondary indicator" elements.

Cost-benefit analyses of spectroscopic oil analysis programs, with the acronym "SOAP," have been demonstrated in many applications to be very significant. Continuously and heavily used plant, such as diesel express trains, where daily oil sampling and analysis gives an immediate cost-benefit ratio of 10:1 in direct costs and 20:1 for indirect costs when service reliability benefits are included.

The analytical methods for wear metals have generally moved to inductively coupled plasma (ICP) atomic emission systems. A small sample is automatically extracted from a sampling bottle, diluted with kerosene, and sprayed into the ICP analyzer plasma torch at 6000–8000°C. The very high temperature of the plasma excites the metal particulates to high energies, which emit light of a characteristic atomic wavelength. Duplicates (or more) are readily programmed. The emission from each metal present is detected and reported. The cost of additional wear element detection is marginal once the ICP system is set up.

The ready availability of duplicate sample determinations and insertion of calibration standards gives a high level of quality control as precision, accuracy, and reproducibility to the final results. The analytical data generated by the ICP system is readily handled, quantified, and then placed into a file for that engine

TABLE 2.1   Wear Elements in Lubricants and Their Sources

	Source
Major Elements	
Aluminum	Primary component of piston alloys, also bearings, washers/shims and casings of accessories. From corrosion of engine blocks, fittings, and attachments.
Chromium	Used as a hard(er) coating to reduce wear, indicates wear of chromium plating on engine bores, shafts, piston ring faces, some bearings and seals.
Copper	With zinc in brass alloys and tin in bronze alloy wearing components, copper present in journal, thrust, and turbocharger bearings, also cam, rocker, gear, valve, and small-end bushings. Also, fabricate oil cooler cores.
Iron	Still a major, massive component of engines, gearboxes, and hydraulic systems. Lubricant contact through cast bores, cylinder liners, piston ring packs, valve guides, rolling element bearings, chains, and gears. Difficult decision given by wearing component increased trace levels of iron.
Lead	In bearings, solder joints as "lead/tin alloy" and also seals.
Molybdenum	A wear reduction coating on first piston ring faces for some diesel engines.
Nickel	From valves, turbine blades, turbocharger cam plates, and bearings.
Silver	Alloys in bearings, bearing cages, and bushings for diesel engine small ends, turbochargers, and rolling element bearing applications in gas turbines.
Tin	Common alloy in bearings with aluminum, bronze, and brass fittings, seals, and also in cooler matrix solder.
Titanium	Top end of market, gas turbine bearing hubs, turbine blades, and compressors.
Zinc	With lead and tin in common alloys such as brass and also some seals.
Minor Elements	
Antimony	May be used in bearing alloys.
Boron	Borates used as cooling system anticorrosion agents, presence in lubricant and hydraulic fluids shows leak in cooling system matrix.
Magnesium	Increasingly used as an alloy with aluminum for accessories and casings.
Manganese	From corrosion of manganese steel alloys, occasionally in valves.
Sodium	Usually sodium borate as cooling system anticorrosion agent. Increasing trace presence in fluids shows leak in cooling system matrix, marine applications indicate ingress of coolant sea water.
Silicon	Piston wear. As silica, indicates road dust ingress, particularly damaging as hard particulate, which causes high levels of wear, shows air filter and breather system failure, particularly mining and deserts.

system, which can then be compared with previous results. This is concentration level "trending" in its simplest form. The overall effect is to give a high throughput of high quality analyses at low cost. While the automated sampling ICP multiple element system has a high capital cost of $150–200 k ($300–400 k) each, the high sample throughput can cut the unit cost per sample down to 50 p ($1). An atomic absorption (AA) apparatus can be used instead of the ICP system but suffers from the disadvantage of only determining one element per analysis from the nature of this method. The older emission system of an electric discharge between either still or rotating ("Rotrode") carbon electrodes is still used but the advantages of the ICP

TABLE 2.2   Some Indicative Combinations of Wear Elements

Elements	Indicative Cause
Sodium and boron	Coolant leakage into lubricant, as through head gasket failure
Lead and copper	Main or big-end bearings
Copper, silver, and iron	Turbocharger bearings
Chromium and iron	Piston rings
Silver, copper, and lead	Small-end bush
Iron and copper	Oil pump wear

TABLE 2.3   Corrective Levels for Lubricant Deterioration

Deterioration Level	Action
Normal	Within average, no action
Alert	Within average $\pm 2\sigma$, action $\rightarrow$ increase sampling frequency
Urgent	Within average $\pm 4\sigma$, action $\rightarrow$ maintenance needed, can be deferred
Hazardous	Beyond average $\pm 4\sigma$, action $\rightarrow$ immediate maintenance, no deferral or trend in analysis >60% average
Dangerous	Trend in analyses up to 90% of alert level, action $\rightarrow$ shutdown/recall immediately/immediate urgent maintenance

system for high throughput of samples are gradually displacing it. The ICP spectroscopic technique and oil samples are brought together as a condition monitoring system.

It is meaningful to analyze trends in the wear element test data, which monitors the deterioration of the oil condition. Absolute and rate of change data concentration values can be used to assess the deterioration of a lubricant or hydraulic fluid — the ideal scheme, with regular sampling, servicing, and replenishment at preprogrammed intervals. It is rare for this regularity to hold; the reality is that sampling/servicing and replenishment of fluids occur irregularly and this must be adjusted numerically in the trend data. From these "trending analyses," element concentration indicators can be developed by various statistical methods using system failure modes to set individual wear metal levels at which corrective or remedial measures must be taken for the deterioration of the lubricant, such as in Table 2.3.

2.3.3   Particulates and Ash in Lubricants

The accurate measurement of particulates and ash in a lubricant sample is very important in assessing its deterioration because the excessive build-up of soot, dirt, or particulates in general can prevent the normal protective function of that lubricant. The term "particulates" covers a wide range, including insoluble matter, sediments, and trace metals as very fine diameter particulate. Larger metal particles such as metal flakes and spalled debris are not covered, these being covered by separate analyses and filtration.

2.3.3.1   Dirt and Particulates in Lubricants

Controlling the cleanliness of any lubricant or hydraulic system as it deteriorated with use was very important in the past and will be even more important in the future because of the following reasons:

- System reliability is increasingly important and a major contributor to equipment failure is particulate contamination in the system operating fluid.
- Systems perform at higher energy levels for longer periods and maintained to be "cleaner" so as to deliver that performance.
- Equipment tolerances are decreasing for high precision components ($\sim$5 μm clearance or less) and in automotive and hydraulic components they are increasingly common. Smaller particulates, for example, 2 μm dependent upon its nature, can agglomerate and clog sensitive components such as control and servo valves.
- For automotive applications, two trends lead to increased particulate levels:

    Exhaust gas recirculation, for environmental exhaust emission reduction, primarily for NOx, having the additional beneficial effect for emissions of depositing particulate into the lubricant rather than being emitted. However, this creates a problem of enhanced particulate levels for the lubricant.

    Strong consumer pressure for increased service intervals, already up to and beyond 50,000 mi (80,000 km) for trucks and 30,000 mi ($\sim$50,000 km, or every 2,000 years) for some new 2005 light vehicles.

Lubricant must last longer and yet meet enhanced performance standards. Enhanced levels of particulate are now envisaged, well above 1%, up to 2 or 3%, a steep challenge for the lubricant to remain effective under these conditions.

2.3.3.2   Useful Definitions

"Particulates" and "dirt" are descriptions that require a more precise description and definition, as follows:

- "Particulates" are small, up to 15 μm maximum, either carbonaceous, inorganic compounds or fine metal particles, where the metal particulates result from "rubbing wear."
- "Dirt" is road dirt ingested by faulty induction air filters, poor seals or defective/absent air breather components; the parts that survive are usually hard particles such as silicates (from sand, etc.).
- "Metal debris" is comprised of larger metal flakes or spalled particulates resulting from catastrophic micro-failures or incipient major failures such as parts of gear teeth being separated.

Hydraulic fluids develop haze or very light deposits over a considerable time of their service life; petrol/gasoline engines develop black particulates slowly over their service life, while diesel engines rapidly develop black particulates. The operating limit of circulating lubricant filters is in the range 10 to 15 μm, whereas it has been shown that removal of the "larger," >10 m, particles from a circulating lubricant system can reduce catastrophic bearing failures by 25%. Further, for hydraulic systems it is claimed that 80% of failures can be avoided if particulates >5 μm are removed by filtration. While not going to these levels of filtration, higher levels of filter efficiency are now incorporated into new designs. This must happen to meet the enhanced levels of filtration required over the enhanced periods of service operation. However, one problem is that the enhanced levels of filtration can remove the small, fine, metal particulate, which is used for wear data and trend analysis.

2.3.3.3   Particulate Analyses

There are a number of methods available for the measurement of soot in lubricants. These measurement methods can be grouped into three categories, as where the particulate is:

1. Removed from the liquid, then oxidized while measuring the mass loss.
2. Separated by addition of solvents to the lubricant sample and the precipitated mass measured.
3. Measured within the neat, or diluted, lubricant sample for absorbance, scatter, or obscuration at a given wavelength.

2.3.3.4   The Enhanced Thermogravimetric Analysis, ASTM D5967 Appendix 4 (Colloquially Known as the "Detroit Diesel Soot Test")

Total particulate in a degraded oil sample is determined by thermogravimetric analysis (TGA), where 20 mg of oil in a pan on one arm of an electronic balance is heated under a programmed temperature furnace environment in a nitrogen atmosphere. Differentiation is made between carbonaceous and incombustible ash by increasing the temperature and changing to an oxygen atmosphere. A 20-mg sample is larger than normal but is necessary because the final objective, the soot content, will be less than 1 mg. The temperature environment is held at 50°C for 1 min, raised to 550°C at a rate of 100°C/min, maintained isothermally for 1 min, and then raised to 650°C at 20°C/min. The method considers the residual sample at this stage to be composed of soot and incombustible material with liquid hydrocarbons removed. The atmosphere is then switched to oxygen and the furnace temperature raised to 750°C at 20°C/min and maintained for a stable weight for at least 5 min. The changes in weights at different temperatures and atmospheres are due to soot being the difference in weight between 650°C in nitrogen and 750°C in oxygen. The residual material is incombustible ash and metallic residues, assuming that all of the remaining lubricant base stock is driven off and oxidized at the higher temperatures under oxidizing conditions.

2.3.3.5   Optical Particulate Measurements

A very desirable feature in particulate measurement is a linear relationship between particulate signal, by light absorption or scattering, and particulate concentration. This relationship generally holds as a linear relationship of a certain slope up to ~1.5% particulate concentration, followed by a linear relationship

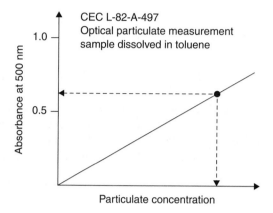

FIGURE 2.4   CEC L-82-A-497 calibration plot for particulate determination in degraded lubricant by dilution in toluene.

with a higher slope at higher particulate levels. While an overall linear relationship is very desirable, the major problem is the change in relationship between signal, however derived, and particulate concentration in the region of 1.5% concentration. Two methods measure sample particulate concentrations, one infrared by direct sample absorption and one in the visible by dilution in toluene.

The visible method, capillary electrophoresis chromatography (CEC) L-82-A-497, "Optical Particulate Measurement," dilutes the degraded oil sample in toluene, a solvent, which disperses all of the particulate, and then measures the absorbance of the diluted solution at 500 nm in a spectrophotometer. Standardization uses a lubricant or hydraulic fluid sample of known pentane insolubles content to construct a calibration curve (Figure 2.4).

The method is quick, repeatable, and accurate, provided that the sample disperses well and does not cause light scattering, which will add to the apparent OD. This method was adopted by the CEC to measure soot developed in lubricant samples from the Peugeot XUD11BTE engine test and uses 0.1 g of oil sample in known aliquots of toluene. The solvent aliquot volume is increased to bring the OD within an acceptable range. The OD plot for lubricant samples dispersed in toluene and measured at 500 nm should be linear with a high correlation coefficient. The only drawback is that some additives or degradation products may cause light scattering and an incorrect result.

2.3.3.6   Infrared Measurements at 2000 cm^{-1}

Soot does not absorb in any specific region of the infrared region but as small particulate scatters the incident radiation in a nonphotometric manner. Theoretically, light scattering of a spherical, uniform diameter, particulate is proportional to the fourth power of the wave number. From this, the background scatter in the infrared spectrum of a used lubricant containing particulate should decrease across the infrared region from 4000 down to 400 cm^{-1}. Background scatter does decrease but not as much as predicted by theory, probably because the particulate is not monodisperse and certainly not spherical. The chosen measurement point is 2000 cm^{-1} because there are no absorbing groups present in lubricants. Increase in lubricant absorption at 2000 cm^{-1} with engine run time is mainly dependent upon the mass of soot particulate present, with second order effects due to the effective particulate size and shape, and therefore is somewhat dependent upon engine type. High levels of soot particulate give high absorbance levels and inaccuracies in spectrophotometry, which can be overcome by using thinner path length cells. The results are in absorbance and need calibration for percentage soot. The advantage of the method is that it is a direct measurement on the sample, without the effects of adding solvents and the like, and that it arises from infrared measurements, which could be undertaken for another set of measurements in any case. The disadvantage of the method is that the sample spectra need to be the difference spectra, that is, the difference between the engine test run samples and the original, fresh oil, which may not always be available.

2.3.3.7 Particle Size Distribution

A more fundamental view of the nature of particulates in degraded lubricant and hydraulic fluids is the distribution of particulate sizes. This can be done continuously by light scattering or discontinuously by a range of physical filters. The latter is self-explanatory but the first needs explanation. Particles suspended in a medium scatter incident light at an angle dependent upon particle size and also upon the wavelength of the incident light. The second is simplified by using a monochromatic source such as a laser. The particles are assumed to be spherical, a very broad-brush approximation. A variable correction factor is needed for the nonspherical nature of the particulates, such as a rodlike nature with a defined length/width ratio. Particulate light scattering optics uses a collimated laser light source, usually a He/Ne (red) laser, expanded by a lens into a broad beam, which diffuses the sample cell. The light is dispersed/scattered by the suspended particulate in the sample cell and then collected by a similar second lens and focused onto a detection plate. The detection plate samples the intensity of the scattered light at a large number of points and transforms into a particle size distribution by suitable software. The resolution of the method depends upon the spatial discrimination of the detector plate.

Particle size distributions for a range of samples from engine runs using a range of related lubricant formulations show that these particulate distributions are interdependent, the smallest particulate size distribution leading to the successive growth of the larger particle size distributions. The interdependence of these particulate distributions measures the effectiveness of dispersants, for the particulate can successively agglomerate from the initial size of around $0.1~\mu$m diameter to 1 to 7 to 35 μm and then larger diameters. If the dispersant within the lubricant is not degraded, then the agglomeration process will be stopped or reduced.

2.3.3.8 Particulates in Hydraulic Fluids

Hydraulic fluid cleanliness is crucial to the continued operation of hydraulic systems, avoiding component damage and failure. The level of cleanliness is many orders of magnitude down (better) from that accepted for lubricants. Instead of values of mass particulate, the emphasis for hydraulic fluids is on the number of particulates in the range of 2 to 15 μm, a range correlated to the probability of component problems. With this stimulus, several methods of electronic particle number counting have been developed, based upon the following:

- Light absorption
- Flow decay
- Mesh obscuration

These methods are continuous and easy to use; their main problem is the large amount of data that they generate for the size and number of particulates but without reference to the composition of those particulates. Wear metal or chemical analytical data is required to properly understand the complete picture of particulate composition in hydraulic fluids. A fundamental problem is the lack of suitable, repeatable reference standards. When used for equipment monitoring, it is very important that the response of the counter has a high particle size correlation with the size of particles, which cause damage to the fine tolerance components of the system. A 5 μm diameter was regarded as the lower limit of damaging particles until recently, but this is now reduced to 2 μm as an indicator of potential damaging conditions, approaching the limit of discrimination between two such particles.

One type of mesh obscuration particle counter uses three successive micro-screens of 15, 5, and 2 μm pore size (Figure 2.5). Laminar fluid flow through this array of screens generates pressure drops, caused by oversized particles partially blocking the respective pore size filter, recorded by differential pressure transducers. Count data from hydraulic samples is statistically derived through correlation with data from a calibration standard. This counter is effective for most oils of different levels of obscuration (light-black) and is relatively insensitive to other counter-indicators such as entrained water and air in degraded lubricant samples.

Another method of electronic particle size counting uses the blocking behavior of a particle size distribution in a degraded lubricant sample passing through a single, monosized micro-sieve of either

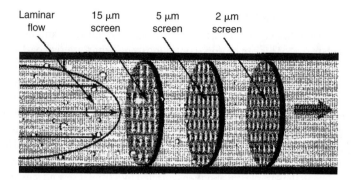

FIGURE 2.5   The principle of the mesh obscuration particle counter. (From *Machinery Oil Analysis — Methods, Automation and Benefits*, 2nd ed., Larry A. Toms, Coastal Skills Training, Virginia Beach, VA, 1998. With permission.)

15, 10, or 5 μm pore size (Figure 2.6). A correlation is assumed between the particle size distribution of an unknown sample and that of a standard. The measured parameter is the differential flow across the micro-screen, which converts flow decay measurements to an ISO cleanliness code.

An optical particle counting method uses a path of collimated light passed through a hydraulic oil sample and then detected by an electrical sensor. When a translucent sample passes through the sample, then a change in electrical signal occurs. This is analyzed against a calibration standard to generate a particle size and count database, linked to an ISO cleanliness value. The output values of the light absorption particle counter are badly affected by the following factors:

- The opacity of the fluid raising the background value to the level that the instrument no longer works, overcome by sample dilution with a clear fluid.
- Entrained air bubbles within the sample are counted as particles, which confuse the system, and are removed by ultrasonics and vacuum treatment.
- Water contamination is more difficult to deal with, causing increased light scattering. But significant levels or water, such as >0.1 or >0.2% levels, will fail the oil anyway.

The continued monitoring of particle cleanliness in hydraulic fluids within systems is a very important process to maintain the integrity and performance of complex hydraulic systems.

2.3.3.9   Ash Content

The "sulfated ash" content of a lubricant is an important property and can be included under particulates in degraded lubricants. It gives a meaningful indication of the detergent additive content and is useful

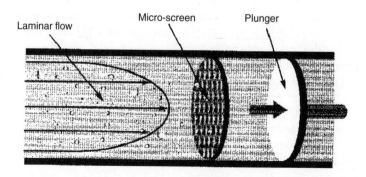

FIGURE 2.6   The principle of the flow decay particle counter. (From *Machinery Oil Analysis — Methods, Automation and Benefits*, 2nd ed., Larry A. Toms, Coastal Skills Training, Virginia Beach, VA, 1998. With permission.)

as a control test in the oil blending process. While it is a property only normally used for new formulations, results for degraded lubricants have considerable interference from both wear metals and other contaminants.

The problem with sulfated ash arises from inorganic compound deposits in the ring zone and on the piston crown. The problem becomes very important when extensive deposits build up on the piston crown from low/medium power level operation, such as for a taxi engine in town. However, when such an engine is used at extended higher energy power levels, such as extended motorway journeys, the deposits on the piston crown become very hot, retaining heat and glow. They can become so hot that they melt part of the piston crown to the extent of penetration, that is, a hole, causing catastrophic deterioration of the engine, which is the downside of sulfated ash content.

The upside of metallic detergent inclusion into lubricant formulations is their ability to reduce the deposition of carbonaceous substances and sludges in the ring zone and piston crown. The essence of the problem is to balance the level of metallic soap sulfonate in the original formulation and the amount of sulfated ash that results. Sulfated ash is a major contribution to the overall formation of ash, contributing to crown land deposits above the piston rings, valve seat deposits (and thus leakage through seat burning), and combustion chamber deposits. These deposits cause pre-ignition of the gasoline/air mixture, leading to a decreased fuel octane rating for the same engine called octane rating decrease (ORD). It is beneficial to reduce the impact of this effect by minimizing ash deposits. Ash formed from lubricants can also contribute to diesel engine particulate emissions.

Recalling that the sulfated ash content is important for new lubricants, the simplest test is the ASTM D842 Ash Test where the ash content of a lubricant is determined as a weighed sample, to constant weight, of oil burned for 10 min at 800°C. The mass measured is that of the incombustible solids, be they wear metals or other incombustibles such as fine metallic particles or silicaceous dust. The ASTM D874 Ash Test is an improved ASTM D842 method in that the oil sample is combusted until the carbon residue and metallic ash is left. Sulfuric acid is added, the sample is reheated and weighed to constant values. The last stage converts any zinc sulfate to zinc oxide.

The sulfated ash tests indicate the concentration of the metal-based additives in fresh lubricant blends. Problems arise from (i) any phosphorus present forming pyrophospates of variable composition, giving higher and more variable results and (ii) magnesium sulfate being variably converted to its oxide. Carefully conducted, the sulfated test gives a reasonable measure of additive metals present in a lubricant formulation. The weight of metal present can be converted to the expected sulfated ash content by the conversion factors given below:

To Estimate Sulfated Ash Content from Metal Content:

Metal Conversion Factor	Metal % to Sulfated Ash
Zinc	1.25
Sodium	3.1
Magnesium	4.5
Calcium	3.4
Barium	1.7

If the lubricant has been formulated with magnesium-based detergents or boron-based dispersants, then these methods of sulfated ash are unreliable. The sulfated ash test is also unreliable for used lubricants, due to the following reasons:

- The presence of incombustible contaminants.
- Additives will be degraded during service life and are thus changed chemically but the constituents will continue to appear in the ash residue at the same concentrations as for the new oil.
- A trend toward ashless detergents, which undermines the relevance of the sulfated ash test as a measure of detergent in a formulation.

It is important to check the sulfated ash method against reference blends wherever possible.

2.3.4   Acidity and Base Reserve

Determining the alkaline reserve or acid content of a degraded lubricant fluid should be straightforward by analogy to acid/base titrations in water. But this is the simplistic point that causes so many problems with determining "base" and "acid" numbers in degraded lubricant and hydraulic fluids. To thoroughly understand "base number," an appreciation is needed to determine the following:

- How it arises
- How it has been, and is currently, measured
- The problems of those analyses
- What this means for lubricant use/extended use and condition monitoring

While the idea of a "number" is simplistic and therefore appealing, the reality is complex and we need to look at the points made above, in order.

2.3.4.1   The Need for Base Number Measurement

The need to measure the base number in some form as a property of a lubricant/degraded lubricant arises from the acidic products formed during the service life of that lubricant. The acid formation process can be rapid or slow, according to the stress that the lubricant is exposed to. The emphasis must be on the effect that the "service life" of the lubricant involves, in terms of either high temperature and pressure or over a short and intense, or a very long-term and less severe, service interval.

The starting position is that most lubricant base fluids have some, maybe greater or lesser, basic properties that neutralize acidic components introduced into them. As the performance requirements of lubricants developed, it became evident that the naturally occurring anti-acidic properties of unmodified base stocks were not sufficient to prevent lubricant and hydraulic oils becoming acidic and corroding the components of the system. The development of detergent additives had two effects:

- The organic nature of the additives themselves had an additional, but marginal, anti-acid contribution.
- However, more importantly, the detergent additives had the ability to solubilize as inverse micelles alkaline, inorganic material such as calcium oxide/carbonate or the corresponding magnesium salts (much less used). These compounds react with acidic products formed in the lubricant to produce neutral salts, which bind the acidity as an innocuous product. Barium compounds are not used now because of toxicity problems.

2.3.4.2   Sources of Acidity-Induced Degradation

Acidity in lubricants arises from two sources:

- The (declining) sulfur content of fuels, forming sulfur oxides, primarily sulfur dioxide (SO_2).
- The reaction ("fixation") of atmospheric nitrogen by reaction with atmospheric oxygen in the high temperatures, 2000 to 3000°C, of the combustion flame front, forming nitrogen oxides such as nitric oxide (NO) and nitrogen dioxide (NO_2), primarily the former, which then slowly oxidizes to the dioxide.

Sulfur and nitrogen dioxides (SO_2 and NO_2) dissolve in any water present to give the mineral acids of sulfurous/sulfuric and nitrous/nitric acids. The two forms of each acid are given because the dioxides initially dissolve in water to give the first, weaker, acid and then oxidize to the stronger, second acid.

Organic acids are formed by the partial oxidation of hydrocarbons. Normally, hydrocarbon oxidation is considered as going through to complete combustion with water and carbon dioxide as the final products. But combustion/thermal degradation can be partial, with hydrocarbon end groups forming carbonyl groups to make aldehydes, ketones, and carboxylic acids, the last as:

$$R-C=O$$
$$|$$
$$OH$$

Organic acids are not normally regarded as strong acids; acetic acid has a dissociation constant in water of 1.8×10^{-5} at 298 K and is regarded as a weak acid, the prime constituent of cooking vinegar. But various R-group substituents can increase the dissociation to make the acid stronger, such as for trichloroacetic acid. Two points to particularly consider for the strength of organic acids:

1. Acid dissociation constants increase with temperature; the higher the temperature, the stronger the acid.
2. The value given is for acetic acid in water. Acid:base interactions and equilibria are considerably different in other solvents, often making organic acids stronger.

Applying these to organic acids in degraded lubricants, the lubricant is a drastically different solvent to water, which also operates at high temperatures. As an example of the strength of organic acids, the railways originally lubricated their steam engine cylinders with animal fats before hydrocarbon oils were available. The high steam temperatures within the cylinders degraded the fats into their constituent organic acids, which corroded the metals present, particularly the nonferrous metals such as copper, lead, zinc, and so on.

The acidity generated within a degraded lubricant during its service life is a mixture of inorganic strong acids and weaker organic acids. This mixture is one of the causes of the analytical problems in determining the acidity of both the acids, and the remaining alkaline reserve added to neutralize that acidity, in a lubricant formulation. This is the need to determine the base number in a lubricant, both new and used. It is a standard analytical measurement for degraded lubricants.

2.3.4.3 Measurement of Base Number

An acid is normally associated with the bitter, corrosive, sometimes fuming in their concentrated form, properties of the mineral acids, classically sulfuric, nitric, hydrochloric, and phosphorous acids. There are others but these are the common mineral acids. Their common property is the ability to donate/give a proton (H^+) to a base. Sulfuric acid then becomes an anion, such as sulfate (SO_4^{2-}), nitrate (NO_3^-), chloride (Cl^-), or phosphate (PO_4^{3-}).

The common bases as alkalis, such as sodium hydroxide, caustic soda, potassium, and ammonium hydroxides are strong bases with sodium carbonate as a mild alkali or weak base. Again, as for the acids, there are many others but these are the commonly used alkalis. The common feature of alkalis is the hydroxide group (OH^-), which accepts the proton from the acid to form water (H_2O).

Aqueous acids and bases in equal amounts neutralize each other to form a neutral salt and water, as in the standard neutralization of hydrochloric acid by sodium hydroxide:

$$HCl + NaOH \rightarrow NaCl + H_2O$$

Whichever way this is done, by adding acid to alkali or the reverse, for equal amounts of acid and alkali, the end result is a neutral solution of pH 7. If the strength of one of the solutions is accurately known, then the concentration of the other solution can be calculated — basic chemical laboratory work. Neutralization is shown by an indicator with different colors in acid or alkaline solution, neutralization being shown by a color balance between the two forms. Litmus is one example of a neutralization indicator, being blue in alkaline and red in acid solution. Progress of acid/base titrations can equally be followed by other methods, such as:

- The pH electrode combined with the standard calomel electrode to follow either the solution pH or the potential difference in millivolts (mV) between the electrodes.
- The electrical conductivity of the solution between two platinum plate electrodes, because both the proton, H^+, and the hydroxide ion, OH^-, have high conductivities relative to other ions and both H^+ and OH^- are at a minimum at the end point, pH 7.

From these fundamental considerations, if the alkaline reserve (base number) of a degraded lubricant is a base, then it should be possible to titrate it against a standard acid solution to determine how much base is present; that is, the basis of base number determination, transferred over from water-based acid/alkali

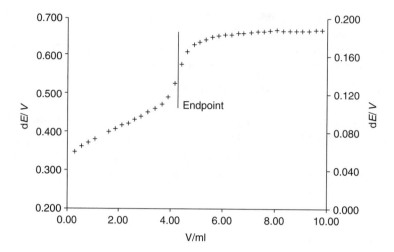

FIGURE 2.7   mV vs. volume plot for the titration of new/slightly degraded lubricants by the IP 177/ASTM D664 method.

neutralizations to the analysis of new and degraded lubricants in a variety of organic solvent mixtures. Many acids have been used to titrate the alkaline reserve in a lubricant but they give different values, particularly for heavily used samples.

2.3.4.4   IP 177/ASTM D664 — Base Number by Hydrochloric Acid Titration

This is a joint method developed by the Institute of Petroleum in the United Kingdom and ASTM in the United States and was the earliest method for measuring the base content of a new or degraded lubricant or hydraulic fluids. It is still preferred by some operators and has essentially been reintroduced by the IP 400 method; see Section 2.3.4.7 later, with the same solvent and acid titration system but with a different detection system.

The solvent for the titration of the lubricant/hydraulic sample must dissolve the sample and be compatible with the titrating acid. In this case, it is a mixture of toluene, isopropyl alcohol, and a very small amount of water. The acid is dissolved in alcohol and the two solvents are completely miscible. The progress of the neutralization reaction is followed using a combination of a glass electrode and the standard calomel electrode, a standard nonaqueous solvent analytical procedure. The signal used is the potential difference between the electrodes expressed as mV.

The neutralization works well for new and slightly used lubricants. The mV difference signal varies as a sharp sigmoidal form when mV is plotted against acid titration volume (Figure 2.7). The neutralization endpoint is at the mid-point of the sharp rise, as indicated. There is no problem with the analysis for new and lightly degraded samples, the neutralization curve is sharp, and the endpoint is clear.

Problems arise as more extensively degraded lubricants are analyzed. The clear form of the neutralization curve slowly degrades with increased degradation of the lubricant sample until its form is lost and there is no clear endpoint (Figure 2.8). A procedure is suggested where an endpoint value to work to is used instead, but this is an unsatisfactory solution.

There are several strong arguments against the use of the IP 177/ASTM D664 method for base number:

- The hydrochloric acid has an acid strength in the solvents used in this method, which only reacts with, and therefore determines "strong alkalinity," >pH 11, in the lubricant sample. It does not determine "mild alkalinity," up to pH 11, although it is not clear whether this is a crucial difference.
- The method has poor reproducibility, although this is improved by using the replacement ASTM D4739 method, which uses a very slow potentiometric titration, 15 min/1 ml acid reagent added — an extremely slow method.

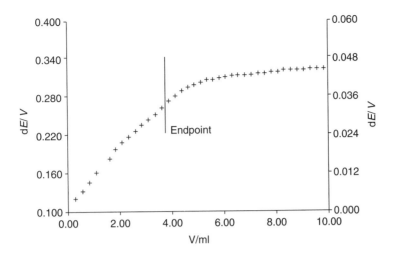

FIGURE 2.8 mV vs. volume plot for the titration of heavily degraded lubricants by the IP 177/ASTM D664 method.

- The sensitivity and fragility of the electrodes is important, the glass electrode is particularly fragile. Replacement glass electrodes must always be available, "conditioned" in the reaction solvent and ready for use. Another problem is that the electrode surfaces are gradually fouled by carbonaceous particulate in degraded lubricant samples and the electrode must be replaced.
- This method is not unique as against the others, but all base number methods use chemicals with various forms of hazards, which are expensive to dispose of. The formal method uses a large test sample, 20 g, in 120 cm³ of solvent, the volume of which is increased by the ensuing titration.

The test results are presented as milligrams of potassium hydroxide per gram sample equivalent. When applied to analyze successively degraded lubricant samples from engine bench or field tests, the IP 177/ASTM D664 base number method results tend to decline quickly in the initial stages of the test and then decline more slowly in contrast to results from other methods (Figure 2.9). It is generally held that a lubricant with a base number approaching a value of 2 should be replaced. Therefore, a base number of 2 for a degraded sample shows that its alkaline reserve equates to 2 mg potassium hydroxide per gram

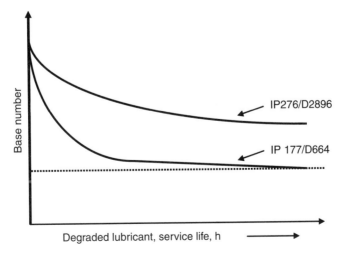

FIGURE 2.9 Base number degradation values for same successive lubricant samples by IP 177/D664 and IP 276/D2896.

of sample. While the titration uses hydrochloric acid, this is related to its equivalent as potassium hydroxide. To sum up, the IP 177/ASTM D664 method suffers from the following:

- Poor reproducibility, particularly for heavily degraded samples
- Lack of clarity in what it means
- Fragile apparatus
- Requiring large sample masses and solvent volumes

2.3.4.5   IP 276/ASTM D2896 Base Number by Perchloric Acid Titration

This method is really a modification of the previous IP 177/ASTM D664 method, arising from the perception that changing the titrating acid from hydrochloric to the stronger perchloric ($HClO_4$) will react with both strong and mild alkalinity in lubricant samples. It is argued that the results using perchloric acid reflect the total additive content of the formulation. To accommodate the change in acid, the solvent must be modified as well and is a mixture of chlorobenzene and glacial acetic acid. The detection method is the same as for IP 177/ASTM D664, a combination of the glass electrode and the standard calomel electrode. The titration is the same and the plot of mV against acid volume has the same sigmoidal shape as given in Figure 2.7.

Unfortunately, the method suffers from the same problems for heavily degraded samples, the plot then becoming indistinct with no clear endpoint, as in Figure 2.8. In this case, reproducibility is as poor as for the IP 177/ASTM D664 method. In this case, the method suggests a "back titration" with a much poorer range of reproducibility and repeatability. When this method is used to analyze degraded lubricant samples from engine bench or field tests, the IP 276/ASTM D2896 method base number results decline slowly throughout the test, in contrast to the results for the same samples analyzed using the IP 177/ASTM D664, as set out in Figure 2.6. There is no sharp decline in the initial stages of the test. There is a clear difference in results from the same samples between the IP 276/ASTM D2896 and the IP 177/ASTM D664 methods. As before, it is generally held that a lubricant with a base number approaching a value of 2 should be replaced. The test results have the same values as for the IP 177/ASTM D664 method. The solvents and chemicals used in IP 276/ASTM D2896 are even more hazardous and difficult/expensive to dispose of than those used in the preceding IP 177/ASTM D664 method. The following summarizes the IP 276/ASTM D2896 method:

- It is a modification of the previous IP 177/ASTM D664.
- It gives generally higher base number values, said to reflect the total, strong, and mild together, alkalinity present in a lubricant formulation.
- It has the same problem of an indistinct endpoint for heavily used samples.
- The solvents and chemicals used are hazardous and difficult/expensive to dispose of.

2.3.4.6   ASTM D974 — Base Number by Color Indicator

This method is worth noting but is now used relatively little. The method is very similar to IP 177/ASTM D664 method but uses a naphtholbenzein indicator color change to determine the neutralization endpoint. The results are expressed in the same way as IP 177/D664.

2.3.4.7   IP 400 — Base Number by Conductimetric Titration

IP 400 is relatively recent (there is no equivalent ASTM method) and directly addresses the problems of the previous methods. Chemically, it is identical to IP 177/ASTM D664 but the crucial difference is that it uses a conductimetric detection method to follow the progress of the neutralization reaction. It measures the resistance or its inverse, the conductivity, of a solution between two platinum plates rigidly held in a glass tube, shown both as a diagram and picture in Figure 2.10. The plates are typically 10 mm square, welded to platinum wires, which exit through the wall of the glass tube to external connections. The conductimetric probes are very robust and work as well when bright metal or when coated with carbonaceous particulate. The only problem with electrode contamination occurs when the carbonaceous

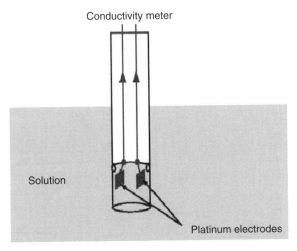

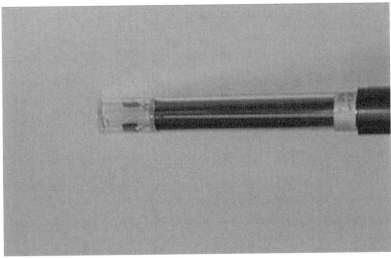

FIGURE 2.10   Diagram and picture of conductimetric cell.

particulate coats the wall of the glass probe containing the electrodes sufficient enough to cause an electrical short circuit.

The conductimetric cell does not need to be a special model. Excellent results can be obtained using standard cells as used in initial physical chemistry laboratory experiments. Special cells are only constructed for automated systems, which use small volumes of sample, solvent, and titrating acid solution.

The conductivity of the solvent plus sample is low, of the order of 2 μS (microsiemens) and increases linearly as the titration proceeds (Figure 2.7). At the endpoint, the gradient of the linear plot changes sharply. The endpoint is determined by the intersection of two linear plots, as shown in Figure 2.11.

The crucial point about IP 400 is that the quality of the endpoint does not change with the sample condition, either new, lightly used, or heavily used, as shown in Figure 2.11. The intersection of the linear sections moves to lower values as the alkaline reserves of the samples reduce. Reproducibility is good, within the limits set for IP 276/ASTM D2896, with no deterioration with sample use, as shown. The test results are very close to those obtained using the IP 177/ASTM D664 method under its best conditions, which is not surprising as the chemistry is the same. Smaller volumes are used, the sample weight specified by the IP 400 procedure is 5 g but very good reproducibility and repeatability have been achieved down to 0.1 g for small volume and unique samples. The titration is relatively quick compared to the previous

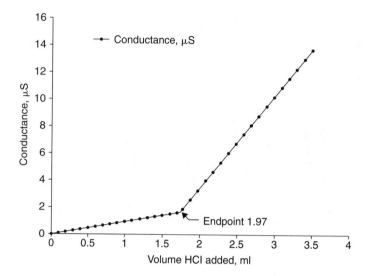

FIGURE 2.11   Conductivity vs. titration volume plot for IP 400.

potentiometric methods, for during the titration the solution conductivity stabilizes as soon as the added aliquot of acid is thoroughly mixed. The IP 400 procedure is simple and straightforward. This demonstrates that the problems of the previous base number methods are associated with the following:

- The potentiometric electrodes and their physical reactions.
- The potential difference titrations of a number of substances in the used lubricant samples against their total conductivity.

2.3.4.8   Precision of Base Number Determinations

The precision of these determinations has the following two forms, from cooperative test programs carried out between participating laboratories:

1. *Repeatability*, by the same operator and the same laboratory
2. *Reproducibility*, by different operators and laboratories

The format of precision is interestingly different from that normally encountered. It is set as a requirement that the results on the same sample should not vary by more than the stated limit values more than 19 cases out of 20, an interesting approach to a 95% confidence limit.

	Repeatability	Reproducibility
For IP 177/ASTM D664, base number		
By manual methods	7 mg	20 mg
Automatic methods	6 mg	28 mg

	Repeatability	Reproducibility
For IP 276/ASTM D2896, base number		
New lubricants	3%	7%
Used lubricants	24%	32% with back titration
For IP 400 base number		
New and used lubricants	$0.17x1/2$	$0.31x1/2$
where x is the average of the results		

Note that there is no distinction in precision between new and used lubricant samples for the IP 400 conductimetric method for base number determination.

2.3.4.9 Fourier Transform Infrared Spectroscopy Methods

Fourier transform infrared spectroscopy (FTIR) has been applied to tne analysis of degraded lubricant and hydraulic fluids. The method is not direct in the sense of reading a value from a scale. The analysis is conducted indirectly by first obtaining various parameters derived from the difference spectra between the time sample and the original lubricant. Multivariate analysis and principal component regression (PCR) are then applied to these parameters to determine the base number. The overall process is now well established as a technique for measuring used lubricant properties. It has considerable potential as a nonwet, relatively "dry" method, which does not need to use hazardous laboratory chemicals.

2.3.4.10 Summary for Base Number Measurements

For the reasons developed above, the base number value for a degraded lubricant is not a straightforward measurement. Any quoted values are not absolute and must be related to the method used to determine that value. The problems of the IP 177/ASTM D2896 and IP 276/ASTM D2896 methods lie with:

- Repeatability/reproducibility difficulties introduced by the potentiometric electrode reactions.
- The interpretative differences seen between "strong" and "weak" alkalinity.

The ASTM D974 colorimetric method is rarely used. The IP400 method is much better for base method measurement because of its clarity of endpoint, which is sustained for new, somewhat degraded, and heavily degraded lubricant samples. The FTIR difference analyses combined with chemometric statistical analyses can predict base number and show considerable promise with very substantial reductions in the use of solvents and reactants.

2.3.4.11 Sources and Effects of Acidity

In addition to acidity caused by combustion of inorganic compounds to give mineral acids, as described in Section 2.3.4.2, and also the oxidative degradation of hydrocarbon fuels and lubricants, hydraulic fluids will also degrade through localized high temperatures.

Localized high thermal stress on a hydraulic fluid will, in due course but over a considerably longer period than for lubricants, cause thermal degradation and oxidation. These conditions will cause the physical properties of the hydraulic fluid to go outside its specification and it must be replaced.

Acidity in degraded lubricant and hydraulic fluids corrodes system components. Corrosion combined with erosion gives enhanced wear rates, particularly in systems with mixed metals in contact by electrochemical effects. Corrosion can also generate solid debris within the system leading to clogging of tubes, filters, and obstruction of system operation. In collaboration with water, corrosion leads to rust formation. The areas that are prone to acidity attack are (i) bearing corrosion and (ii) cam and tappet corrosion and rusting.

It may be surprising that acidity and alkalinity can exist together in a used lubricant. It is again helpful to go back to the explanation given in Section 2.3.4.2. The effect of hydrocarbons as solvents with their low permittivities (dielectric constants) has the effect of giving a greater range of acidity and also alkalinity to the components present in the system. Acid-base interactions can range from (i) complex formation, AHB- of acid, AH, and base, B-, together, with the acid proton shared between the acid and base, and (ii) to the full transfer of the proton from the acid to the alkali as normally understood by "neutralization." In the former case, the substance can be both acid and basic (alkaline). Therefore, base and acid numbers can coexist in the same system. Generally, in a new automotive lubricant sample, the base number will be high, of the order of 6–10 KOH units and the acid number can be of the order of 0.5–1.0 units.

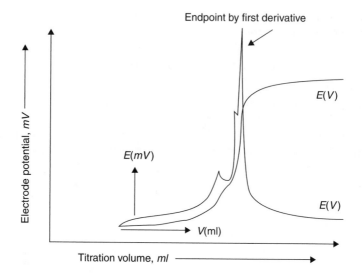

FIGURE 2.12   Acid number determination by the IP 177/ASTM D664 method, mV vs. volume plot and first derivative plot.

Acid number is the corollary of base number but has not been subject to the same level of controversy as described for base number. Only recently has there been difficulty with the acid numbers of synthetic ester lubricants used in gas turbine engines, addressed by a sampling and analytical error (SAE) method. Developments in applying conductimetric methods to the determination of acid number are also discussed as part of a method to determine base and acid numbers sequentially for the same sample in the same apparatus.

2.3.4.12   Acid Number Determination by IP 177/ASTM D664

This method is directly analogous to the base number determination described previously in Section 2.3.4.4. The solvents for the sample are the same: a mixture of toluene, isopropyl alcohol, and water with the titrant being potassium hydroxide in alcohol. The method follows the neutralization of the sample solution by alcoholic alkali by using the glass and standard calomel electrode pair, giving a millivolt potential difference between the electrodes against titration volume, V. The form of the titration is again a sigmoidal curve, with the endpoint at the change of gradient, the point of inflection, at the center of the sigmoidal plot. The endpoint is more clearly shown by the first derivative, $d(mV)/dV$ plotted against V (Figure 2.12), which is an appropriate repeat of Figure 2.7.

The test results are presented as milligrams of potassium hydroxide per gram sample equivalent, the same as for base number. The limits of repeatability and reproducibility are the same as for the determination of base number, as can be seen in the following table, as a requirement that the results on the same sample should not vary by more than the stated limit values of more than 19 cases out of 20.

	Repeatability	Reproducibility
By manual methods	7 mg	20 mg
Automatic methods	6 mg	28 mg

2.3.4.13   D974 — Acid Number by Color Indicator

Again, it is worth noting this method but it is used relatively rarely. It is very similar to IP 177/ASTM D664 but instead of a potentiometric method, it uses the color change of an indicator, naphtholbenzein, to determine the neutralization endpoint. The results are expressed in the same way, in milligram KOH per gram of sample.

2.3.4.14 Sampling and Analytical Error Determination of Acid Number in (Gas Turbine) Synthetic Ester-Based Lubricant

Gas turbine lubricants are subjected to high temperatures in the center of the engine. But these lubricants are not exposed directly to combustion gases, as in a reciprocating engine. The high temperature within the central engine bearings causes breakdown of the esters to give acids and polyhydric alcohols and their degradation products. The organic acids give acidity to the ester lubricants, which cause corrosion to the engine unless controlled. It is important that this acidity is controlled for current gas turbines as future engines will operate at even higher internal temperatures and cause even more acidity. The results for acid number of the SAE method can be addressed by scrupulous attention to experimental detail in IP 177/ASTM D664.

2.3.4.15 Simultaneous Conductimetric Determination of Base and Acid Numbers

The conductimetric base number determination of IP 400 involves dissolving the lubricant sample in a toluene/isopropyl alcohol/water solvent and then titrating that solution with an alcoholic solution of hydrochloric acid, to give the well-known plot of Figure 2.13, Conductivity vs. Titration Volume.

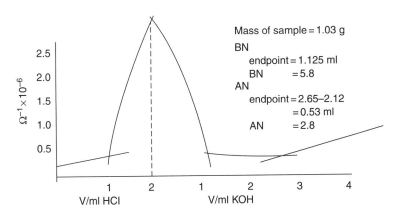

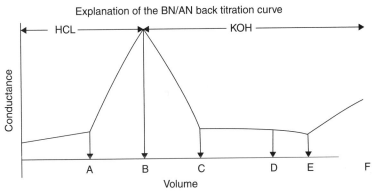

The titration curve is seen to pass through a maximum of six different regions. The region up to point B represents the addition of HCl and from point B to point F represents the addition of KOH.

FIGURE 2.13 Sequential conductimetric acid and base titration of lubricant sample.

The sector A-B-C in Figure 2.13 is the base number titration, exactly the same as an IP 177/ASTM D664 titration for base number. This plot can be reversed by the addition of alcoholic alkali, which gives an almost exact symmetrically reversed plot, sector C-D-E. Further addition of alkali then titrates the original acidic content of the lubricant sample, sector E-F-G as the acid number titration. This method gives results within the IP and ASTM limits for repeatability.

2.3.4.16 Relationship Between Acid and Base Numbers of Degraded Lubricants

The relationship between the acid and base numbers for a degraded lubricant sample were developed for higher sulfur fuels and previous additive packages. Thus, the general rules, which were that if the acid number rose to be greater than the declining base number, "crossing over," then this was a condemning limit for the lubricant charge. Further, if the base number declined below a value of 2, then this was a separate condemning limit for the lubricant charge.

However, the gradual move to "low" and "lower" sulfur fuels for diesel fuel and, separately, modern additive packages can extend a system's lubricant charge life. The condemning limits for degraded lubricants have changed considerably.

2.3.5   Water Content

Water commonly contaminates machinery lubricant and hydraulic systems; its presence reduces the load-carrying ability of a lubricant and increases wear. In addition, it promotes oxidation and corrosion. For synthetic polyol esters, water degrades the base stock back to its component acid and polyol. Maximum "safe" levels of water are usually taken to be 0.1–0.2%, higher for engines, lower for machinery and hydraulic systems. Water contamination of engine lubricant and hydraulic systems commonly arises from the following:

- Combustion water, recalling that hydrocarbon combustion gives carbon dioxide and water as products. Some of the water passes into the crankcase as "blowby" down the side of the piston and condenses at the lower temperatures of that region of the engine.
- Condensation of water in engines or hydraulic systems on standing or condensation into fuel tanks/hydraulic fluid reservoirs when operating at low/very low ambient temperatures.
- Leakage into the fluids from cooling systems, such as circulating cooling water in engines by gasket failure, or leakage within the matrix of a heat exchanger. Almost all heat exchangers leak to some extent; acceptable ones leak very, very little but the leakage rate eventually increases with corrosion to become significant.
- Water in lubricants degrades their formulation by absorbing acid gases to form strong acids. The presence of water in formulations can cause the additive package to precipitate out ("drop out") as a severe form of degradation, which leaves the base oil only to lubricate the system.

There are various methods to determine water in hydrocarbons and also lubricant and hydraulic fluids for the following reason:

- The different nature of relatively pure hydrocarbons, as fuels, and lubricant and hydraulic fluids as complex formulations.
- The different nature of the physical methods used to determine water in these fluids.
- The varying nature of water at different concentrations. Water is a very complex physical substance for which complete models have yet to be accepted. Many models have been proposed for the physical properties of water but it is clear that "bulk" water, as a large polymeric but transient structure, has different physical properties from smaller groups of water molecules or, indeed, individual water molecules. Indeed, the infrared spectrum of very dilute water in organic solvents is a relatively

narrow band centered on one frequency, whereas higher concentrations have appreciably wider bands at a shifted frequency.

From these considerations, the various viewpoints of the methods used to measure the water content of lubricant and hydraulic fluids can be appreciated.

2.3.5.1 The IP 74/ASTM D95 --- Water in Petroleum Products and Bituminous Materials by Distillation (the "Dean and Stark" method)

The IP 74/ASTM D95 ("Dean and Stark Method") for the determination of water in hydrocarbon fluids is a "total" method, rather gross and sensitive up to the 12% level. The method selectively distils water from petroleum products to separate and measure it using an organic solvent. It is an applied steam reflux distillation, which separates and concentrates the condensed water into a separate, calibrated, test tube (Figure 2.14).

One problem in measuring the volume of water is complete separation of the water and hydrocarbon in the calibration test tube, which can be clear (complete) or hazy (incomplete), dependent upon the nature of the fluids and additive components present. The glassware apparatus for the Dean and Stark distillation is shown in Figure 2.11. Note that the calibrated test tube in the system is positioned such that the water evaporated from the hydrocarbon fluid sample is collected and measured.

The Dean and Stark method can be seen as a "total" water determination method as it collects all of the water from the sample that can be volatilized. Its limitation is that it uses an equilibrium water distribution between the (sample + organic solvent) and the (water + organic solvent); thus, almost all of the water is removed to the measurement calibrated test tube. 100 g of oil sample is continuously distilled/refluxed with ~100 cm^3 of xylene, an aromatic solvent immiscible with water. The procedure is continued for 1.5 to 2 h to ensure that all of the water has been transferred. The percentage of water present in the sample

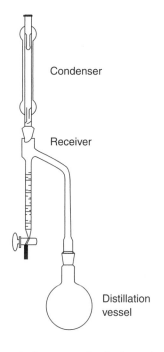

FIGURE 2.14 The Dean and Stark method apparatus for the determination of water in hydrocarbon fluids.

is expressed as the volume of water in the graduated test tube multiplied by 100% and divided by the mass of oil sample.

The method is direct with an unequivocal measurement of water but has the following disadvantages:

- Lack of sensitivity
- Occasional problems of measuring the water content because of incomplete separation of water/xylene in the measuring tube
- The time of measurement, upward of 1.5 to 2 h per sample
- Personnel intensive

2.3.5.2   IP 356/ASTM D1744 — Determination by Karl Fischer Titration

The Karl Fischer method of water determination is frequently discussed and results from it are often quoted. It uses the reaction of water with iodine and sulfur dioxide in a pyridine/methanol solution, which is unpleasant to use. Iodine in a methanol/chloroform solution is an alternative reagent.

The reagent reacts with hydroxyl groups, –OH, mainly in water but also in other hydroxylic compounds such as glycol, CH_2OH–CH_2OH, and depolarizes an electrode. The resulting potentiometric change is used to determine the endpoint of the titration and thus calculate the concentration of water in the oil sample. While the Karl Fischer method might be used to determine the water content of a formulated lubricant or hydraulic sample, it has never been approved for this purpose.

The method was originally developed to determine the concentration of water in crude oil and can be used to determine water in fuels. When used to determine water in new and degraded formulated lubricants and hydraulic fluids, the method overdetermines a "water response" because the reagent not only reacts with water but also some of the additives present. This is a problem because the Karl Fischer response for a new oil can give a blank value of 2%, mainly from the additive pack. But a failure limit for water in internal combustion engines is typically set at 0.2% or lower; thus, the failure limit is an order of magnitude less than the blank value. Worse, however, is the problem of the oil additives degrading during service life, which may form unknown compounds that may or may not react with the Karl Fischer reagent. The blank value is now in doubt for used samples. It can be estimated but this leaves a possibly large margin of error. Therefore, the Karl Fischer titration method for the determination of water in new and degraded formulated lubricant and hydraulic fluids is fraught with difficulty.

Some variations have been tried, such as gently sparging the oil sample with dry nitrogen and thus blowing the water content as vapor over into a Karl Fischer titration. This takes a long time to complete and it is uncertain at what point, if complete at all, "all" of the water has been transferred for measurement. The Karl Fischer determination of water in formulated new/degraded lubricant and hydraulic fluids is unsuitable because of reactions with additives. When additives are absent, then the Karl Fischer method is a sound method to determine the water content of "pure" hydrocarbon fluids, such as base oils.

2.3.5.3   Water Content by FTIR Spectrophotometry

The O–H group in water has a strong, broad, and distinctive infrared absorption from 3150 to 3500 cm^{-1}, centered on 3400 cm^{-1}. The absorption band is broad because the O–H group is hydrogen bonded for groups of water molecules. As the concentration of water decreases, it becomes less hydrogen bonded or exists more as smaller groups of bonded molecules or even individual molecules, and its molar absorption increases. Therefore, the calibration curve tends to be nonlinear at lower concentrations. A representative set of FTIR spectra for different levels of water contamination of hydrocarbon oils is given in Figure 2.15 and inspection demonstrates the nonlinearity of the water absorbance in the 0.0 to 0.2% concentration range.

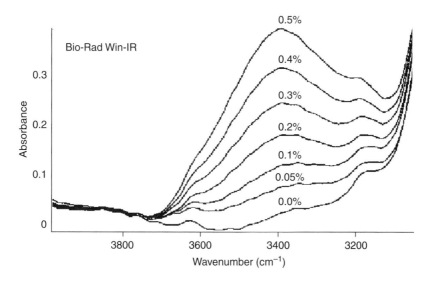

FIGURE 2.15 FTIR spectra of hydrocarbon fluids degraded with water. (From *Machinery Oil Analysis — Methods, Automation and Benefits*, 2nd ed., Larry A. Toms, Coastal Skills Training, Virginia Beach, VA, 1998. Courtesy Bio-Rad Laboratories.)

The method works well except for formulations using polyol ester base oils or with high dispersant/detergent additive levels. In the first case, the problem arises from the polyol ester infrared absorption in the previously used 3150–3500 cm^{-1} region, centered on 3400 cm^{-1}. Subject to detailed baseline corrections, the 3595 to 3700 cm^{-1} region is used instead to determine water contamination in these oil samples. This corresponds to a singly bonded O–H group. For high detergent/dispersant lubricant samples, the hydroxyl absorption band is not seen but a background increase in absorption occurs between 3000 and 4000 cm^{-1}. This effect is nonlinear and must be calibrated with standard solutions. It is separate from baseline shifts due to soot and particulates, which are unlikely to be present in this type of lubricant formulation. These two effects point to the main limitation of the FTIR method, which essentially reduces to the need to know the nature of the fresh, unused lubricant. This means that the FTIR method cannot be applied universally and will give errors occasionally, when samples of oils based on polyol esters or formulations contain high levels of dispersants/detergents. Other than this limitation, the FTIR method is very useful and shows great potential for the rapid and accurate determination of water degradation in many new and used lubricant formulations.

2.4   Minor Methods of Investigating Lubricant Degradation

Description of the following methods as "minor" means that they are only used in particular and individual circumstances to investigate the degradation of lubricants and hydraulic fluids. They do not form part of a routine investigation of degraded lubricant samples.

2.4.1   Density, "Gravity," or "Specific Gravity"

Density of a lubricant sample is also referred to as its "specific, or API, gravity" and has little value as a measure of the degraded lubricant's fitness for purpose. The determination of lubricant sample density is now readily measured to three significant figures using vibrating tube detectors, a much shorter and accurate procedure than the density bottle method or glass hydrometer, ASTM D1298.

But the density information gained has little importance because the density of a degraded lubricant should be close to that of the original material. Changes in density show contamination by a solvent, such as fuel dilution, a different product inadvertently added, or a build-up of foreign material. The differences are nevertheless small and, as an example, fuel dilution needs to be extensive to see a significant change in density.

2.4.2 Flash Point of Degraded Lubricant

Flash point determination of lubricant samples can now be considered more readily due to automated instruments being readily available. Both manual and automated methods are based upon the Pensky-Martins method, as in ASTM D93 for diesel lubricants. The method brings together considerations of volatility, combustion limits, and ignition temperatures to give a useful measure of great utility.

Flash point values of degraded samples rarely increase. If they do, a higher viscosity fluid has been inadvertently added. Much more likely is a decrease in flash point for a degraded lubricant sample caused primarily by fuel dilution resulting from cold/low temperature engine operation. Thermal decomposition of the base oil under extended power operation may also generate lighter fractions which reduce the sample flashpoint.

Reduced flashpoints of degraded diesel lubricants due to fuel dilution would normally be associated with a decreased viscosity value and a crosscheck should be done for this. The quantitative extent of fuel dilution is usually nonlinear with respect to flashpoint and should be measured by either gas chromatogrphy or FTIR methods. Various method procedures exist for increasing accuracy, such as the Cleveland open cup, ASTM D92, the Pensky-Martins closed cup, ASTM D93, and the Setaflash small scale closed cup, D3828. Flash points for degraded petrol/gasoline fuel dilution in degraded lubricants are measured by ASTM D322.

2.4.3 Foaming of Lubricants

Lubricant foam has a low load carrying ability. Excessive foam build-up in a reservoir or sump will rapidly lead to excessive wear and catastrophic failure of the system. Too high a level of lubricant in an engine sump, by overfilling or miscalibration of the level indicator (dipstick), causes the crankshaft and connecting rod big-end caps to whip up the lubricant into an all-pervading foam and rapid damage ensues. Air leaks into the oil flow or an open drop from a supply pipe into a hydraulic fluid reservoir can generate foam. Operationally, engines should not be overfilled, the level indicator correct, leaks stopped, and supply pipes extended to deliver return lubricant below the normal liquid surface level in a reservoir.

While base oils have little foaming tendency, modern lubricant formulations contain many additive substances such as detergents, which can enhance their tendency to foam. Surface active additives will also increase the foaming tendency of a formulation. ASTM D892 measures the foaming tendency of a hydrocarbon fluid but is much more relevant to the fresh, unused material under laboratory conditions than degraded samples in operating systems.

Resolving a used lubricant foaming problem should be treated with great care, but fortunately it is relatively rare. Foaming of the new formulation is controlled by the addition of liquid silicone polymers, which reduces the surface tension at the contact points of the foam cells. This allows the lubricant to drain away and the foam to subside. However, formulating the optimum silicone concentration requires extensive work as too little or too much silicone additive increases the foaming tendency of the formulation. Adding an antifoam silicone liquid *in situ* to a foaming, degraded lubricant or hydraulic fluid should be approached very carefully and incrementally.

2.4.4   System Corrosion ("Rusting") with Degraded Lubricants

Lubricant formulations contain rust inhibitors and a system which is maintained well, with a maximum water content of 0.1% for engines and 0.01% for other systems, should not have rust problems. Corrosion inhibitors are needed for systems and vehicles that are used intermittently, such as military vehicles in prolonged storage or vehicles delivered from one side of the world to the other. Repeated sequences of cold engine starts and very short drive distances up to dealer delivery causes condensation of water in the engine and ensuing corrosion of ferrous parts. The contaminants building up in a degraded lubricant system can adversely affect the action of the rust inhibitor present by competitive adsorption at metal surfaces.

The ASTM D665 rust method applies to lubricant formulations. If rust corrosion is either suspected or present by discoloration, then degraded samples should be sent for testing.

2.4.5   Demulsibility and Interfacial Tension of Degraded Lubricants

The demulsibility characteristic of a lubricant is its ability to separate from water when emulsions are formed in a system. While the test is performed for new lubricants, the build-up of trace contaminants may reduce the separation from water in emulsions for degraded samples, hence the term "demulsibility." Testing degraded lubricants for demulsibility in the laboratory may not be indicative of that sample's performance in operating systems. ASTM D1401 is the demulsibility test for turbine lubricants and ASTM D2711 is for medium- and high-viscosity lubricants. A slightly different procedure of ASTM D2711 is used for extreme pressure (EP) lubricants.

Interfacial tension (IFT) is a measure of the surface energy of a fluid against a solid surface or an immiscible standard fluid. Additives contribute to that surface energy and a decrease in interfacial tension indicates that these additives are being deactivated or removed in some way, or depleted by oxidation. A decrease in interfacial tension is an early indication of oxidation before changes are noticed in acid number or viscosity. Alternatively, the circulating lubricant is collecting certain compounds in the system added as rust inhibitors, which have polar structures.

Interfacial tension measurements of degraded lubricants are useful for rust- and oxidation-inhibited turbine and transformer oils by ASTM D971. The results should not be interpreted on their own but related and compared to changes in other measurements of the system, particularly viscosity and acid number.

2.4.6   Instrumental Analytical Techniques

The spectroscopic, chromatographic, and x-ray analytical techniques represented by FTIR, gas and liquid chromatography (GC/LC), and x-ray diffraction (XRD/XRF) are increasingly used to investigate degraded lubricants. The long-term trend is for the cost of the instruments to decrease and their resolution to increase with enhanced information technologies.

FTIR is increasingly used to analyze degraded samples, particularly for sequential samples compared to new, unused samples of the same lubricant. Selected regions of the infrared region are used to follow particular aspects of sample degradation. The method is given additional power through the use of multivariate data analysis. Toms describes the application of FTIR to the analysis of degraded samples.

Chromatography, particularly liquid chromatography, may be used to analyze additives in lubricant formulations. CEC has very high resolution of additives and can follow their depletion in successive degraded samples. Gel permeation chromatography (GPC) can follow the degradation or scission of polymer chain lengths and therefore mean molecular weights of additives such as VIIs, dispersants, and other polymeric additives.

X-ray diffraction is mainly used for quality control and to identify unknown deposits; of more importance is XRF and x-ray absorption fine structure (XAFS), used to identify the elements in compounds, liquids, and solids found in operating systems.

These instrumental techniques will have increasing importance in the analytical investigation of degraded lubricant samples. An important issue, already begun, is to bring these instrumental analytical techniques and their specific application developments into the set of standard analytical methods for the lubricant and hydraulic fluid manufacture and service use industries.

2.5   Case Studies of Degraded Lubricants

2.5.1   A Degraded Lubricant Sample from a Heavy-Duty Diesel Engine

	New	Used	Standard	Change
Appearance	Mid-brown, transparent	Black, opaque ←		
Odor	Mild	Diesel, slightly burnt ←		
Viscosity at 40°C, mm²/sec	71.31	61.82	D445	−13.3%
at 100°C	11.71	10.58		−9.7%
Viscosity index	160	162		+1.25%
Acid number, mg/g/KOH	2.8	4.5	D664	+1.7 units
Base number, mg/g/KOH	9.6	4.5	D4739	−5.1 units
Water	Nil	Nil	Crackle test	
Percentage of soot	0	1.2%	TGA ←	
Percentage of fuel dilution	0	4%	GC ←	
ICP Elements, mg/kg or ppm			D5185	
P	350	437		+24.9%
Zn	400	602		+50.5%
Ca	1100	1267		+15.2%
Ba	<1	<1		
B	129	90		−43.3%
Mg	10	15		
Na	<1	4		
Fe	<1	9		
Al	<1	<1		
Cr	1	<1		
Cu	<1	2		
Pb	<1	8		
Sn	<1	4		
Ni	<1	<1		
Mo	<1	<1		
Si	6	25		

The arrows conjoin the results for appearance, odor, soot, and fuel dilution to show how qualitative direct observation is supported by instrumental analyses.

Overall Comment and Judgement: This degraded lubricant is still fit for service despite the identical acid and base numbers. Once used as a "condemning limit," the crossing of acid and base numbers no longer means the end of a lubricant charge. Low sulfur fuels and modem additives can greatly extend the life of a lubricant charge in a system.

The viscosity decreases at both temperatures are just within condemning limits and need to be followed carefully in future sampling. The additive elements P, Zn, and C have increased, probably due to volatilization of base oil and the associated decrease in volume. The increase in silicon is considerable and is queried for a maintenance check — is the air filter allowing dust into the engine system?

Overall, this oil can remain in the engine system as still being "fit for service" and able to protect it but will probably require replacement at the next sampling interval.

2.5.2 A Degraded Grease Sample

	New	Used	Standard	Comment
Drop point, °C	181	143	D2265	Substantial decrease ←
Penetration test				
Unworked	310	450	D217	
Worked	319	459		Softened ←
Consistency, NGLI	1	000	D217	Oil Separation?
Appearance	Bright	Dull		
Texture	Smooth	Coarse ←		
Color	Mid-brown	Black ←		
Odor	Very mild	Unpleasant, oxidized/burnt ←		
Water (% by weight)	0.0	0.18	D95	
Ash (% by weight)	2.7	4.2		
Insolubles (% by weight)	0.06	2.6	D128	Wear metal debris and ingress of dirt
Changes in wear elements by ICP Pb		Iron — major increase Al — minor increase Pb — minor decrease Si — traces detected		
Soap elements, mg/kg or ppm				
Ca	3000	5900 +97%		
Na	100	400 +300%		
K	<10	<10		
Li	13,000	6100 — 53%		
Pb	16,500	7300 — 56%		
FTIR analysis	Normal soap	(By microscope FTIR) Separated regions ← of partially oxidized hydrocarbon oil, ← lithium, and calcium soap strand regions		

Overall Comment and Judgement: The analyses of this degraded grease sample indicate that it is no longer fit for service and ought to be replaced immediately. Further consideration is warranted if it is the right grease for this purpose (see below).

The first seven observations, from "Drop point" down to "Odor" indicate to an experienced operator that the grease is failing in its purpose. The grease is beginning to separate into a heterogenous gel, from a smooth to a coarse consistency, shown by its decreased NGLI classification. The microscope FTIR of the oil areas shows oxidation, possibly from overheating. This is borne out by the unpleasant/burnt smell and change to black in color from mid-brown, indicating overheating.

The ash and insolubles values indicate that the grease is both collecting wear debris and picking up dirt. It is failing to protect against wear, shown by the increased Fe levels, and dirt is being ingested, from the increased Na and trace Si, possibly through faulty seals — a maintenance issue.

The soap elements show contrary trends, calcium and sodium have increased, possibly concentrated by loss of oil content from overheating. In contrast, the lithium soap thickener and lead EP compounds are decreasing, reflected in the decreased lead found in the spectroscopic analyses.

Overall, there are conflicting indicators in this set of analyses. Clearly, this grease formulation is not doing well in this application. An experienced operator examining this analysis, and reviewing previous analyses, might well consider if this grease is the correct grade for the job. Further technical advice is needed and should be sought.

2.5.3   A Degraded Lubricant Sample from a Gas-Fueled Engine

	New	Used	Standard	
		Comment ←		
Appearance	Clear	Reasonably clear ←		
Odor	Mild	Slightly acrid/burnt		
Viscosity at 40°C, mm²/sec	125.6	163	D445	+29.8%
at 100°C	13.3	15.6		+17.35
Viscosity index	101	97		−4.0%
Flash point, °C	>200	>200	D92	
Acid number, mg/g/KOH	0.92	1.78	D664	+0.86 unit
Base number, mg/g/KOH	5.8	2.6	D2896	−3.2 units
UIUts				
Water	Nil	Nil	Hotplate and crackle test	
Insolubles	0	0.1%	D893	
FTIR oxidation	Nil	Slight ←		
Nitration	Nil	Slight ←		
ICP Elements, mg/kg or ppm			D5185	
P	500	470		
Zn	560	510	No major changes for any	
Ca	2500	2100		
Ba	<1	<1		
Mg	370	365		
Na	<1	123		
Fe	<1	55		
Al	<1	25		
Cr	<1	<1		
Cu	<1	33		
Pb	<1	5		
Sn	<1	2		
Ni	<1	6		
Mo	<1	3		
Si	5	25		

Comment: This degraded lubricant is still acceptable for service. Although thickened, its viscosity index has declined slightly. Base number has depleted and should be monitored carefully but has not decreased below the acid number value. It is interesting that the elemental data results have no major changes, the additive element concentration levels are stable, in contrast to the heavy duty diesel sample analyses presented in Case Study 2.5.1. This may be due to a combination of a concentration effect, from volatility of the lubricant formulation, and that ICP elemental analysis does not discriminate between active and inactive ("spent") species. Note that the slightly different analytical parameters used for the overall analyses of degraded heavy duty diesel lubricants samples and those for a gas engine degraded sample.

2.5.4 A Degraded Hydraulic Fluid

	New	Used	Standard
Appearance	Bright	Dark brown ←	
	Clear	Not clear	Extensive degradation ←
Color	2.9	7.5 by dilution	D1500
Viscosity at 40°C, mm^2/sec	58 7.3	70	D445
at 100°C	7.3	8.8	
Acid number, mg/gKOH	1.0	2.1	D664
Water	Nil	Nil	Crackle test
Percentage of insolubles, gravimetric,	15	270	Substantial ←
(as mg/l) increase			increase
n-Pentane	0	2.5	D893A
Toluene	0	0.5	
Resins, by difference		2.0	Substantial increase ←
ICP elements, mg/kg or ppm			D5185
Additive elements			
P	550	390	→ −29.9%
Zn	560	402	→ −28.2%
Wear Metal Changes			
Fe	<1	52 ←	
Cu	1	17	
Na	<1	21	
Pb	<1	8	
Cr	<1	3	
Sn	<1	2	

Overall Comment and Judgement: Hydraulic fluids normally remain much cleaner and less degraded than automotive lubricants. This degraded sample is, however, the exception and has several independent condemning features. The service age of this sample is unknown.

The color has changed from bright/clear to dark brown/not clear. Color measurement on the degraded sample can only be done by dilution and the "not clear" report indicates extensive suspended particulates.

The acid number is concerning; to have increased by 1.1 units from the (normally) low value of 1.0 when usual practice sees a decrease indicates that oxidation is occurring. The insolubles by pentane/toluene comparison show a substantial production of resins, again related to oxidation, paralleled by the increase in particulates determined by gravimetry.

The viscosities at both 40 and 100°C have increased by over 20%. The additive elements are declining in a uniform manner and the wear elements show an increase, mainly in iron, supported by increases in other minor wear elements.

This hydraulic fluid has come to the end of its effective life, possibly beyond that, and needs rapid replacement for the continued mechanical health of the machine it operates. Overall, the sampling interval for fluid analyses should be reviewed to consider a shorter time. Or a regular sampling and analytical program should be set up for this machine.

2.5.5 Overview of Degraded Lubricant Analyses

Review of the four examples given shows that each suite of analyses has been specifically targeted for the particular lubricant's service application. A complete suite of analyses is neither needed nor cost-effective, only those that give an insight into the extent of degradation of the lubricant, that is, its "condition are neeeded."

The use of conjoining arrows in each set of analyses shows how various effects are connected together by causal factors. Therefore, decisions on the extent of a lubricant's degradation and subsequent action should be broadly based to cover as many independent variables as possible.

Degradation of lubricants will always occur but this can be controlled and contained by a suitably frequent program of sampling and appropriate analyses. With increasing reliability of machines occurring and expected, the monitored degradation of the lubricants for these machines is now essential.

Bibliography

ASTM (www.ASTM.com) and IP (energyinst.org.uk) method numbers have been given in text and are readily accessed from the appropriate websites. Note that although the IP website is derived from the former Institute of Petroleum, hence "IP," this Institute merged to form the Energy Institute in 2004.

There are many publications concerning aspects of lubricant degradation among other associated topics and a list would be extremely long. Very good introductions to lubricants and their degradation are contained in:

1. *Machinery Oil Analysis — Methods, Automation and Benefits*, 2nd ed., Larry A. Toms, Coastal Skills Training, Virginia Beach, VA, 1998.
2. *Automotive Lubricants Reference Book*, A. Caines and R. Haycock, 1st ed. published by SAE, 1996 and 2nd ed. by Mechanical Engineering Publications Ltd, London and Bury St Edmunds, UK, 1996.

These books as well as the ASTM and IP websites give an excellent introduction to studying the degradation of lubricants in their various formulations.

References

[1] S.B. Saville, F.D. Gainey, S.D. Cupples, M.F. Fox, and D.J. Picken, A Study of Lubricant Condition in the Piston Ring Zone of Single Cylinder Diesel Engines Under Typical Operating Conditions, SAE 881586 (1988).

3

Lubricant Properties and Test Methods

3.1   Introduction

Proper lubrication of mechanical machinery is essential for the high reliability and maximum usable life of the machine. Thus, associated lubricating systems and fluids must be properly selected and managed. Lubricating oils and greases are formulated for a wide range of industrial, marine, and commercial applications and consequently are required to fulfill a wide range of specific properties and functions. It follows that the range of test methods utilized to quantify these characteristics is also broad.

Standard test methods for lubricants have evolved from a variety of sources including: the National Lubricating Grease Institute (NLGI) [1]; American Petroleum Institute (API) [2]; Society of Automotive Engineers (SAE) [3]; Coordinating Research Council, Inc. (CRC) [4]; Federal agencies (Army, Navy, Air Force); instrument manufacturers; and original equipment manufacturers (OEM) [5–8]. Most of the test methods utilized across industrial fields are described in the International ASTM [9,10] "Standards on Petroleum Products and Lubricants" Volume 05. A number of these methods are also joint standards with the API, Institute of Petroleum (IP) [11], Deutsches Institut fur Normung e.V. (DIN) [12], American National Standards Institute (ANSI) [13], and International Standards Organization (ISO) [14].

In this chapter, the test methods presented are primarily those most frequently used from the array documented by the International ASTM Standards on Petroleum Products and Lubricants. To facilitate coverage of the wide variety of test types, two primary testing categories [15] are presented:

- Performance tests that define and quantify critical fluid properties, performance parameters, and specifications. These methods are generally utilized for new-oil specification, manufacturing quality control, and user acceptance for specific machinery applications.
- Condition tests that measure operational parameters such as the accrual of fluid contamination and degradation characteristics. These test methods are generally utilized for in-service oil monitoring.

The performance tests are subdivided into oil and grease applicability and then into the test methods specific to the respective lubricant properties or characteristics associated with the category. It is not the intent to cover all test methodology in detail, but rather provide a brief description of the most popular methods and practices with focus on the (1) property or characteristic parameter being measured, (2) the test methods and apparatus employed, and (3) the significance of the test and its results. If operational details are desired, reference should be made to the original source of the tests, that is, ASTM, etc. For ease in locating specific tests, they are presented alphabetically in each category.

3.2 Lubricant Performance Tests

A modern lubricant is expected to operate efficiently over a wide range of temperatures and stresses while in contact with machinery metals, many different chemical compounds, and other debris resulting from combustion or other processes. Oil performance tests have been designed by the petroleum industry and OEM to define the performance or properties of oil products required for proper operation in each class of machinery application and for product manufacturing quality assurance and control (QA/QC). In addition, machinery owners often utilize selected performance parameters to determine the suitability of a specific new-oil product for use in a specific machine and to verify bulk product specifications upon delivery.

3.2.1 Lubricating Oil Tests

3.2.1.1 Additive Metals in Unused Oils — ASTM D4628, ASTM D4927, ASTM D4951, ASTM D6443, ASTM D6481

Most lubricating oils contain additive compounds that contain metals. These include detergents, dispersants, corrosion inhibitors, antioxidants, antiwear, and extreme-pressure agents. In unused fluids, the concentration of metals associated with additive compounds is a rough measure of the additive concentration.

The ASTM D6443 determines the level of calcium, chlorine, copper, magnesium, phosphorus, sulfur, and zinc in unused lubricating oils by wavelength dispersive x-ray fluorescence spectrometry. Matrix effects (interelement interferences) are reduced by regression software. Results are reported in mass percent. High-concentration additives may be diluted to produce values within the calibration range of the instrument. Alternatively, the following methods may be utilized for determining additive metals in unused lubricating oils. Barium, calcium, phosphorus, sulfur, and zinc may be determined by ASTM D4927 wavelength dispersive x-ray fluorescence spectrometry. Phosphorus, sulfur, calcium, and zinc may be determined by the ASTM D6481 test utilizing an energy dispersive x-ray fluorescence spectrometer. Barium, calcium, magnesium, and zinc may be determined by ASTM D4628 test utilizing atomic absorption spectrometry. Barium, boron, calcium, copper, magnesium, phosphorus, sulfur, and zinc may be determined by utilizing the ASTM D4951 atomic emission spectrometry by an inductively coupled plasma spectrometer. Test methods for determining the level of chlorine and sulfur in lubricating oils are given below.

3.2.1.1.1 Comments
The tests for determining additive metals are applicable only to unused lubricants and are not appropriate for evaluating the remaining additive concentration in in-service oil, as the metal component of the expended additive may remain present in the oil. In addition to these instrument methods, there are a number of single element methods such as the ASTM D1091, which determines the value of phosphorus.

3.2.1.2 Aniline Point — ASTM D611

Hydrocarbon oils have a tendency to cause swelling of rubber parts such as elastomer hoses, seals, or gaskets. The aniline solubility temperature is a useful method to determine the compatibility of a lubricant with rubber parts. In the ASTM D611 test, specified volumes of the chemical aniline (a benzene derivative) and the oil sample are placed in a tube and mechanically mixed. The mixture is heated until the two phases become miscible. The mixture is then cooled at a controlled rate. The test endpoint is reached when the mixture separates into two phases. The temperature at the moment of phase separation is recorded as the aniline point for the oil sample. The measurement exhibits an inverse relationship in that the lower the aniline point, the greater the tendency for rubber swelling. Oils that are high in aromatics will have the highest solvency to aniline, and thus the lowest aniline point. Paraffinic oils have the least solvency and the highest aniline point. Napthenic, cycloparaffin, and olefin based oils exhibit an intermediate solvency to aniline and thus have intermediate aniline points.

3.2.1.2.1 Comments
The main significance of the aniline point test is the determination of the swelling tendency of a given organic elastomer from exposure to the oil being tested. In this regard, the test method is only applicable for temperatures that fall between the freezing and boiling points of the sample mixture.

Applicability: hydraulic fluids; steam, hydro, and industrial gas turbine oils; and electrical insulation oils. International equivalent standards include DIN 51775, IP 2, and ISO 2977.

3.2.1.3 Ash Tendency Test — ASTM D482, ASTM D874
Modern internal combustion engines require lubricant formulations that reduce the formation of carbonaceous deposits and sludge. The most common agents for deposit reduction are metallic detergent and dispersant additives containing compounds of calcium, magnesium, potassium, zinc, sodium, and barium. A rough measure of the level of these compounds in new-oil blends is provided by the ash content or sulfated ash content test. The simplest measurement is the ASTM D482 test where the ash content is the quantity or mass of incombustible solids left over after burning a specified quantity of oil for 10 min at a temperature of 800°C. The ash will include metallic additives and any contaminants containing metals. The ASTM D874 test is an improved method for determining the ash content of lubricating oils. In this method, the sample is burned until only the carbon residue and metallic ash remain. The residue is treated with sulfuric acid, reheated, and weighed. The sulfated ash test indicates the concentration of the metal-based additives in new-oil blends.

3.2.1.3.1 Comments
For oils containing a magnesium-based detergent and a boron-based dispersant, the sulfated ash measurement will be unreliable. The ash test is also unreliable for used-oil measurements. In this case, the ash test results will also include contaminant compounds and wear metals in addition to the desired additive compounds. The test does not distinguish between consumed and remaining usable additives — the measured additive compounds will include both reserve (not-yet-used) and consumed additive constituents, generating a concentration similar to new-oil. The trend toward ashless additive compounds renders this test a dubious measurement of detergent level.

Applicability: The ASTM D482 is utilized for petroleum fuels, crudes, and lubricating oils. The ASTM D874 is generally utilized for unused lubricating oils.

International equivalent standards include (D482) IP 4 and ISO 6245 and (D874) IP 163 and ISO 3987.

3.2.1.4 ASTM Color — ASTM D1500
The color of a lubricant provides no direct indication of its lubricating properties; however, a change in color does signify a change in chemistry or the presence of a contaminant. Lubricant color assessment is normally based on transmitted light, and the common color scales used for color interpretation are based on this principle. The basic assumption of the color test is that oil color can be related to deterioration or degree of refining and that any color change is due to deterioration or contamination. New-oil color is determined by the ASTM D1500 colorimeter method. The test compares the transmitted light of an oil with a set of standard glass color slides under controlled conditions. Note that new-oils complying with specific legislation (e.g., special tax exemptions) may have added dyes to indicate the special classification or status.

3.2.1.4.1 Comments

The ASTM color test is only applicable to mineral oils that do not have dyes or other materials that change the oil's natural color. The test is a rough measure of water and oxidation by-products in inhibited steam and hydro turbine lubricating oils.

International equivalent standards include IP 196, ISO 2049, and DIN 51578.

3.2.1.5 Carbon Residue Characteristics --- ASTM Dl89, ASTM D524

When oil is subjected to evaporation or is exposed to atmospheric air for long periods, oxidation of the nonvolatile constituents of the oil results in a carbonaceous residue. Because other oil ingredients present in lubricating oils often produce non-carbonaceous deposits, it is desirable to determine the ash content as well in order to establish the true value of the carbon content.

The Ramsbottom (D524) and Conradson (D189) tests are used to determine carbon residue characteristics. In the more common Conradson test, the oil sample is heated in a porcelain crucible placed within a special steel crucible for a prescribed period of time. The increase in weight of the porcelain crucible indicates the amount of carbon residue. In the Ramsbottom test, the sample is heated at a specific temperature in a glass bottle with a small opening until all the volatile ingredients have evaporated and "coking" and "cracking" have further decomposed the oil. The weight of the remaining material is the carbon residue.

3.2.1.5.1 Comments

The degree of carbon residue is of secondary importance for well-refined, quality-controlled lubricants and so the carbon residue test is not frequently performed. In addition, the test is of little value for synthetic lubricants with their exceptional thermal stability. In automotive applications, engine design, fuel selection, and operating conditions are as important in carbon deposition as the intrinsic carbon residue content of the oil. Consequently, the test is generally performed during the selection of hydrocarbon oils for such applications as heat-treating, air compressors, and high-temperature bearings.

International equivalent standards include (Dl89) IP 113, DIN 51551, and ISO 6615 and (D524) IP 14 and ISO 4262.

3.2.1.6 Chlorine Content --- ASTM D808

Chlorine compounds are deleterious to the environment and significant efforts have been undertaken to reduce the amount of these compounds in machinery fluids. Thus, in those machinery applications where it may be desirable to determine the presence of a chlorinated compound or contaminant, the following test may be utilized. The D808 method utilizes a pressure vessel that provides a gravimetric determination of chlorine content. The method is suitable for either new or used oils as well as greases. The presence of other halogen compounds will obscure the results.

3.2.1.6.1 Comments

These test methods only determine the amount of chlorine present and do not indicate fluid condition or effectiveness in a particular application. Note that chlorine is an oxidizer. For most machinery fluid applications, chlorine compounds are contaminants and their presence in any concentration will be considered harmful to machinery operations. For any machinery fluid still utilizing a chlorine-based additive, the optimum concentration must be monitored: too much may produce excessive corrosion, while an insufficient quantity may limit the fluid's effectiveness.

3.2.1.7 Copper Corrosion Resistance --- ASTM D130

Many lubricated machines are comprised of parts containing copper alloys. It is therefore essential that copper components be properly lubricated and that the oil be noncorrosive to metallic copper. Most petroleum crudes contain sulfur compounds, some of which are corrosive to copper. During the refining of high quality lubricants, most of the corrosive compounds of sulfur are removed. However, some lubricant types are blended with sulfur-based emulsifying or extreme-pressure additives, which can also corrode copper surfaces. The degree of protection from copper corrosion provided by a lubricant is determined from the ASTM D130 test. In this test, a specified volume of oil is placed in a covered beaker at a temperature of 125°C. A polished copper-strip specimen is immersed into the oil for a specified period (usually 2 h) after which time it is checked for the degree of tarnishing or corrosion. The degree of corrosion is determined by comparison of the copper specimen with a set of ASTM (tarnish) standards.

3.2.1.7.1 *Comments*

The copper corrosion resistance test method is suitable for hydrocarbon products having a Reid vapor pressure (D323-99a) no greater than 18 psi. The test will provide a rough measure of the corrosion prevention characteristics for oil utilized in machinery to lubricate native bronze and brass parts. The test is also suitable for determining the best candidate from a series of potential oil products. The test is not suitable for determining the overall corrosion prevention characteristics of lubricating oils.

International equivalent standards include ISO 2160, IP 154, and DIN 51811.

3.2.1.8 **Density and Specific Gravity — ASTM D1217, ASTM D1298**

Density, a numerical expression of the mass-to-volume relationship of a fluid, is sometimes used as a measure of lubricant composition consistency or production uniformity. Density is also used as a rough indicator of hydrocarbon type or volatility. In the petroleum industry, density is usually expressed as kilograms per cubic meter (kg/m^3) at 15°C. Kilograms per liter (kg/l) is also used.

The ASTM D1217 test determines density and specific gravity of pure hydrocarbons or petroleum distillates that boil between 90 and 110°C. The test utilizes the Bingham pycnometer at the specified test temperatures of 20 and 25°C.

The ASTM D1298 test determines the density, specific gravity, and API gravity values of petroleum and non-petroleum mixtures. The test is performed in a controlled environment and results are read off a glass hydrometer at convenient temperatures, and corrected to the standard temperature, 15.56°C (60°F), by means of the "petroleum measurement tables" issued by ASTM International and IP. The API gravity reading is determined from the hydrometer reading by the formula:

$$\text{API Gravity} = \frac{141.5}{\text{sp. gr. } 60/60°F} - 131.5$$

3.2.1.8.1 *Comments*

Density and specific gravity are generally utilized for establishing weight and volume factors for shipment, storage, delivery, and quality assurance. The American Petroleum Institute and ASTM provide API gravity and temperature corrections conversion tables that include density and specific gravity. Note that paraffinic oils exhibit lower specific gravity than base oils that contain naphthenic or asphaltic components. Specific gravity measurements used in combination with other measurements (ASTM D2501) provide some indication of the hydrocarbon composition of oils. Such determinations, however, should be limited to oils that are predominately hydrocarbons.

International equivalent standards include (D1298) IP 160, API 2547, DIN 51757, and ISO 650.

3.2.1.9 **Emulsibility, Demulsibility Characteristics — ASTM D1401, ASTM D2711**

A highly refined petroleum lubricant resists the tendency to form emulsion mixtures with water and will generally phase-separate upon standing. However, in pressurized circulating lubrication systems, the mechanical action of the pump and other components can cause oil and contaminating water to form an emulsion. Moreover, the system flow rate may be high enough to prevent sufficient standing time in the reservoir to allow phase separation to occur. The water separability or demulsibility property of a lubricant is determined from the ASTM D1401 test. In this test, 40 ml of oil sample and 40 ml of distilled water are placed in a graduated cylinder and stirred for 5 min at 54°C to create an emulsion mixture. The degree of separability is indicated by the time required for the oil/water phase separation to take place. The progress of the oil/water phase separation is measured at 5-min intervals. The D2711 test is utilized for medium- and high-viscosity lubricating oils.

3.2.1.9.1   Comments

Oil emulsions make poor lubricants. Persistent emulsions will increase component wear, oil oxidation, and sludge and varnish deposits that can in turn clog filters, reduce cooling efficiency, and promote corrosion. Bearing, gear, and hydraulic systems that are prone to water contamination require good lubricant demulsibility to prevent severe wear damage to moving parts. Oils utilized for these components should be verified for oil–water separability before recommendation for service.

International equivalent standards include (D1401) IP 19, DIN 51599, and ISO 6624 and (D2711) DIN 51353.

3.2.1.10   Extreme-Pressure Properties of Lubricants — ASTM D2782, ASTM D2783, ASTM D3233, ASTM D6121, ASTM D6425

One of the most important attributes of lubricating oil is the ultimate load that can be sustained without seizure or scoring of the lubricated sliding surfaces. The "seizure" condition relates to welding or fusion of metal asperities on the rubbing test pieces, while the "scoring" characterizes the nature (furrowed scar) of the seizure.

ASTM D3233 utilizes a Falex pin and vee block apparatus to determine the load carrying properties of fluid lubricants. The Falex device utilizes a pin rotating against a pair of v-shaped blocks immersed in the lubricant sample. The tester rotates the steel pin at 290 rpm. The load is continuously increased by a ratchet mechanism in 250 lbf (1112 N) steps until the end of the test. In both cases, the result is the load at which pin failure occurs.

ASTM D2782 utilizes the Timken apparatus to determine the extreme pressure of lubricating fluids. In this method, the test machine utilizes a steel test cup rotating at 800 rpm against a steel block lubricated with the test sample. Addition or removal of weights varies the load between the cup and the block. The test results are indicated by the maximum weight at which no scoring occurs.

ASTM D2783 utilizes the Falex four-ball test machine to determine the wear properties of extreme pressure lubricants. The four-ball tester employs a steel ball rotating at 1760 rpm against three stationary steel balls lubricated with the test sample. Addition or removal of weights is used to vary the load between the rotating and stationary balls. The test results are reported as the dimensions of scar marks made on the ball surfaces. Interpretation of the scarring up to seizure determines the load/wear index and the weld point.

The ASTM D6121 test, commonly referred to as the L-37, evaluates the load-carrying ability of extreme-pressure (EP) lubricants under low-speed and high-torque conditions common in final hypoid drive axle applications. The final axle assembly is run in for a specified period at prescribed load, speed, and temperature after which the test phase is operated for 24 h at a prescribed load, speed, and temperature. The bearings and gears are evaluated for wear, corrosion, and deposits in accordance with CRC manual 17.

The ASTM D6425 standard test for measuring friction and wear properties of EP lubricating oils utilizes the SRV test machine. This test determines the oil's coefficient of friction and its ability to resist wear in machinery applications subjected to high-frequency, linear oscillation motion. The SRV machine utilizes a test ball oscillated at constant frequency, stroke amplitude, and load against a test disk. The result is determined from the mean wear scar diameter.

3.2.1.10.1   Comments

Friction machine tests are a proven means for assessing the load carrying capacity of lubricants and the influence of various antiwear, antiscuffing, and extreme-pressure additives used in bearing and gear lubricants. Since these tests are intended to evaluate relatively low-viscosity oils for circulating systems, they are generally not suitable for evaluating compounded fatty lubricants used for open or hypoid gears. In addition, due to the subjective nature of the results, precise conclusions are not possible. However, these tests are useful screening tools and when supplemented with other test methods that provide antiwear and corrosion characteristics, type of agent, etc., a better interpretation of the relative overall lubricant performance may be made. For precise data, antiwear and extreme-pressure characteristics measurements should be performed on the specific bearings and gears and these should be operated under the same load, speed, temperatures, etc. that are representative of the intended machinery application.

International equivalent standards include (D3233) IP 241, (D2782) IP 240, (D2783) IP 239, and (D6425) DIN 51834.

3.2.1.11 Flash and Fire Points — ASTM D56, ASTM D92, ASTM D93, ASTM D1310, ASTM D3828, ASTM D6450

The flash and fire points of a lubricant are measures of the fluid's volatility and flammability. The flash point refers to the minimum temperature at which there is sufficient vapor to cause a flash of the vapor/air mixture in the presence of an open flame. The fire point is the minimum temperature at which the production of vapor is sufficient to maintain combustion. When selecting a lubricant for a given application, the main concern for the lubrication engineer is the assessment of its potential for explosion or fire under the anticipated conditions of operation or storage. Consideration must also be given to the potential for contamination by more volatile fluids, such as the dilution of an engine lubricant by fuel, which greatly increases the potential for damage due to explosion or fire. The flash and fire points of a lubricant are determined from several tests including the following.

ASTM D56 flash point by "Tag closed cup" measures the flash point of liquids with a viscosity below 5.5 cSt (mm^2/sec) at 40°C (104°F) and a flash point below 93°C (200°F). The specimen is placed in a closed cup and heated at a slow constant rate. At specified intervals an ignition source is lowered into the cup. The lowest temperature, corrected to standard barometric pressure, at which the vapor above the sample ignites, is taken as the flash point.

ASTM D92 flash point by "Cleveland open cup" measures the flash and fire point temperatures of all oil products except those with a flash point below 79°C. The test cup is filled with a specified amount of sample and the temperature is raised. At specified intervals an open flame is passed over the cup. The lowest temperature, corrected to standard barometric pressure, at which the vapor above the sample ignites, is taken as the flash point.

ASTM D93 flash point by "Pensky-Martens closed cup" measures the flash point temperature of oil products with improved accuracy. The test cup is filled with a specified amount of sample and closed. The sample temperature is raised and at specified intervals an open flame is lowered through a shutter into the cup. The lowest temperature, corrected to standard barometric pressure, at which the vapor above the sample ignites, is taken as the flash point.

ASTM D1310 flash point and fire point by "Tag open cup" measures the flash point and fire point temperatures of liquids having flash points between 0 and 325°F (−18 and 165°C) and fire points up to 325°F.

ASTM D3828 flash point by "small scale closed cup" measures the flash point temperature of oil products with improved accuracy. It can measure the flash point or the occurrence of a flash at a specific temperature. The test cup is filled with a specified amount of sample and closed. The sample temperature is raised and at specified intervals an open flame is lowered through a shutter into the cup. The lowest temperature, corrected to standard barometric pressure, at which the vapor above the sample ignites, is taken as the flash point.

ASTM D6450 flash point by "continuously closed cup" (CCCFP) test utilizes a 1-ml test sample placed into a closed but unsealed cup. Air is injected into the cup during the test. This test determines the flash point of fluids between 10 and 250°C. The flash point is the temperature reading at the moment the flash-induced pressure increase is sensed.

3.2.1.11.1 Comments

The flash point of an oil product is a rough measure of the fluid's volatility or lower explosive limit corresponding to a vapor pressure of about 2 to 5 torr. The measure is an important property when selecting fluids for high-temperature machinery applications. In addition, the flash point temperature is useful as a quality control measure where a change in it can be related to a commensurate change in fluid volatility due to product changes or the presence of ingress contamination by more or less volatile fluids.

International equivalent standards include (D56) IP 304, (D92) IP 36 and ISO 2592, (D93) DIN 51758, ISO 2719, and IP 34, and (D3828) IP 303 and ISO 3679.

3.2.1.12   Foaming Characteristics — ASTM D892, ASTM D3427, ASTM D6082

A reliable lubricant should release entrained air or other gas and resist foaming. Excessive foaming is detrimental to the operation of most machinery fluid systems. Foam can fill the internal spaces such as a separator or reservoir resulting in poor system efficiency or failure; cause a vapor block in filters resulting in oil starvation; and cause excessive wear of lubricated parts due to the poor load-carrying ability of entrained gas (air). Excessive foaming can result in oil loss due to the overflow of the reservoir through vents and create maintenance problems.

The ASTM D892 test for foaming characteristics provides an empirical rating of the foaming characteristics of a lubricant sample under specified temperature conditions. A metered volume of dry air is blown through a diffuser immersed in the sample for a period of 5 and 10 min. The foaming tendency is determined as the foam volume (milliliters) at the end of the blowing period. The foaming stability is determined from the settling period for three or more sequences. Alternately, for oil temperatures above 93.5 to 150°C, the similar ASTM D6082 procedure is used.

The ASTM D3427 air release properties test determines the ability of a lubricant to release entrained air at a controlled temperature. In this procedure, compressed air is blown through the sample for 7 min and the time taken to release the entrained air is recorded. The volume of entrained air remaining is determined from sample density measurements with a density balance.

3.2.1.12.1   Comments

Foaming is attributed to air entrainment due to mechanical working of the oil during machine operation. In addition, the presence of water and surface active materials in the oil such as rust preventatives, detergents, etc. can cause foaming. Foaming may be controlled to some extent with the use of additives. Since these materials can increase the tendency of the oil to entrain air, the "optimum" amount of additive for the oil application must be determined.

International equivalent standards include (D892) ISO 6247, DIN 51566, and IP 146, (D3427) DIN 51381, IP 313, and ISO 9120.

3.2.1.13   Hydrolytic Stability Characteristics — ASTM D2619

A reliable lubricant should resist the tendency to hydrolyze at machinery operating temperatures when in the presence of water and copper components. Poor hydrolytic stability gives rise to the formation of acidic by-products and insolubles, which in turn results in deposits of varnish and sludge and chemical leaching of copper and other machinery metals. This property is of particular importance for equipment utilizing ester-based lubricating or hydraulic fluids. The ASTM D2619 hydrolytic stability characteristics test determines the ability of a lubricant to resist the tendency to hydolyze (acidify) in the presence of water and machinery metals. The test employs a vessel containing 75 g of oil, 25 g of water, and a copper specimen. The vessel is closed and rotated for a period of 48 h at a temperature of 95°C (200°F). The hydrolytic stability value is determined from measurements of the resulting insolubles level, acid number, viscosity increase, and copper loss.

3.2.1.13.1   Comments

The ASTM D2619 test is useful in the evaluation of a variety of potential hydraulic fluids. In particular, the effectiveness of hydrolytic stability inhibitors in synthetic fluids (i.e., phosphate, silicate esters) can be evaluated and compared. As with other similar tests, the most reliable information is interpreted from test results using actual application experience.

3.2.1.14   Interfacial Tension — ASTM D971, ASTM D2285

Lubricating oils, as a result of oxidation or modification by additives, display differences in the degree of wetting, spreading, and boundary lubrication on various surfaces. Such differences are indicated by interfacial tension (IFT) values. The measurement of the strength of the boundary film (interfacial tension) employs a special device (tensiometer) which presents a planar ring of platinum wire, supported by a precision spring, to the interface between the test oil and water. The force required to detach the ring from the surface of the liquid with the higher surface tension (usually the water) is a measure of the IFT. The test is extremely sensitive and requires practice and control to obtain reproducible results.

3.2.1.14.1   Comments

Strong mutual attraction between oil surface molecules is responsible for surface tension. When materials such as oxidation by-products are present, the film strength (surface tension) of the fluid is modified. The measurement of change in interfacial tension can compliment other performance tests such as neutralization number as a means of detecting excessive fluid oxidation, and life-limiting values may be applied. Note that a rigorous chemical analysis method such as Fourier transform infrared (FT-IR) is preferable for analyzing the degree and type of oxidation by-products present in used-oils. IFT measurement on new oils, with the exception of electrical insulating oils, is generally not appropriate.

International equivalent standards include (D971) ISO 304 and DIN 53914 and (D2285) ISO 9101.

3.2.1.15   Low-Pressure Volatility of Lubricants — ASTM D2715

Machinery operating in reduced atmospheric pressure or vacuum conditions requires lubricant products with low volatility and decomposition rates. The ASTM D2715 encompasses the necessary apparatus and procedure to determine evaporation and decomposition rates of lubricating materials. In the D2715 procedure, a prescribed quantity of the lubricant is exposed in a vacuum thermal balance device. As the evaporated material collects on a condensing surface, the decreasing weight of the original sample is recorded as a function of time. With additional instrumentation, it is possible to obtain quantitative evaporation data along with the identity of the volatile and decomposition by-products. The procedure is repeated at increasing temperature levels and the evaporation rate(s) determined for each temperature. The vapor pressure may also be determined when the molecular weight of the test lubricant is known.

3.2.1.15.1   Comments

Determination of evaporation rates under prescribed conditions may provide a basis for the estimation of an approximate useful life if attrition of the lubricant can be attributed to evaporative loss. Data interpretation will be more reliable when the other factors that influence low-pressure performance of lubricants are also evaluated. This test is limited to speciality low-vapor pressure liquid and solid lubricants.

3.2.1.16   Neutralization Number (Acid/Base Number) — ASTM D664, ASTM D974, ASTM D2896, ASTM D3339

New lubricant products will normally exhibit acidic or alkaline characteristics, depending on their intended function, additive mix, manufacturing process, and the presence of contaminants or degradation by-products formed during service. For example, internal combustion engine lubricants are formulated to exhibit an alkaline reserve, while turbine lubricants tend to be somewhat acidic. The degree of acidity or alkalinity of a lubricant is derived from its neutralization number.

The ASTM D664 acid number by potentiometric titration may be used to determine the change in relative acidity of a lubricant regardless of color or other properties. In this test, the sample is dissolved into a mixture of toluene, isopropyl alcohol, and water and titrated with alcoholic potassium hydroxide. The meter readings are plotted against the respective volumes of the titrating reagent utilized. The acid number is indicated as the quantity of base, expressed in milligrams of potassium hydroxide per gram of sample, required to titrate the sample from its initial value to that of a calibration standard.

The ASTM D974 acid or base number by color indicator titration is a simple qualitative method for determining acid or base number. In this test, the sample is dissolved into a mixture of toluene, isopropyl alcohol, and water and titrated with an alcoholic base or acid solution until the endpoint is indicated by a color change of the added naphtholbenzein solution. The acid number (AN) is determined as the quantity in milligrams of potassium hydroxide (KOH) required to neutralize all the acidic by-products in 1 g of lubricant sample. The base number (BN) "alkalinity reserve" is determined as the quantity of acid, expressed in equivalent milligrams of KOH, required to neutralize all basic constituents in the lubricant sample.

The ASTM D2896 base number by potentiometric titration test may be used to determine the change in relative base number of a lubricant regardless of color or other properties. In this test, 20 g of oil sample is dissolved into 120 ml of titration reagent glacial acetic acid and chlorobenzene (or mixed xylenes) solution which is titrated with perchloric acid. The potentiometric readings are plotted against the respective volumes of titrating solution. The base number is calculated from the quantity of acid needed to titrate the solution, expressed in milligrams of potassium hydroxide per gram equivalent.

For testing smaller samples, utilize the ASTM D3339 method. This test is especially intended where the sample volume is too small to allow accurate analysis by D974 or D664.

3.2.1.16.1 *Comments*
The D664 and D2896 potentiometric methods are more versatile and provide more reproducible data than the D974 colorimetric method. These methods may also be utilized to analyze dark oils and oils with weak-acid, strong-acid, weak-base, or strong-base constituents even though the dissociation constant of a strong acid or base may be up to 10^3 greater than the next strongest component. The constituents that contribute to overall oil acidity include phenolic compounds, resins, esters, inorganic and organic acids, heavy metal salts, and salts of polybasic acids. The constituents that contribute to alkalinity include organic and inorganic bases, salts of polyacidic bases, amino compounds, etc. Some amphoteric additives such as inhibitors, detergents, salts of heavy metal, soaps, and other fillers can produce either an acid or alkaline response. An increase in used-oil acidity is generally due to the presence of oxidation by-products.

Neutralization number is useful for production quality control, specification, and the purchasing of lubricants. It is also used in combination with other tests, for example, interfacial tension, to ascertain changes that occur in service under oxidizing conditions. When used to monitor lubricant condition in systems such as steam turbines, hydraulic systems, and transformer and heat transfer units that are normally free of interfering contaminants, neutralization number may be used to indicate when the oil must be changed, restored, or reclaimed. Note that rigorous chemical analysis methods such as FT-IR will provide better data granularity on the individual constituents contributing to changes in an oil's acid or base number.

International equivalent standards include (D664) IP 177, DIN 51558, and ISO 6618 and (D974) IP 139 and ISO 6618.

3.2.1.17 Oxidation Stability — ASTM D943, ASTM D2272, ASTM D2893, ASTM D5763, ASTM D5846, ASTM D6514
All oil products will begin to oxidize and degrade from the moment of use. Exposure to atmospheric oxygen promotes the formation of acidic by-products that degrade into sludge and varnish deposits. Good oxidation stability is one of the key requirements of a lubricant and an important factor in the estimation of remaining useful service life. Oxidative by-products are generally characterized as insoluble resins, varnishes, and sludges, or soluble organic acids and peroxides. The usable service life of in-service oil tends to be inversely proportional to the level of these contaminants. These tests are generally utilized to determine the oil's oxidation stability and provide an indication of the remaining usable life.

The ASTM D943 lubricant oxidation characteristics test was developed to determine the oxidation stability of petroleum lubricants in the presence of oxygen, water, copper, and iron. A volume of oil is reacted with oxygen in the presence of water and a catalyst of iron and copper at a temperature of 95°C for a period of 500 to 1000 h. Samples are removed periodically during the test period for analysis of acidic content. The ASTM D664 acid number test is typically performed to determine sample acidity.

The oxidation characteristics test is discontinued when the AN reaches 2.0 mg KOH/g or above. Note that some new oils may exhibit AN of up to 1.5 mg KOH/g. The D943 test is of limited applicability for these oils.

ASTM D2272 — rotating pressure vessel oxidation test (RPVOT) is a relatively short duration test procedure and generally preferred over other methods. This test utilizes a rotating pressure vessel in which the oil sample, water, oxygen, and a copper catalyst coil is placed. The vessel is charged with oxygen to 90 psi and is rotated at 100 rpm in a temperature-controlled bath. The test is timed until a specific drop in oxygen pressure signals the endpoint. The remaining oil life is reported as minutes of bomb life.

The D2893 test was developed to determine the oxidation stability of extreme pressure lubricants. In this test, dry air is bubbled at a rate of 10 l per hour through 300 ml of oil at a temperature of 95°C for a period of 312 h. Samples are removed periodically during the test period for analysis of viscosity and precipitation number.

ASTM D5763 test determines the oxidation and thermal stability characteristics of gear oils using the universal glassware apparatus. In this test, 100 g of oil is exposed to a continuous flow of dry air at 120°C for 312 h. The result is determined from an evaluation of the total sludge, viscosity change, and oil loss.

ASTM D5846 evaluates the oxidation stability of petroleum-based hydraulic and turbine lubricating oils using the universal oxidation test apparatus. The oil is exposed to air at 135°C until the acid number increases by 0.5 mg KOH/g over the new, untested oil or until insoluble deposits are observed.

The ASTM D6514 evaluates the high-temperature oxidation stability and deposit-forming tendency of inhibited steam and gas turbine lubricating oils in the presence of air, copper, and iron metals. The test is performed for 96 h at 155°C after which the AN, viscosity, and sludge level are reported.

3.2.1.17.1 Comments

The soluble by-products of oxidation tend to increase oil acidity and viscosity and promote corrosion of metal parts. The rate of oxidation is basically a function of time. However, high operating temperature and the presence of contaminants such as water and catalysts such as copper accelerate the oxidation process. In general, the oxidation rate of petroleum oil will double for every 10°C (18°F) increase above the nominal operating temperature of the machine's fluid system.

The main problem with these test methods is that the associated test procedure does not duplicate equivalent in-service conditions. The presence of oxidation by-products and some additives (acting as catalysts) will reduce the effectiveness of these tests for in-service oils, generally causing the oxidation stability results to read low. Applying an oil purifier to remove oxidation by-products before sampling will generally improve oxidation stability test results.

Oxidation resistance is critical in steam turbine oils because of the seriousness of turbine bearing failures. Other applications such as gear, transformer, hydraulic, heat transfer, and gas turbine systems also require lubricants with excellent oxidation stability. Although direct data comparisons are not always reliable, oxidation stability test results form the basis for comparison of the relative life of competitive lubricant products as well as new and in-service oils.

International equivalent standards include (D943) IP 157, DIN 51587, and ISO 4263.

3.2.1.18 Pentane Insolubles of Lubricating Oils — ASTM D893, ASTM D4055

When the oil is diluted by a prescribed volume of pentane, insolubles such as resins, dirt, soot, and metals tend to drop out of suspension. These precipitates are collectively called pentane insolubles. A high level of pentane insolubles indicates a high level of oxidation or contamination of the lubricant. Further treatment with toluene will release additional precipitates collectively referred to as toluene insolubles. A high level of toluene insolubles indicates a high level of external contamination such as soot or dirt. The difference between pentane insolubles and toluene insolubles indicates the level of oxidation by-products in the oil. The ASTM D893 centrifuge or D4055 micro-filtration methods may be utilized to determine the degree of pentane and toluene insolubles in lubricating oils. These methods also determine the pentane insolubles of high-detergency engine oils, where a coagulant is used to release insolubles held in suspension by the detergent/dispersant additive.

In the ASTM D893 method, pentane insolubles are determined by placing 10 g of sample into a centrifuge tube, filling the tube to the 100-ml level with pentane, and centrifuging. The precipitate is washed twice with pentane, dried, and weighed to the nearest 0.1 mg. The percent pentane insolubles is calculated from the before and after weight of the centrifuge tubes. In addition, the pentane insolubles residue can be further treated with toluene to dissolve and separate out the resins formed during the oil oxidation process. In this procedure, 10 g of sample is placed into a centrifuge tube, filled to the 100-ml level with pentane, and centrifuged. The precipitate is washed twice with pentane, once with a toluene alcohol solution, once with toluene, dried, and weighed to the nearest 0.1 mg. The residue left over after this test is referred to as toluene insolubles.

In the ASTM D4055 method, pentane insolubles in lubricating oils are determined by membrane filter analysis. This procedure utilizes a submicron filter membrane as a means to extract the insolubles precipitate. The filter membrane is cleaned and weighed to the nearest 0.1 mg. One gram of sample is placed into a volumetric flask and filled to the 100-ml level with pentane. The sample solution is filtered and the membrane is re-weighed to the nearest 0.1 mg. The percent pentane insolubles is calculated from the before and after weight of the membrane filter.

3.2.1.18.1   Comments

Pentane and toluene insolubles are generally utilized to indicate the amount of soot and unburned fuel by-products in used diesel engine oils. It is usually preferable to utilize a more rigorous chemical analysis method such as FT-IR, which will provide individual results for soot and acidic by-products in used diesel engine oils.

3.2.1.19   Pour and Cloud Point — ASTM D97, ASTM D2500, ASTM D5771, ASTM D5772, ASTM D5773, ASTM D5949, ASTM D5950, ASTM D5985, ASTM D6749

Many types of machines must be started while cold and the ability of a lubricant to flow and lubricate properly during (and just after) start-up is an important factor in the selection of the lubricant. A general classification of oil response to low temperature conditions may be determined by two temperature measurements: pour point and cloud point. When an oil product is cooled sufficiently, a point is reached when it will no longer flow under the influence of gravity. This is referred to as the pour point. At this temperature, the lubricant cannot perform its primary functions. However, agitation by a pump will break down the wax structure and allow paraffinic type oils to be pumped at temperatures below their pour point. Naphthenic oils contain little or no wax and their pour point is reached by virtue of the increase in viscosity due to the low temperature. Naphthenic oils cannot be pumped readily near their pour point.

Just prior to reaching the pour point, a paraffinic lubricant becomes cloudy as a result of the crystallization of waxy constituents. These materials crystallize in a honeycomb structure at low temperatures. The temperature at which a cloudy appearance is first observed is referred to as the cloud point. At this temperature, most oils will not perform effectively enough for reliable machine operation. Under these conditions, very viscous oil or waxy constituents may accumulate and plug oil passages or filters. Note that the cloud point should not be confused with cloudiness or color changes due to high lubricant stress or contamination.

In the ASTM D97 pour point test, the oil is first heated to ensure solution of all ingredients and elimination of any influence of past thermal treatment. The oil is then cooled at a specified rate, and checked at temperature increments of 3°C. The test apparatus removes the sample vessel from its cooling jacket and tilts it at a 90° angle as prescribed by the D97 test method until no flow is observed for a 5-sec interval. The pour point result is then reported as 3°C higher than the temperature at which the sample ceased to flow.

In the ASTM 2500 cloud point test, the oil is first heated to ensure solution of all ingredients. In this test, the oil's temperature is measured in 1°C temperature changes. The instrument typically utilizes a coaxial fiber optic sensor positioned above the test sample to determine and record temperature changes. The temperature indicated at the initial appearance of crystallization signifies the cloud point.

ASTM D5771 (stepped cooling method), D5772 (linear cooling rate method), and D5773 (constant cooling rate method) tests determine cloud point of petroleum products by utilizing automatic systems for sample cooling and optical sensors to detect the cloud point. During the controlled cooling process, the optical sensor monitors the transparency of the sample for a change in crystalline structure. The temperature at which wax crystals are sensed is taken as the cloud point. The test apparatus may be operated over a temperature range of -40 to $+49°C$ with a resolution of $0.1°C$.

ASTM D5949 standard test for determining pour point of petroleum products utilizes the automatic pressure pulsing apparatus. This test applies a controlled burst of nitrogen gas onto the sample surface while the sample is being cooled. The instrument detects sample movement over a temperature range of -57 to $+51°C$ with an optical sensor.

ASTM D5950 standard test for determining pour point of petroleum products utilizes the automatic tilt apparatus. This instrument automatically tilts the test lubricant fixture during cooling and detects movement of the surface of the test specimen with an optical sensor. The test may be operated over a temperature range of -57 to $+51°C$. This test method is not intended for use with crude oils.

ASTM D5985 standard test for determining pour point of petroleum products utilizes the rotational method. This test determines pour point via an automatic instrument that continuously rotates the test specimen against a suspended detection device during the cooling process. The no-flow point is detected by a change in the fluid viscosity or crystal structure that is sufficient to impede flow. The test may be operated over a temperature range of -57 to $+51°C$. This test method is not intended for use with crude oils.

ASTM D6749 standard test for determining pour point of petroleum products utilizes the automatic air pressure tester. This test determines pour point by an automatic application of positive air pressure onto the sample surface during the cooling process. The lowest temperature at which surface deformation of the sample is observed is the pour point. The test may be operated over a temperature range of -57 to $+51°C$. This test method is not intended for use with crude oils.

3.2.1.19.1 Comments

The pour point depends to a great extent on the type of crude from which the oil is made. Naphthenic oils exhibit naturally lower pour points. These materials, however, also tend to thicken more rapidly as the temperature lowers. Paraffinic base oils, on the other hand, contain waxy materials that solidify and become insoluble at temperatures near or slightly below the pour point. In these cases, flow may occur if the oil is subjected to mechanical shear or agitation. However, paraffinic oils must have sufficient flow characteristics to be able to be drawn into the pump. In contrast, naphthenic oil, although free of waxy structures, will not respond to shear forces when the pour point is reached. Again, pumpability will suffer unless the pour point can be depressed. The pour point of some paraffinic base oils may be improved by using refining processes or formulation with pour depressant additives.

The cloud point of an oil is the temperature at which a distinct cloudiness or haze is discerned in the bottom of the test jar as paraffin wax and other materials, normally soluble, begin to separate. Since the test is a visual observation, the measurement is limited to oils that are transparent. Operating capillary or wick-fed oil systems at the cloud point will restrict lubrication effectiveness. In this case, wax particles or any fine dispersion contained in the oil will retard or may even prevent the flow of lubricant.

International equivalent standards include (D97) IP 15 and ISO 3016.

3.2.1.20 Precipitation Number --- ASTM D91

Semi-refined petroleum oils or black oils often contain naphtha insolubles, which are referred to as asphaltic materials. When determining the suitability of an oil product for use as a basestock lubricant, the asphaltic contamination level must be determined. In addition, in-service lubricants accumulate soluble and insoluble contaminants and the by-products of oil oxidation. These materials are detrimental to oil performance and long service life. Contaminants that are insoluble in oil may be separated and quantified by centrifuging the sample. However, soluble contaminants must be precipitated out of an oil sample by treatment with a solvent.

The ASTM D91 precipitation number test method determines the asphaltic residues remaining in black oils and steam cylinder stock after refining. These residues are largely insoluble in paraffinic naphtha and, therefore, they may be readily extracted. Asphaltic residues are, generally, deleterious to in-service lubricants as they exhibit lower thermal and oxidative resistance, thus acting as a catalyst to increase oil oxidation. The ASTM D91 method defines the precipitation number of a lubricant as the quantity in milliliters of precipitate or sediment formed when 10% by volume of oil and 90% by volume of naphtha solvent are centrifuged under prescribed conditions.

3.2.1.20.1   Comments
The D91 precipitation test may be applied to any oil but is primarily intended for steam cylinder and black oils. The test may be used to show the presence of foreign matter in the oil. Since the D91 test measures total solids that are insoluble in naphtha, separate analysis methods are required to quantify individual insolubles and any naphtha soluble materials that may be present.

International equivalent standards include IP 75.

3.2.1.21   Refractive Index — ASTM D1218, ASTM D1747
Refractive index is a consistent fluid property and therefore a useful technique for checking fluid uniformity. The property is also useful in characterizing basestocks for fluids that have equivalent molecular weights. The refractive index of oil increases with increase in molecular weight, from paraffins to naphthenes to aromatics.

Refractive index is defined as the ratio of the velocity (of a specified wavelength) of light in air to that in the test fluid. In the case of process oils and other petroleum products, the D-line of sodium (5893 Å/589 nm) is the most frequently used wavelength. To obtain the refractive index, a drop of sample is placed on the measuring prism face and the light source is adjusted in line with the refractometer telescope while the sample is uniformly lighted. After alignment, the scale is read and the value converted to refractive index at the appropriate temperature. The test temperature and light source are also reported with the refractive index reading.

The ASTM D1218 test method determines the refractive index of transparent and light-colored hydrocarbons that have a refractive index ranging from 1.33 to 1.50, and at temperatures from 20 to 30°C. The test utilizes a high-resolution refractometer.

The ASTM D1747 test determines the refractive index of viscous materials and low melting point solids that have a refractive index ranging from 1.33 and 1.60. These materials are tested at temperatures between 80 and 100°C.

3.2.1.21.1   Comments
In conjunction with viscosity and specific gravity, refractive index may be used for hydrocarbon type identification.

3.2.1.22   Rust Prevention Characteristics — ASTM D665, ASTM D1748, ASTM D3603, ASTM D6557
Iron and steel alloys are the major metals used for fabricating machinery. Any contact between the native metal with air and moisture will cause rusting. It is therefore necessary to protect internal machinery surfaces from contact with dissolved or free moisture that may be carried along by the lubricant. In addition to surface damage, rust particulates will promote oil oxidation, cause abrasion of wear surfaces, clog oil passages and filters, and damage sensitive components such as servo-valves, etc.

The ASTM D665 test utilizes a vertical steel rod immersed in a mixture of 300 ml of test oil and 30 ml of water (distilled or sea). The mixture is heated to 60°C (140°F) and stirred continuously during the test. After a period of 4 h the rod is examined for rusting. At the end of the time period, the test is reported as pass or fail depending on the observance of any rust.

The ASTM D1748 test measures the ability of metal preservatives to prevent rusting of steel under conditions of high humidity. Polished steel panels are dipped in the test oil, allowed to drain, and suspended in a humidity cabinet at 48.9°C (120°F) for a specified test period. The size and number of rust spots on the test panel is interpreted as the result. A pass is taken to be no more than three spots of rust, where none is greater than 1 mm in diameter.

In the ASTM D3603 method, a polished steel cylinder is immersed into a prepared sample, consisting of 275 ml of the test lubricant and 25 ml of distilled water for a 6-h period. At the end of the time period, the test is reported as pass or fail depending on the appearance of any rust.

In addition to the above methods, the ASTM D6557 test may be used to evaluate the rust preventive characteristics of automotive engine oils. This test utilizes a series of test tubes containing the sample and a steel ball. As the tubes are mechanically shaken, air and an acidic solution (to provide a corrosive environment) are continuously applied to each tube for an 18-h period. The steel balls are imaged by an optical system to quantify the rust prevention capability of each sample oil. This standard method was designed to replace the ASTM D5844.

3.2.1.22.1 Comments

The main significance of the rust prevention test is the determination of the ability of a given lubricant to maintain a reliable lubricant film and deny exposure of the metal surface to entrained and atmospheric oxygen.

Applicability: hydraulic fluids; steam, hydro, and industrial gas turbine oils; and electrical insulation oils.

International equivalent standards include (D665) ISO 7120, DIN 51566, and IP 146 and (D3603) NACE TM-01-72.

3.2.1.23 Saponification (Sap) Number ASTM D94

Many lubricants have fatty compounds added to increase their fluid film strength and improve water emulsibility characteristics. The fatty material displays a strong affinity for metal surfaces and enables oil to combine physically with the water instead of being displaced by it. The saponification (sap) number indicates the amount of these compounds (i.e., the degree of compounding). The sap number of a lubricant can be determined from two basic titration methods, one colorimetric and the other potentiometric.

In the ASTM D94 sap number test, a specific quality of KOH is added to a specific quantity of sample and heated. In the reaction that follows, the fatty compounds are converted to a soap or saponified. Excess KOH is then neutralized by titrating with hydrochloric acid. The sap number is the quantity (milligrams) of KOH that was consumed in the reaction with the oil.

3.2.1.23.1 Comments

In practice, the sap number is usually considered together with the neutralization number to determine the relative levels of acidic and fatty compounds in the lubricant. Note that the test results of new oils will vary depending on the level of acidic additives or other constituents in the formulation. In used oils, the sap value tends to increase in conjunction with the neutralization number. In addition, the nature and degree of contaminants will further distort used-oil test results. Considerable experience is necessary to properly interpret the results and meaning of sap number data.

International equivalent standards include IP 136.

3.2.1.24 Sulfur Content — ASTM D129, ASTM D1266, ASTM D1552, ASTM D2622, ASTM D4294, ASTM D5453

All crude oils contain sulfur in elemental or various compound forms. Some sulfur compounds are acidic and corrosive, others have functional uses, for example, as naturally occurring antioxidants. During the lubricant refining process, the nonharmful sulfur compounds are retained to increase oxidation resistance. Additional sulfur compounds may be added to improve oxidation resistance or extreme-pressure performance for a particular application. The concentration of active, combined, and total sulfur content of lubricating oils may be determined from the ASTM DI29 sulfur content by pressure vessel oxidation method, the ASTM D1266 sulfur by titration method, the ASTM D1552 high-temperature method, the

ASTM D2622 or D4295 x-ray fluorescence (XRF) spectroscopy methods, and the ASTM D5453 total sulfur by ultraviolet fluorescence method.

The ASTM D129 sulfur by pressure vessel oxidation technique is generally applicable to all lubricating oils for measuring the total sulfur content. The procedure involves the ignition and combustion of a small sample of oil under pressurized oxygen. The products of combustion are collected and the sulfur is precipitated and weighed as barium sulfate.

For measurement of sulfur levels from 0.01 to 0.4 mass%, the ASTM D1266 method is used. In this test, the sample is burned in a furnace and the combustion gases are titrated.

The ASTM D1552 high-temperature method determines the concentration of sulfur in lubricating oils to 0.005% from the reaction of a known amount of potassium iodide with the sulfur dioxide given off by burning the oil sample in a high-temperature oxygen stream.

The ASTM D2622 sulfur by XRF spectroscopy method determines the concentration of sulfur in lubricating oils by x-ray fluorescence spectroscopy. Modern high-power wavelength and energy dispersive XRF instruments can measure sulfur concentrations of 0.005% and are not affected by the presence of dissolved metals. The test utilizes an automated spectrometer system and requires much less sample preparation and analysis time than other sulfur determining methods. Alternatively, the ASTM D4294 sulfur by XRF spectroscopy may be used to determine the total sulfur content from 0.05 to 5 mass %.

The ASTM D5453 test method determines total sulfur in liquid hydrocarbons that boil from approximately 25 to 400°C, with viscosity between 0.2 and 20 cSt, at room temperature. The sulfur is oxidized to sulfur dioxide, which absorbs energy from ultraviolet (UV) light. The amount of sulfur is derived from the resulting UV fluorescence.

3.2.1.24.1   Comments

The undesirable consequences of sulfur compounds in oil are corrosion of metal surfaces, particularly copper. However, in machinery applications utilizing hypoid gears or metal cutting tools, the greater extreme pressure and increased load-carrying ability of the lubricant formulation effectively offsets the corrosion problem. With the exception of "active" sulfur containing lubricants, sulfur content of general-purpose lubricants is usually limited by machinery operating specifications.

International equivalent standards include (D129) IP 61, (D4294) ISO 8754, and (D5453) ISO 20846.

3.2.1.25   Thermal Stability — ASTM D2070, ASTM D5579, ASTM D5704

Thermal stability refers to the resistance to oil degradation or property change due to thermal stress and is characteristically important for machinery oils.

The thermal stability of hydraulic fluids is determined by the ASTM D2070 test. The test fluid is heated to 135°C for 168 h in a beaker containing copper and iron rods. At the end of the test, the rods are rated visually for discoloration. Additionally, the change in viscosity of the oil, the amount of sludge formed in the oil and the weight loss of the copper rod can be determined.

The ASTM D5579 test determines the thermal stability of heavy-duty manual transmission lubricants utilizing a cyclic durability test stand. The thermal stability performance of the lubricant is determined from a specified number of shifting cycles that can be performed without failure of synchronization when the transmission is operated while continuously cycling between high and low range. The results are determined from a visual inspection of the transmission parts.

The ASTM D5704 test determines thermal and oxidative stability of lubricating oils used in manual transmissions and final drive axles. The test, commonly referred to as the L-60-1 test, utilizes a heated gearcase containing two spur gears, a test bearing, and a copper catalyst. The lubricant is heated to a specified temperature and operated for 50 h at prescribed loads and speeds during which air is bubbled through it. Thermal stability is determined from viscosity, insolubles, and gearcase cleanliness.

3.2.1.25.1 Comments

These tests are useful for determining oil deterioration characteristics under operating conditions and are suitable for determining additives and base oils for thermal degradation tendencies.

International equivalent standards include (D5704) STP512A L-60-1.

3.2.1.26 Viscosity and Viscosity Index --- ASTM D445, ASTM D2270, ASTM D2983, ASTM D4683, ASTM D5481, ASTM D6080, ASTM D6616, ASTM D6821

Generally speaking, viscosity is the measure of a lubricant's internal friction or resistance to flow or the ratio of shear stress to shear rate. In a given machine, the optimum thin film thickness separating the moving surfaces can be maintained only if the operational viscosity range is correct. Consequently, viscosity is probably the most important property of a lubricant specification. Changes in the viscosity of in-service oil can indicate ingress contamination or fluid degradation. Contamination of oil by fuel, light oils, or solvents will lower its viscosity level. Oxidation and combustion by-products tend to thicken the oil, a condition indicated by an increase in viscosity level.

Dynamic (absolute) viscosity is the measure of the tangential force required to shear one parallel plane of a fluid over another. The force is proportional to the fluid viscosity, the planar area being sheared, and the rate at which the adjacent parallel planes are being forced to slide over each other, divided by the thickness of the film. At any given temperature, the dynamic viscosity is related to force, fluid density, film thickness, and area by the following formulae:

$$\text{Dynamic Viscosity} = (\text{Force/Area}) \times (\text{Film Thickness/Velocity})$$

$$= \text{Kinematic Viscosity} \times \text{Density}$$

The force/area parameter is reduced to units of pressure — Pascal (Pa). The film thickness/velocity parameter is reduced to units of time — seconds (sec). Thus, the dynamic viscosity is expressed as Pascal-seconds or Poise (P). For most lubricant applications, Pascal-seconds and Poise are inconveniently large units and dynamic viscosity is generally reported in milliPascal-seconds (mP) or centipoise (cP).

The absolute viscosity is generally measured by a rotating spindle or ball-type viscometer such as the Brookfield or tapered bearing simulator (TBS). Note that consistent measurement results will depend on reproducibility of the oil type and measurement conditions. Multigrade oil will exhibit a decrease in viscosity as the shear rate increases; thus, absolute viscosity values will change with different instrument types unless exact shear rate and temperature conditions can be duplicated. The Brookfield viscometer measures dynamic viscosity under low shear rate conditions. To ensure effective lubrication of high shear-rate machines, lubricating oils should also be tested with a high shear-rate viscometer such as the TBS. The TBS viscometer measures absolute viscosity under high and very high shear-rate conditions and allows the evaluation of monograde and multigrade oils at or near the shear-rates of modern high-performance reciprocating machines.

The ASTM D445 kinematic viscosity test is the most popular measurement used to determine the nominal viscosity of machinery oils. The unit of measurement is the Stoke and is equal to one square centimeter per second. For most lubricant applications, the Stoke is an inconveniently large unit. A smaller unit, the centistoke (cSt) — equal to one square millimeter per second — is preferred. The D445 test is usually performed at a lubricant temperature of 40 or 100°C to standardize the results obtained and allow comparison among different users. Note: Always report the oil temperature when performing kinematic viscosity. The ASTM D445 test determines the kinematic viscosity of liquid lubricants by measuring the time taken for a specific volume of the liquid to flow through a calibrated glass capillary viscometer under specified driving head (gravity) and temperature conditions. The method is applicable both to transparent and opaque fluids. The kinematic viscosity is the product of the time of flow and the calibration factor of the instrument.

The ASTM D2983 utilizes the Brookfield rotary viscometer. The Brookfield instrument measures dynamic viscosity as a function of the resistance to the rotational force of a metal spindle immersed into the oil. This test quantifies dynamic viscosity under low-temperature, low shear rate conditions. Note: Always report lubricant temperature when performing viscosity tests. The test reports viscosity readings in centipoise (cP). The D2983 test procedure was developed to determine the viscosity performance of crankcase, gear, industrial, and hydraulic oils over the temperature range of +5 to −40°C. Note that a variation of this method is widely used to determine the viscosity of new and used lubricating and hydraulic oils. In this case, the test is generally performed at ambient temperature. Performing the test at ambient temperature produces ambiguous results. Utilizing an immersion bath to control the sample temperature during the test will improve results. Alternatively, the ASTM D6080 may be used to determine the viscosity of synthetic and petroleum hydraulic fluids.

The ASTM D5481 determines the apparent viscosity at high-temperature and high-shear rate (HTHS) utilizing a multi-cell capillary viscometer. This viscometer is instrumented for pressure, temperature, and timing. The test is conducted at a temperature of 150°C and a shear rate corresponding to an apparent shear rate at the viscometer tube wall of 1.4 million reciprocal seconds (1.4×10^6 sec^{-1}). The viscosity is determined from calibrations previously established for Newtonian oils over a range of 2 to 5 mPaS at 150°C. Note that the results determined by this method may not correspond with viscosity measurements generated by other instruments.

The ASTM D6616 viscosity at high shear rate by TBS viscometer utilizes a tapered rotor that rotates at 3600 rpm in a close fitting stator containing the oil under test. The TBS measures viscosity as a function of the resistance to the rotational force of tapered journal and plain bearing set operating under a prescribed load. The shear rates are applicable to high-speed reciprocating machinery. High shear-rate viscosity measurements may also be performed by the ASTM D4683 using the TBS.

ASTM D6821 determines the viscosity for driveline lubricants (gear oils and automatic transmission fluids) utilizing a constant shear stress viscometer at temperatures from −40 to 10°C. The viscosity is determined by applying a prescribed torque and measuring the resulting rotational speed after a controlled temperature regime. The result is reported as milliPascal seconds (mPa/sec).

Viscosity index (VI) is defined as the extent to which oil resists viscosity change due to temperature change. The higher the viscosity index, the less an oil's viscosity will change as a result of temperature change. Viscosity index data is generally used to assess the viscosity performance of oil over the normal operating temperature range of a machine including its start-up and shutdown conditions. The viscosity of most machinery fluids will tend to decrease as the fluid temperature increases. The viscosity index will change with oil chemistry. However, the viscosity of all lubricants does not respond to temperature changes in a consistent manner. Monograde oil (Newtonian) will generally exhibit a constant viscosity/temperature at all (except very high) shear rates. For monograde oils, the viscosity will decrease as the temperature rises, and vice versa. However, multigrade oil is a blend of light basestock oil and a thick viscosity index improver (VII) agent. Multigrade oil is a non-Newtonian fluid and its viscosity will not be as affected by temperature change as will be similar viscosity monograde oil.

The ASTM D2270 procedure determines viscosity index from kinematic viscosity (ASTMD445) readings taken at 40 and 100°C.

3.2.1.26.1 *Comments*

Each machinery application will require a specific viscosity grade lubricant depending on operating temperature, load, and speed conditions. In general, low speed, high-load, or high-temperature machines will utilize higher viscosity oil than machines that are lightly loaded or operate at higher speeds or temperatures. Thus, care is essential in selecting the correct lubricant for each specific application. First, the viscosity results determined by each ASTM approved method will generally not correspond to the viscosity reading of another. Since oil manufacturers may utilize an instrument of choice, care is required when comparing the viscosity of different products. Second, not all oil manufacturers' brands will exhibit the same viscosity for a given oil type or grade when the same viscosity measurement method is utilized. Table 3.1 is a compilation of three manufacturers' oils by grade. The table compares the range of viscosity for ISO, SAE, and AGMA oil products. The table also indicates the acceptable minimum and maximum viscosity range

TABLE 3.1 Range of Viscosity for ISO, SAE, and AGMA Oil Products

Viscosity at 40°C	Low limit	High limit	ISO R&O, AW, EP	SAE monograde	SAE multigrade	AGMA R&O	AGMA EP
2.2	1.98	2.42	ISO 2				
3.2	2.88	3.52	ISO 3				
4.8	4.14	5.06	ISO 5				
6.8	6.12	7.48	ISO 7				
10	9.0	11.0	ISO 10				
15	13.5	16.5	ISO 15				
22	19.8	24.2	ISO 22				
32	28.8	35.2	ISO 32	SAE 5			
46	41.4	50.6	ISO 46	SAE 10		#1	
50	45.0	55.0			SAE 0W30		
68	61.2	74.8	ISO 68	SAE 20	SAE 5W30	#2	2EP
68	61.2	74.8			SAE 10W30		
68	61.2	74.8			SAE 20W 20		
85	76.5	93.5		SAE 30	SAE 5W40		
100	90	110	ISO 100	SAE 30	SAE 5W50	#3	3EP
100	90	110			SAE 10W40		
100	90	110			SAE 15W40		
100	90	110			SAE 75W90		
115	103	126		SAE 30	SAE 5W50		
115	103	126		SAE 40	SAE 10W40		
115	103	126			SAE 15W40		
130	117	143		SAE 30	SAE 20W40		
130	117	143		SAE 40	SAE 75W90		
150	135	165	ISO 150	SAE 40	SAE 80W90	#4	4EP
150	135	165		SAE 50	SAE 20W50		
180	162	198		SAE 40	SAE 20W50		
220	198	242	ISO 220	SAE 50	SAE 80W140	#5	5EP
220	198	242		SAE 90			
320	288	352	ISO 320	SAE 60	SAE 85W140	#6	6EP
460	414	506	ISO 460	SAE 140		#7	7EP
680	612	748	ISO 680			#8	8EP
1000	900	1100	ISO 1000	SAE 140		#8A	8AEP
1500	1350	1650	ISO 1500	SAE 250		#9	9EP
2200	1980	2420	ISO 2200				
3200	2880	3520				#10	10EP

for each ISO oil grade. Note that the viscosity for SAE types such as SAE 30 or SAE 15W40 can vary by more than the 10% limit imposed on ISO types. Therefore, determining the actual viscosity of SAE and AGMA oils before use is an important consideration if maximum lubrication reliability is to be obtained.

International equivalent standards include (D445) IP 71, DIN 51566, and ISO 3104, and (D2270) IP 226 and ISO 2909.

3.2.1.27 Wear Prevention/Load-Carrying Properties — ASTM D2670, ASTM D2882, ASTM D4172, ASTM D5182

Load-carrying ability is generally a function of the oil's basestock lubricity and the antifriction and antistick characteristics provided by additives. The complex relationships between the oil and the various additives require different test methods and apparatus depending on the oil blend and intended machinery application.

ASTM D2670 wear properties of fluid lubricants by Falex pin and vee block determines the wear properties for fluid lubricants by means of the Falex pin-on-vee-block machine operating under prescribed conditions. This test determines the load carrying ability and wear properties of fluid lubricants by the degree of wear of a steel pin (journal) rotating at 290 rpm against a pair of v-shaped blocks immersed in the lubricant sample for a period of 15 min. The load on the pin is ratcheted to a prescribed load for the duration of the test. The wear scars on the pin and vee blocks are interpreted to determine the test fluid's wear properties.

ASTM D2882 wear properties of hydraulic fluids by constant volume pump determines the degree of wear on pump components by high-pressure hydraulic fluids. The test utilizes a rotary vane pump that pumps the test fluid through a standardized circuit at 2000 psi for 100 h. The results are determined from the weight loss of the pump cam ring and vanes.

ASTM D4172 wear preventive characteristics of lubricating fluids determines the antiwear characteristics of fluid lubricants utilizing the four-ball wear test machine. The tester utilizes a steel ball rotating at 1200 rpm against three stationary steel balls lubricated with the sample. The sample is temperature controlled at 75°C and the load between the upper and lower balls is set at 147 or 392 N. The test is run for 60 min and the results are determined from the scar diameters worn in the three lower balls.

ASTM D5182 standard test method determines the scuffing load capacity of oils used to lubricate hardened steel gears. The test utilizes the Forschungstelle fur Zahnrader und Getriebebau (research site for gears and transmissions) visual method, commonly referred to as the FZG visual method. The test is primarily used to assess the resistance to scuffing of low-additive oils such as industrial gear oils and transmission and hydraulic fluids. In this test, the FZG gear machine is operated at constant speed for a prescribed period at increasing loads until failure criteria is reached. The FZG method is not appropriate for evaluating high EP oils as the load capacity of these oils generally exceeds the capacity of the test rig.

3.2.1.27.1 Comments

Gear tests are a proven means for assessing the load-carrying capacity of lubricants and the influence of various antiwear, antiscuffing, and extreme-pressure additives. Since these tests are intended to evaluate relatively low-viscosity oils for circulating systems, they are not generally suitable for evaluating compounded fatty lubricants used for open or hypoid gears. In addition to the above methods, the WAM high-speed load capacity test method has been submitted to SAE E-34C for approval in AIR4978 "Temporary Method for Assessing the Load Carrying Capacity of Aircraft Propulsion System Lubricating Oils." The test utilizes the WAM1, WAM3, or WAM4 machines to provide Ryder-like load capacity data of gas turbine and gearbox oils.

International equivalent standards include (D2882) IP 281 and DIN 51389, (D5182) DIN 51354, IP 334, and IP 166.

3.2.2 Semisolid Grease Lubricant Tests

3.2.2.1 Apparent Viscosity of Greases — ASTM D1092

The apparent viscosity of grease is defined as the ratio of shearing stress to rate of shear calculated from Poiseuilles' equation for viscosity using capillary tubes. The ratio varies with shear rate that is proportional to the linear velocity of flow divided by the capillary radius. Rheological properties of grease such as apparent viscosity can be determined with a rather complex apparatus, the principal elements of which are a pressure cylinder containing a floating piston and attached capillary flow tubes. A calibrated hydraulic pumping system and temperature-controlled test chamber complete the test setup.

In ASTM D1092, the chamber temperature is controlled over the range of −54 to 38°C. The procedure employs eight capillary tubes and two pumping speeds to determine the apparent viscosity at 16 different shear rates. Because of grease thickness, measurements are limited to ranges of 25 to 100,000 P at 0.1 sec, and 1 to 100 P at 15,000 sec. At very low temperatures, the shear rate range may require reduction because of the force needed to move the grease through the smaller capillaries.

3.2.2.1.1 Comments

Apparent viscosity of grease is of general use in considering flow problems occurring in the distribution and dispensation of grease from a central source through a network. Volume rate of flow, pressure drop in the system, etc. can be predicted and adjustments made to accommodate the lubrication needs of specific systems and various operating conditions. Apparent viscosity depends on the type of oil and the type and amount of thickener used in the grease.

3.2.2.2 Cone Penetration Test — ASTM D217, ASTM D1403

Grease consistency describes the relative softness or hardness of grease. Grease consistency depends on the type and viscosity of the base oil and the type and proportion of the thickener used in the formulation of the grease. The NLGI classifies lubricating grease according to the ASTM D217 cone penetration test. The D217 test instrument measures grease penetration depth at a temperature of 25°C, for four categories of grease preparation:

1. *Unworked sample:* The grease sample is subjected to minimum disturbance in its transfer to the grease test vessel.
2. *Worked sample:* The grease sample is subjected to the shearing action of 60 double strokes in a standard grease worker.
3. *Prolonged worked sample:* The grease sample is subjected to the shearing action of 60 double strokes at a temperature of 15 to 30°C, followed by an additional 60 double strokes at a temperature of 25°C (standard grease worker).
4. *Block penetration sample:* This test indicates the penetration at 25°C on the freshly prepared face of a cube of grease that is sufficiently hard to maintain its shape.

The results of the ASTM D217 cone penetration test are indicated by the depth (in tens of millimeters) up to which the cone descends into the grease sample. Alternatively, the cone penetration number of grease products may be determined with one-quarter and one-half scale instruments using the ASTM D1403 standard method.

3.2.2.2.1 Comments

The higher the consistency number, the harder the grease product, as indicated by the lower depth penetration distance. Harder greases have less mobility and will tend to stay in place longer; however, flow and pumpability characteristics may suffer. Greases that are too soft can increase rotational friction due to churning. Consequently, it is important that grease consistency is sufficient to maintain a reliable lubricant film for the machinery application. Most general-purpose greases tend to have an NLGI consistency of one or two. Table 3.2 indicates the standard NLGI consistency grades as determined by the ASTM D217 cone penetration test.

When the cone penetration test is performed at a temperature of 0°C on the same grease type, the sample will generally be one to two NLGI consistency numbers higher (harder). Conversely, grease will generally indicate one NLGI consistency number lower (softer) when the test is performed on the same grease at 43°C.

International equivalent standards include (D217) IP 50 and ISO 2137.

3.2.2.3 Copper Corrosion Resistance — ASTM D4048

Copper alloys are utilized in many greased components and it is essential that these parts are properly lubricated by the grease and that the grease is noncorrosive to metallic copper. Since grease is essentially thickened lubricating oil that may contain compounds that are corrosive to copper, it is desirable to test grease products for their corrosion tendency.

TABLE 3.2   NLGI Consistency Grades Determined by the ASTM D217 Cone

NLGI consistency number	Worked penetration range (tens of millimeters)
000	445 to 475
00	400 to 430
0	355 to 385
1	310 to 340
2	265 to 295
3	220 to 250
4	175 to 205
5	130 to 160
6	85 to 115

In the ASTM D4048 test, a polished copper strip specimen is immersed into a specified volume of grease at 100°C for a period of 24 h after which it is compared with the copper strip corrosion standard to determine the degree of tarnishing or corrosion.

3.2.2.3.1   *Comments*

The copper corrosive tendency results have some value in providing a product-to-product or batch-to-batch assessment of the compositional quality, consistency, and uniformity of the grease.

International equivalent standards include (D4048) IP 112, ISO 2160, and DIN 51811.

3.2.2.4   Dropping Point of Grease — ASTM D566, ASTM D2265

The dropping point is defined as the temperature where grease passes from a semisolid to a liquid state. The dropping point temperature determines the maximum service temperature range for grease.

The ASTM D566 test procedure determines the dropping point of grease under controlled conditions. The test apparatus includes a test cell that is immersed in a 400-ml Pyrex™ bath for heating at the prescribed rate. An integral heater, temperature controller, and stirrer maintain uniform control of the temperature and its rate of rise up to 288°C. The test sample is heated incrementally until the dropping point is observed. The temperature at which a liquid drop first falls from the sample cup is reported as the dropping point of the sample. The ASTM D2265 standard test method may be utilized to determine grease dropping point over a wider temperature range up to 316°C. This test is usually performed on high-temperature grease products.

3.2.2.4.1   *Comments*

Grease must not soften and flow during normal operating temperatures. Note that grease consistency and firmness will change when the machinery warms up from ambient to operating temperature. The dropping-point property is useful for checking the quality and uniformity of a specific manufacturer's product and for comparing various grease brands for a specific application. Do not estimate consistency or dropping point by look-and-feel observations. Grease will be much softer in summer than in winter and also when the operating temperature is above ambient. Pay particular attention to automatic grease dispensers that must operate over wide temperature ranges.

Since most grease products are semi-solid mixtures of oil and one or more kinds of soap or thickening agents, they do not usually exhibit well-defined melting points. Therefore, the dropping point only approximates this characteristic and has limited significance in terms of service performance. Many other factors (i.e., design, speeds, loads, evaporation losses, thermal cycles, etc.) greatly influence useful operating temperature of the grease and, therefore, must be recognized in lubricant selection for a particular application.

International equivalent standards include (D566) IP 132, ISO 2176, and DIN 51806, and (D2265) IP 132, DIN 51806 and ISO 2176.

3.2.2.5 Extreme-Pressure Properties of Grease — ASTM D2509, ASTM D2596, ASTM D5706

One of the most important attributes of grease is the maximum load it can sustain without seizure or scoring of the lubricated parts.

The ASTM D2509 utilizes the Timken wear test machine for rating the load-carrying and EP properties of grease lubricants. The Timken apparatus employs a hardened steel ring rotating against a flat steel block that is lubricated by the test lubricant. The maximum load applied without scoring is the Timken OK load (pass). The minimum load that causes scoring is reported as the score load.

The ASTM D2596 test utilizes the Falex four-ball test apparatus where four half inch diameter steel balls (AISI E-52100) are arranged with three balls clamped together so as to cradle and load the top ball. This top ball is then rotated on a vertical axis. The EP property of a grease product is determined from the load-wear index. This property is derived from 10 wear tests of varying loads up to and including the point at which seizure occurs (weld point).

The ASTM D5706 test determines the EP properties of lubricating greases using a high-frequency, linear-oscillation SRV machine. The SRV machine utilizes a steel ball oscillating against a lubricated steel disk. Load is increased in 100 N increments until seizure occurs.

3.2.2.5.1 *Comments*

These tests compare the relative EP characteristics of grease products and are useful in grease product quality control and to compare different grease products for a particular application. Note that there is no general correlation between EP test results and anticipated in-service performance.

International equivalent standards include (D2509) IP 326 and (D2596) DIN 31350.

3.2.2.6 Leakage Tendencies of Wheel Bearing Grease — ASTM D1263, ASTM D4290

Greases used to lubricate automotive wheel bearings must operate under conditions of high speed, high load, and excessive braking forces. Greases subject to oil leakage from mechanical or thermal softening effects can lead to bearing failure.

The ASTM D1263 test method determines the leakage tendencies of wheel bearing greases when tested under prescribed laboratory conditions. The grease is applied to a modified front wheel hub bearing assembly. The hub is rotated at a speed of 660 rpm for 6 h at a temperature of 105°C. The result is determined from the leakage of the grease or oil, or both, in grams and the condition of the bearing surfaces.

The ASTM D4290 accelerated leakage tendency test evaluates the leaking tendency of automotive wheel bearing greases in a modified automotive front wheel hub-spindle-bearings assembly. The D4290 test employs severe conditions including 111 N thrust load, 1000 rpm, and a spindle temperature of 160°C to induce grease deterioration and failure. The test continues for a 20-h period, after which leakage of grease and oil is measured and the bearings are washed and examined for deposits of gum and varnish. The amount of leakage is the test result.

3.2.2.6.1 *Comments*

The test provides a basis for comparing the relative tendencies of different greases to leak from a simulated wheel bearing operation. Since the evaluation is performed under accelerated conditions, it does not equate to actual longtime service nor reproduce actual highway stresses involving load, vibration, and temperature. It is possible, however, to distinguish products having decidedly different leakage characteristics and to judge qualitative changes (i.e., slumping, softening, etc.) in a grease. The value of results is enhanced when used in combination with other stability tests, for example, oxygen bomb stability, roll stability, etc.

3.2.2.7 Life Performance of Grease Products — ASTM D1741, ASTM D3336, ASTM D3337, ASTM D3527

Usable life is an essential consideration when purchasing lubricating grease products. The usable life of grease is related to its ability to resist oxidation and mechanical stresses in the presence of machinery metals and atmospheric oxygen and water.

The ASTM D1741 test describes the measurement of life performance and leakage characteristics under prescribed operating conditions at temperatures up to 125°C. The test apparatus utilizes a cleaned 306 ABEC Class 3 bearing packed with the candidate grease product. The bearing is "run in" at 3500 rpm for about 2 h to generate an equilibrium state and then run to failure. The time to failure is the reported result. In the leakage evaluation test, the test bearing is operated on a 20/4-h on/off cycle until failure is observed.

The ASTM D3336 determines usable life performance employing a high temperature test (up to 371°C). The test utilizes a clean 204 size ball bearing packed with a weighed amount of grease equivalent to 3.2 cm^3. After "working in" the grease by hand rotating the bearing, the system is driven at 10,000 rpm for about 1.5 h to establish equilibrium test conditions and then operated until failure occurs. The time to failure is the reported result.

The ASTM D3337 test employs a small R-4 size, AFBMA Class 7 ball bearing. The method specifies that the bearing is packed $\frac{1}{3}$ full with the grease-under test and immediately installed on the test spindle. The bearing is run in for about 100 revolutions at a speed of less than 200 rpm. The test is conducted at 12,000 rpm until failure. During the test, total test hours, torque-meter tare, torque meter reading, net torque, torque in gram-centimeter, and outer-race bearing temperature are recorded.

The ASTM D3527 life performance test evaluates the high-temperature stability of automotive wheel bearing greases in a modified automotive front wheel hub-spindle-bearing assembly. The D3527 test employs severe conditions (25 lbf [111 N] thrust load, 1000 rpm, 160°C spindle temperature) to induce grease deterioration and failure. The test continues in a 20/4 h on/off cycle until grease breakdown causes the measured drive motor torque to increase past an established endpoint. The number of hours to failure is taken as the test result.

3.2.2.7.1 Comments

The above test methods are screening techniques that may be used for the evaluation and selection of candidate greases for particular applications. Since these tests do not replicate in-service conditions, improved operation in the test rig may not correlate to improved in-service performance. The tests cannot distinguish between minor differences in performance life.

3.2.2.8 Low-Temperature Torque Characteristics — ASTM D1478, ASTM D4693

Greases become more solid-like at very low temperatures and consequently require a greater force to overcome resistance to movement.

The ASTM D1478 test measures the torque (gram centimeters) required to withhold the movement of the outer ring of a 204 size open ball bearing packed with the test grease as the inner ring is rotated at 1 rpm at a prescribed test temperature ranging from −75 to −20°C. The running torque is determined after 10-min rotation of the bearing.

The ASTM D4693 test determines the low-temperature torque of grease-lubricated wheel bearings. The test grease is packed into a specially manufactured, spring-loaded, automotive type wheel bearing assembly. The assembly is heated and cold soaked to −40°C. The spindle is rotated at 1 rpm and the torque required to prevent rotation is measured after a period of 60 sec.

3.2.2.8.1 Comments

As with apparent viscosity, the low-temperature torque displayed by a grease depends on the type and amount of thickener as well as the viscosity characteristics (including temperature coefficient) of the lubricating fluid. Low-temperature torque of greases is of special interest for military, space, and certain applications involving small bearings and relatively limited available torque, for example, mechanical timing devices.

3.2.2.9 Oil Evaporation and Oil Bleed — ASTM D972, ASTM D2595

The proper operation of grease-lubricated components requires the oil base, additives, and thickener to remain consistent during machinery operation. When the ratio of oil to thickener decreases, the grease hardens and lubrication is impaired. Oil separation is the result of two different processes, evaporation and bleed:

1. Oil evaporation refers to the loss of volatile light-ends of the base-oil or additive constituents at high operating temperatures. Oil evaporation causes grease to thicken into a dry paste-like consistency unsuitable for lubrication.
2. Oil bleed refers to the process of oil separation or seeping from grease during operation. This condition is easily recognized by the oily appearance of the greased area and often the formation of little pools of oil. Oil bleed will harden grease, rendering it unsuitable for use.

The ASTM D972 and ASTM D2595 tests determine the evaporation loss tendency of lubricating grease over a wide temperature range. The test apparatus consists of an evaporation cell placed into a temperature-controlled chamber. A helical heating tube wrapped around the cylinder circumference provides the source of a controlled flow of evaporating air as it passes over the surface of the sample and then out through a central stack. Twenty grams of the grease sample is placed in the cell and heated to the desired temperature for a period of 22 h. The D972 test is limited to a temperature up to 149°C. The D2595 augments D972 by utilizing a similar test cell placed into a thermostatically controlled aluminum block oven to determine evaporation losses up to a temperature of 316°C. The sample is weighed before and after the test and the difference is taken as the weight loss due to evaporation.

3.2.2.9.1 Comments

Evaporation losses may include by-products from thermal cracking in addition to egress of volatile ingredients in the grease blend. As these materials evaporate, the grease will also undergo an increase in viscosity. Ultimately, the lubricant will be completely degraded leaving a hard or tarry residue. Note the evaporation test is static and the results cannot be directly interpreted in terms of dynamic in-service conditions.

3.2.2.10 Oil Separation from Lubricating Grease — ASTM D1743, ASTM D6184

Oil loss due to separation causes lubricating grease to thicken and harden, leading to a serious lubrication problem.

The ASTM D1743 standard test method for oil separation from lubricating grease during storage determines the tendency of lubricating grease to separate during storage in a 35-lb pail. The sample is placed on a sieve inside a special test cell and subjected to 0.25 psi (1.72 kPa) air pressure at constant temperature. Any oil that bleeds from the grease during a 24-h period is collected in the cell and weighed.

The ASTM D6184 determines the tendency of oil to separate from lubricating grease at an elevated temperature. The test apparatus consists of a 60-mesh nickel gauze cone with wire handle, mounted inside a 200-ml covered beaker. The grease sample is placed in the wire gauze cone and heated to a test temperature for a period of time. The sample is weighed before and after the test to determine the weight loss from oil separation. The test method is conducted at 100°C for 30 h unless other conditions are required by the grease specification. The test is not suitable for greases having an ASTM D217 penetration greater than 340 (softer than NLGI No. 1 grade).

3.2.2.10.1 Comments

Grease products that are stored for extended periods and products that are intended for use in high-temperature applications should be verified for oil separability before use.

Note the ability of grease to be reliably delivered by a pump is governed by the properties of pumpability and slumpability. Grease pumpability is the ability of a grease product to be pushed through plumbing and nozzles of a powered grease applicator. Slumpability is the ability of a pump to draw grease from a reservoir. Oil separation and subsequent grease hardening will affect the pumpability and slumpability properties of the grease. The pumpability and slumpability of grease products are related to the following factors:

1. Grease products made from highly viscous base oils generally exhibit poor pumpability at low operating temperatures.
2. Grease products with a smooth buttery consistency will generally exhibit good pumpability but poor slumpability.
3. Grease products with a fibrous consistency will generally exhibit good slumpability but poor pumpability.

Choosing the correct grease for use in a powered applicator is essential to effective lubrication of downstream components.

International equivalent standards include (D1743) DIN 51817 and IP 121.

3.2.2.11   Oxidation Stability — ASTM D942

As with oil lubricants, grease products will also oxidize and degrade from the moment of use. Atmospheric oxygen will react with grease constituents and form acidic by-products, which will eventually degrade into sludge and varnish deposits. Since grease lubrication systems generally contain a small quantity of grease, good oxidation stability is a key requirement to a reasonable service life.

The ASTM D942 oxidation stability test determines the oxidative characteristics of grease by means of a pressure drop in oxygen after a prescribed time period has elapsed. The test apparatus includes five tiered shelves that support a number of grease sample dishes. The shelves are placed into a steel pressure vessel and initially pressurized with oxygen to 758 kPa and heated to a temperature of 99°C. At the end of the test period, the pressure drop indicates the amount of oxygen that was consumed by the grease.

3.2.2.11.1   Comments

Changes in the observed oxygen gas pressure are the combined result of gas absorption and reaction processes. A plot of pressure vs. time will normally show a characteristic induction period (gradual decrease) after which the pressure drops sharply, indicating the beginning of the autocatalytic oxidation process and thus the end of usable life. Greases exhibiting lower pressure drops over the prescribed test period will generally have a lower tendency to deteriorate in storage.

International equivalent standards include DIN 51808 and IP 142.

3.2.2.12   Rust Prevention Characteristics — ASTM D1743, ASTM D6138

The ability of grease products to provide a barrier to water and atmospheric oxygen and prevent the rusting of machinery components is an important property.

The ASTM D1743 test determines the corrosion preventive properties of greases using grease-lubricated tapered roller bearings stored under wet conditions at 52°C for 48 h after which they are checked for the degree of tarnishing or corrosion. Prior to the test, the bearings are operated under a light thrust load for 60 sec to distribute the lubricant. This test method is based on CRC Technique L 41, which correlates laboratory results with in-service grease-lubricated aircraft wheel bearings.

The ASTM D6138 corrosion preventive properties test for lubricating greases utilizes a clean lubricated double-row self-aligning ball bearing. The bearing is packed with grease, partially immersed in water, and operated under a prescribed sequence with no load at a speed of 80 rpm for a period of one week after which it is checked for the degree of tarnishing or corrosion (Emcor test).

3.2.2.12.1 Comments

These methods can distinguish between good and poor-performing greases, but are not suitable for ranking manufacturer brands within a specific product category.

International equivalent standards include (D6138) DIN 51802, ISO 11007, and IP 220.

3.2.2.13 Water Wash out Characteristics — ASTM D1264, ASTM D4049

Many grease-lubricated components are normally exposed to environmental moisture. Water contamination generally affects grease reliability by reducing its load-carrying ability due to the formation of water–oil emulsion or by changing its texture, consistency, and/or tackiness.

The ASTM D1264 water washout test evaluates the ability of lubricating grease to adhere to operating bearings when subjected to a water spray. A grease sample is packed into a ball bearing test specimen and subjected to a steady water stream under controlled test conditions. The test result is the percentage of grease (by weight) washed out in a 1-h period, tested at 38 and 79°C.

The ASTM D4049 water washout test evaluates the ability of lubricating grease to adhere to a metal surface when subjected to a water spray. The grease sample is applied to a stainless steel test panel and subjected to a direct water spray at a specified pressure (40 psi), temperature, and time period. The test result is the percentage of grease (by weight) sprayed off after a 5-min period.

3.2.2.13.1 Comments

Response of greases to free water can involve rather complex chemical and physical effects. Some grease products naturally resist water ingress or incorporate it as droplets; other greases absorb water while forming "emulsions," performing normally and otherwise remaining relatively unchanged. The water washout test does not duplicate in-service conditions and results are subjective with relatively poor precision. However, the results are useful for assessing the water washout behavior of greases and comparing candidate greases for a particular application.

International equivalent standards include (D1264) IP 215 and DIN 51807.

3.2.2.14 Wear Prevention/Load-Carrying Properties — ASTM D2266, ASTM D3704, ASTM D4170

Load-carrying ability is generally a function of the oil's basestock lubricity and the antifriction and antistick characteristics provided by additives. The complex relationships between the oil and the various additives require different test methods depending on the oil blend and intended machinery application. Falex, Timken, or similar bearing and gear type wear machines are generally used to determine the wear and antiwear or extreme-pressure additive characteristics of lubricating oils and greases. The ASTM has standardized a number of methods utilizing these wear machines.

The ASTM D2266 wear preventive characteristics of lubricating grease by Falex 4-ball determines the wear characteristics of grease lubricants in sliding steel-on-steel applications. The tester employs a steel ball rotating against three stationary steel balls lubricated with the test sample. The results are reported as dimensions of scar marks made on the ball surface.

The ASTM D3704 wear preventive characteristics of lubricating grease by Falex block-on-ring test machine in oscillating motion determines the wear properties of grease lubricants in oscillating or sliding steel-on-steel applications. The test is used to differentiate between greases with different sliding wear properties. The tester employs a steel ring oscillating against a steel block lubricated with the test sample. The test speed, load, time, angle of oscillation, and specimen finish and hardness are varied to simulate real world machinery conditions. The results are reported as dimensions of scar mark made on the block by the ring oscillation.

The ASTM D4170 fretting wear protection by lubricating greases determines the fretting wear potential and level of protection of greased components and is utilized to differentiate greases with superior antifretting characteristics. The test utilizes two antifriction thrust bearings lubricated with the test sample and operating under prescribed vibratory and load conditions. The test is run at room temperature for 22 h. The result is the weight loss to the upper and lower bearing races.

3.2.2.14.1   *Comments*

These tests are useful screening tools for comparing greases for friction and wear characteristics. However, for precise antiwear results, tests using the specific bearings and/or gears and operated under the load, speed, temperatures, etc. that are representative of the intended machinery application are recommended. Note that fretting is a destructive wear mode caused by low-level oscillatory motion between machinery parts. It frequently occurs in stored, standby, and emergency machinery that are shut down for long periods.

3.2.3   Solid Grease Lubricant Tests

3.2.3.1   Adhesion of Solid Film Lubricants — ASTM D2510

Effectiveness of a solid film lubricant system is dependent on its adhesion to the surfaces to be lubricated. It is important therefore to measure the "bonding" quality of the lubricant to these surfaces.

In ASTM D2510 test, specially prepared aluminum test panels are spray coated to produce a dry film thickness of between 0.005 and 0.013 mm. After measuring the actual dry film thickness deposited, the panels are immersed in a beaker of water (or other fluid) to half the depth of the coating for a 24-h period at room temperature. The panel and coated surface are then dried and the wetted and unwetted surfaces scratched with a stylus. Masking tape placed perpendicular to the parallel scratches is pressed on with a roller. The tape is then stripped abruptly from the surface. Any evidence of film damage or exposure of bare metal is reported as well as any other test conditions or observations that are important to interpretation of the results.

3.2.3.1.1   *Comments*

The procedure provides a means of assessing the effectiveness and continuity of a solid-film lubricant coating process. Since several techniques may be used for applying a particular coating, testing the relative "adhesiveness" of one type of material vs. another should be conducted on a comparable basis for meaningful results. Test conditions may be varied to accommodate specific application requirements.

3.2.3.2   Corrosion of Solid Film Lubricants — ASTM D2649

Since solid film lubricants may contain chlorine, sulfur, and other elements, the corrosiveness of solid film lubricants is a necessary consideration.

In ASTM D2649 test, a solid film lubricant is applied to precleaned aluminum test panels to produce a cured film thickness of between 0.005 and 0.013 mm. A coated panel is then located against an uncoated panel, positioned at a 90° angle in a channel fixture, and placed under a constant load by applying a torque of 2.8 Nm to the nut and bolt clamp. The test assemblies, including blanks, are preheated at 65.6°C for 2 h and then placed in a humidity (95° RH) chamber (49°C). After 500 h, the test units are disassembled and both coated and uncoated surfaces inspected for any pitting, etching, or other evidence of corrosion.

3.2.3.2.1   *Comments*

The method provides a technique for evaluation of corrosion tendencies of applied solid film lubricants. The test, being qualitative, is subject to interpretation. The results can be highly dependent on the procedure, materials and treatments used to obtain the film. Although the evaluation is performed using aluminum, consideration should be given to use of other test metals. It should also be recognized that solid film lubricants generally do not exhibit rust preventive characteristics.

3.2.3.3   Lubricating Qualities of Solid Lubricants — ASTM D1367

Graphite and similar types of solid materials have proven to be useful lubricants alone and when combined with other lubricants such as oils or greases.

The ASTM D1367 is designed to evaluate the lubricating qualities of graphite and also may be used for evaluating other similar solids. To conduct the test, 36 g of the powdered solid are added to 204 g of paraffin base oil having a viscosity of 20.5 to 22.8 cSt at 37.8°C. A cleaned, weighed double-row test ball bearing is mounted on a test drive shaft and the assembly is lowered into a special beaker containing the test mixture. After a 2-h operation at 1750 rpm at room temperature, the test bearing is removed, cleaned, and weighed to determine its weight loss.

3.2.3.3.1 Comments

The test method provides a means for quality control checking the lubricating quality of solid lubricants. Although not a precision method, results may be used for comparative evaluation of different types of solid lubricants used in combination with other lubricating fluids. When the carrier fluid is run separately to establish a "norm," a means for identifying the contribution of the solid lubricant to the lubricating process can be determined. Addition of the "lubricating" solid can cause either an increase or decrease in the wear and load carrying capacity of the fluid lubricant. Possible interactions between additive components in fluids and solid lubricants also may be examined.

3.2.3.4 Thermal Shock Sensitivity of Solid Film Lubricants — ASTM D2511

In applications where extreme temperature variations occur, measuring the capability of a solid film to withstand thermal shock is useful. To perform a test, the solid film lubricant is applied on test panels so as to obtain a finished film thickness of between 0.005 and 0.013 mm that is accurately determined by micrometer measurement. The test panels are first heat soaked at 260°C for 3 h and then immediately cooled to −54°C for 3 h. Subsequently, the test panels are returned to room temperature and examined visually for damage. If the surfaces are intact, additional testing to determine the adhesion of the film (ASTM D2510) may be needed. The same criteria of failure, that is, flakes, cracks, blisters, etc. are evidence of surface coating deficiencies and are noted in the report of the results.

3.2.3.4.1 Comments

As with the adhesion test (ASTM D2510), this evaluation assesses the bond strength of solid film lubricants subjected to extreme thermal stress. The evaluation is more qualitative than quantitative and, hence, is subject to individual judgment.

3.2.3.5 Wear and Load-Carrying Capacity of Solid Lubricants — ASTM D2625, ASTM D2981

Wear characteristics and load-carrying ability of solid film lubricants are important for metal-to-metal sliding applications.

The ASTM D2625 employs the Falex lubricant tester. To determine wear life, two lubricant-coated stationary steel V-blocks are pressed at a specific load against a rotating lubricated steel pin until the solid film is depleted, evidenced by a torque increase of 1.13 Nm above the steady state value. For measuring load-carrying ability, the test is repeated with incremental increases in load until a sharp increase in torque (1.13 Nm) or breakage of the pin occurs. Load-carrying capacity is defined as the highest gage load sustained for a 1-min period. The test equipment is identical to that employed for measurement of the wear properties of fluid lubricants.

The ASTM D2981 test evaluates the wear life of a bonded solid film lubricant under oscillating motion by means of a block-on-ring friction and wear test machine. The test machine comprises a coated steel test ring oscillating against a steel test block. The oscillating speed is 87.5 cpm at 90° arc. The test is run in under a load of 283 kg and then operated at a prescribed load until failure.

3.2.3.5.1 Comments

The ability of the lubricating film to adhere to the applied surface and provide effective separation between moving parts is a prerequisite. These methods offer an effective means of evaluating lubricity and life performance characteristics of various solid lubricants and are useful for comparing the consistency of production lots. Since the test involves the coating of solid film lubricant on test pieces, the results obtained are highly dependent upon the application technique.

3.3   Oil Condition Tests

As all machinery and fluids will begin to deteriorate from the moment of use, fluids should be monitored on a regular basis to determined suitability for continued use. In addition to laboratory analysis, high-value or critical machinery may utilize real-time sensing systems to provide a continuous data stream. The data from both laboratory and real-time sensing are interpreted in conjunction with known machine failure modes, machine usage, configuration, and maintenance data to render an expert opinion on the condition of the machine/oil system [15].

Oil condition testing seeks to determine the level and nature of contaminants. The analysis will comprise methods to detect and quantify the concentration of metals, water, dirt, fuel, insolubles (soot), glycol (antifreeze), degradation by-products, and any other process contaminant that may be key to fluid condition. The primary test methods utilized for oil condition monitoring are presented.

3.3.1   Atomic Emission Spectroscopy --- ASTM D5185, ASTM D6595

Elemental spectroscopy determines the elemental constituents of lubricant fluids by raising their atoms to an excited energy state in a high-temperature source.

The ASTM D5185 inductively coupled plasma (ICP) spectrometry utilizes a radio frequency induction heater to create a hot argon gas plasma torch. The oil sample is atomized and carried by argon gas into the torch. The very high temperature excites metal atoms, which radiate their characteristic emission lines. The emission lines are captured and measured by a digital array sensor or a series of photomultiplier tubes. The ICP can provide parts-per-billion (PPB) sensitivity for compounds and wear metal particulates less than 3 μm in size.

The ASTM D6595 rotary disk emission (RDE) spectrometry is so called because the sample fluid is transported into a high-temperature arc by means of a rotary carbon disc. The disc is immersed into the sample vessel and picks up oil and wear metals as it turns. The arc raises the energy states of the metal atoms in the oil causing them to emit their characteristic emission lines. The lines are measured by the optical system. RDE spectrometers measure particles up to about 10 μm in size at parts-per-million (PPM) sensitivity.

3.3.1.1   Comments

Contaminants and additives that contain metals can be quantified by AES methods. Note that elemental spectrometers are not created equal. Because different instruments utilize different methods of excitation, different excitation energies, and different detection systems, readings do not correlate. If good oil analysis performance is to be achieved, the same instrument type should be utilized to analyze all samples.

3.3.2   Infrared (IR) Spectroscopy --- ASTM (E2412)

Used-oil condition monitoring requires the evaluation of a number of the fluid condition and any contaminants including additives, oxidation by-products, and ingress contaminant materials. Molecular analysis by Fourier transform infrared (FT-IR) spectroscopy provides a reproducible means of determining a wide range of fluid degradation (oxidation, nitration, and sulfation by-products) and contamination (water, fuel, antifreeze, and soot) faults. The FT-IR spectrometer analyzes the chemical structures in an oil sample by evaluating the spectral response of a mid-infrared beam from $\sim$4000 through $\sim$650 cm^{-1} wavelength range.

The beam passes through the oil sample and is altered by the characteristic absorbencies of the various oil and contaminant molecules. A detector and electronic circuit pick up and convert the beam into an audio frequency, which is converted to spectral data by a fast Fourier transform (FFT) software program.

3.3.2.1   Comments

The multiple oil degradation and contamination symptoms generated by in-service oil applications must be individually quantified to determine the cause and thus the remedy. FT-IR provides an excellent means of determining oil contamination and degradation progress as it quantifies the chemical structures (failure mode symptoms) associated with these materials. Note that certain oil additives interfere with IR water detection methods such as demulsifier additives found in hydraulic and steam turbine oils. For these oil types, water concentration may be determined by Karl Fischer titration (KFT).

In addition to benchtop FT-IR analyzers, there are also a number of real-time, IR sensors available. In most cases, IR sensors are miniaturized, ruggedized solid-state versions of traditional IR spectrometers that can be mounted on operating machinery fluid systems. In this regard, they can withstand machinery operating temperatures and vibrations, are autonomous, and eliminate the need for sampling and remote lab analysis.

3.3.3   Particle Counting --- ISO 4406, ISO 11171

Fluid cleanliness is also a key factor for machinery system reliability, especially for hydraulic and precision bearing systems. Insoluble particulates due to dirt ingress and oil degradation will increase with equipment use. These materials will cause silting, gumming and abrasion of machinery parts. In practice, there is a direct relationship between oil cleanliness and component usable life — the cleaner the system, the longer the life of the oil and oil wetted parts. With supplementary filtering and clean oil makeup, the particle count can be expected to maintain a nominal value or "dynamic equilibrium." This "baseline" represents the normal condition of the system. Any increase in particulate counts above the baseline indicates an increase in contamination, regardless of whether the particles are wear metals, ingress dirt, or oil degradation by-products. Electronic particle counting is the preferred method of monitoring system cleanliness. Since the number of particles per milliliter increases dramatically as particle size diminishes, both the size and count data must be interpreted to determine the potential effect on a given machinery/fluid system. The interpretation is further complicated by the fact that even small changes in the count bin sizes have dramatic effect on counts recorded. To overcome this problem, the ISO community established the ISO 4406 and ISO 11171 standards for fluid cleanliness monitoring.

The 4406 standard characterizes fluid cleanliness in increments from 0.01 to 2,500,000 particles per milliliter of sample. The 4406 standard specifies three bin ranges and a series of numerical codes to indicate the count in each bin, where:

- The first code indicates the particle count above 4 μm
- The second code indicates the particle count above 6 μm
- The third code indicates the particle count above 14 μm

The system is open-ended and can expand in either direction, above or below the current codes. In practice, particle count data from a counter is compared to the ISO 4406 table (Table 3.3) for determination of the ISO cleanliness rating for each size range. For example, a sample containing 80–160 particles per milliliter greater than 4 μm, 20–40 particles per milliliter greater than 6 μm, and 5–10 particles per milliliter greater than 14 μm would generate a cleanliness code of 14/12/10.

The ISO 11171 standard defines the requirements for particle counter measurement calibration and data reproducibility. The standard covers flow rates, coincidence error, resolution, and sensor and volume accuracy. It also ensures that all instruments are calibrated to the same National Institute of Standards and Technology (NIST) traceable standards.

3.3.3.1   Comments

For simple hydraulic systems, the particle count will directly correlate with system reliability. However, for machinery systems containing bearings and gears, particle counting does not indicate the nature of the debris and other analysis techniques should be used when high particle counts are observed.

There are a number of particle counters available in bench top and portable versions. In addition, there are a number of online particle-count sensors available for real-time monitoring of equipment fluid systems. The sensors are extensions of the benchtop light-extinction particle counter technology, packaged in a rugged casing to withstand the rigors and high pressures of online machinery fluid systems. The sensors

TABLE 3.3   Determination of the ISO Cleanliness Rating

ISO code	Min. count	Max. count	ISO code	Min. count	Max. count
1	0.01	0.02	15	160	320
2	0.02	0.04	16	320	640
3	0.04	0.08	17	640	1,300
4	0.08	0.16	18	1,300	2,500
5	0.16	0.32	19	2,500	5,000
6	0.32	0.64	20	5,000	10,000
7	0.64	1.3	21	10,000	20,000
8	1.3	2.5	22	20,000	40,000
9	2.5	5	23	40,000	80,000
10	5	10	24	80,000	160,000
11	10	20	25	160,000	320,000
12	20	40	26	320,000	640,000
13	40	80	27	640,000	1,300,000
14	80	160	28	1,300,000	2,500,000

evaluate count data with preset limits and generate automatic alarms for immediate operator attention such as green-light/red-light.

It should be noted that while most instruments have good measurement repeatability, different manufacturers' instruments and sensors will not give the same results for the same sample if the instrument does not meet ISO 11171. In addition, entrained water and air bubbles affect all particle counters. Some models handle air bubbles. However, water bubbles are counted as particles by all models. These problems can result in unreliable counts especially for the smaller particle sizes.

3.3.4   X-Ray Fluorescence Spectroscopy

X-ray fluorescence (XRF) spectroscopy is similar to AES except that x-ray energy rather than heat is used to stimulate the metal atoms in the sample. The x-ray source raises the energy level of the atoms in the sample, resulting in a corresponding release of x-ray energy from the excited atoms. The ASTM D4927, ASTM D6443, and ASTM D6481 are specifically for determining additive metals in unused oils.

3.3.4.1   Comments

XRF spectroscopy can monitor solids, fluids, or dissolved metals in oil (whether naturally occurring, introduced as additives, or resulting from normal wear processes in machinery) and can measure particles of all sizes. In addition to benchtop XRF spectrometers, there are a number of XRF sensors for off- and online elemental analysis. These sensors utilize both flow-through and filter patch technologies.

3.3.5   Water Determination by Karl Fischer Titration --- ASTM D6304

The ASTM D6304 test determines water level by titrating a measured amount of the sample and the Karl Fischer reagent. The reagent reacts with the OH molecules present in water and depolarizes an electrode. The corresponding current change is used to determine the titration endpoint and calculate the concentration value for water present.

3.3.5.1   Comments

Note that many other compounds including some oil additives contain OH molecules. These will be counted as water, skewing the results. Also, the most common interfering materials are mercaptans and sulfides. High-additive package lubricants such as motor oils will generate erroneous results due to additive interference. Consequently, the Karl Fischer test should not be used to evaluate water in crankcase lubricants.

References

[1] National Lubricating Grease Institute (NLGI), www.nlgi.org.

[2] American Petroleum Institute (API), www.api.org.

[3] Society of Automotive Engineers (SAE), www.sae.org.

[4] Coordinating Research Council, Inc. (CRC), www.crcao.com.

[5] Caterpillar Machine Fluids Recommendations, Document SEBU6250-11, Caterpillar Inc. October 1999.

[6] General Motors Maintenance Lubricant Standard LS2 for Industrial Equipment and Machine Tools, Edited by General Motors Corporation, LS2 committee on maintenance lubricant standards, revised 1997.

[7] Special Manual Lubricants Purchase Specifications Approved Products, Document 10-SP-95046 Cincinnati Milacron (Cincinnati Machine) Company, July 1995.

[8] Lubricating Oil, Fuel and Filters Engine Requirements, Detroit Diesel Corporation, 1999, www.detroitdiesel.com.

[9] International ASTM "Standards on Petroleum Products and Lubricants," Volumes 05.01, 0.02, 0.03, and 0.04. www.astm.org.

[10] Rand, Salvatore J., "Significance of Test for Petroleum Products," 7th ed., International ASTM, 2003.

[11] Institute of Petroleum (IP). www.energyinst.org.uk.

[12] Deutsches Institut fur Normung e.V. (DIN), www.normung.din.de.

[13] American National Standards Institute (ANSI), www.ansi.org.

[14] International Standards Organization (ISO), www.iso.org.

[15] Toms, Larry A., *Machinery Oil Analysis — Methods, Automation & Benefits*, 2nd ed., Coastal Skills Training Inc., Virginia Beach, VA, 1998.

4

Contamination Control and Failure Analysis

Jacek Stecki
Department of Mechanical Engineering
Monash University

Definitions

For the purposes of this chapter we use the following definitions.

Cause. The means by which a particular element of the design or process results in a failure mode.

Contaminant. Any foreign or unwanted solid, liquid, or gaseous substance present in the lubricant and working fluid that may have a detrimental effect on the condition of a machine.

Criticality. The criticality rating is the mathematical product of the Severity (S) and Occurrence (O) ratings. Criticality $= (S) \times (O)$. This number is used to place priority on items that require additional quality planning.

Current Controls (design and process). The mechanisms that prevent the cause of the failure mode from occurring, or which detect the failure before it reaches the next operation, subsequent operations, or the end user.

Detection. An assessment of the likelihood that the sensor will detect the cause of the failure mode or the failure mode itself, thus preventing it from reaching the next operation, subsequent operations, or the end user.

Failure Effect. The consequence(s) a failure mode has on the operation, function, or status of an item.

Failure Mode. Sometimes described as categories of failure. A potential failure mode describes the way in which a product or process could fail to perform its desired function (design intent or performance requirements).

Function. Any intended purpose of a product or process. Failure mode and effects analysis (FMEA) functions are best described in verb–noun format within engineering specifications.

Lubricant. Any substance interposed between two surfaces in relative motion for the purpose of reducing friction and/or reducing mechanical wear.

Lubricant/working-fluid-based condition monitoring and diagnostics. Monitoring of the characteristics of the lubricant (liquid or solid) and/or working fluid to determine its current state (diagnosis) and to predict its future state (prognosis).

Nonsolid contamination. Small portions of liquid or gaseous matter (e.g., air, water, chemical substances) present in a liquid lubricant and working fluid, which are the result of ingression into lubrication and hydraulic/pneumatic system.

Occurrence. An assessment of the likelihood that a particular cause will happen and result in the failure mode during the intended life and use of the product.

Particulate-based condition monitoring and diagnostics. The monitoring of the characteristics of solid particulate present in the lubricants and working fluid of a system to determine the current state of the system (diagnosis) and to predict its future state (prognosis).

Risk priority number (RPN). A mathematical product of the numerical Severity (S), Occurrence (O), and Detection (D) ratings. RPN$= (S) \times (O) \times (D)$. This number is used to place priority on items that require additional quality planning.

Severity. An assessment of how serious the effect of the potential failure mode is on the next operation, subsequent operations, or the end user.

Solid contamination (particulate). Small portions of solid metallic and nonmetallic matter present in lubricants and working fluids, which are the result of mechanical wear of surfaces in a machine, electrochemical processes (e.g., electrical discharge), or ingression from the environment.

Symptom (contaminant analysis). A measured or otherwise identified physical, chemical, or morphological characteristic of particulate present in a sample of lubricant/working fluid that provides pertinent information about a particular characteristic of the state of the machine.

Syndrome. A set of symptoms that identifies a certain fault of the machine.

Tribology. The science and technology of interacting surfaces in relative motion, and of related subjects and practices.

Tribology-based condition monitoring and diagnostics. The detection of changes to the tribological characteristics of a system to determine the current state of the system (diagnosis) and to predict its future state (prognosis).

Working fluid. Fluid used as an energy transmission medium (produced from petroleum products or aqueous or organic materials) or for other purposes (e.g., coolant).

4.1   Introduction

In systems employing fluids (produced from petroleum products or aqueous or organic materials) as lubricants or working media, wear particles (metallic wear particles, friction polymers, carbon deposits, etc.) are carried away from the wearing surfaces and circulated throughout the system. In addition, fluids can contain contaminants (undesirable solid, liquid, or gaseous material) that are the result of ingression from the system's environment.

Control of contamination is specially important in hydraulic power and control systems which, due to their advantages over other systems in providing flexible and accurate control of motion and forces, are widely used in automotive, aerospace, machine tools, earthworking machinery, and other areas of engineering. The presence of contamination, resulting from contaminant ingression or from wear in system components, if not controlled, will affect hydraulic system reliability and accuracy of its control actions. Experience shows that the majority of hydraulic systems fail due to damage or operational malfunction caused by the presence of contamination in the system.

The level of fluid contamination in the system is the result of a multitude of factors that interact in a complex way. Some of the more important factors that affect the contamination level are: the rate of contaminant entry to the system from its environment (e.g., dust), rate of wear debris production by system components, filter characteristics (e.g., β-rating [1], dirt holding capacity), type of filtration system (e.g., high-pressure, low-pressure, by-pass), system duty cycle, design of reservoir, fluid loss, and contamination tolerance of system components.

The function of a filtration system is usually defined as removal of contamination from the fluid. No filtration system will remove all contamination from the hydraulic or lubrication system. The best we may expect is that the level of contamination is contained to the quantity below the components' tolerance level. Thus, this function must be redefined as maintaining the contamination level of the fluid below the tolerance level of the system components. Although the above is recognized by most of the users of fluid power systems, in practice, many hydraulic systems have inadequate protection against contamination.

The filtration system is usually the first casualty when the price of a hydraulic system must be competitive. High-quality filters are expensive; thus, to lower the price of the system, the quality of contamination control is lowered by applying smaller size, lower filtration rating filters, or by offering a better contamination control system at an extra cost (this is usually rejected by the customer as in most cases the user has no understanding of filtration requirements). The effects of "cutting corners" when selecting a filtration system will only become apparent later on when system damage or malfunction occurs, forcing modifications to the system, which are usually expensive.

A very common reason for inadequate contamination control is lack of understanding of contamination control technology. Usually the filters are selected by following the equipment manufacturer's recommendations, which specify the required fluid cleanliness level (e.g., in terms of ISO 4406 [2] or NAS 1638 [3] standards), absolute maximum mesh size, and required filter β-rating necessary to maintain correct operation of the various hydraulic components. Filtration system selected on the above basis, without consideration of expected rates of ingressed and system-generated contamination with which the equipment will have to cope and no knowledge of components tolerance to contamination may be totally inadequate in the case of a complicated multi-branch system, consisting of many hydraulic control elements and operating in a dirty environment. The concept of contamination balance, first proposed by Fitch [4], which considers the generation and removal of contamination within the system, provided a mathematical tool to describe interactions between various factors affecting level of contamination in a system and thus led to a better understanding of the contamination characteristics of hydraulic and lubricating systems.

An important step forward was the recognition of the significant economic impact of contamination on the life cycle costs of the plant and that contamination control, both in hydraulic and lubricating systems, must be considered as an integral part of a quality program of the plant. This in turn provided a driving force behind further advances in contamination control theory [5], the development of on-line contamination sensors [6,7], widespread application of on-line monitoring and root-cause and on-condition maintenance, availability of microprocessor based signal and knowledge processing (expert system, fuzzy logic, neural nets), and a marked progress in further development of the filtration technology [8–12].

The design of a contamination control system consists of three major tasks (Figure 4.1):

- Setting of appropriate targets for cleanliness of the system in order to achieve desired life cycle of the system
- Implementing contamination control
- Applying condition monitoring strategies (predictive and/or preemptive) to prevent systems failures

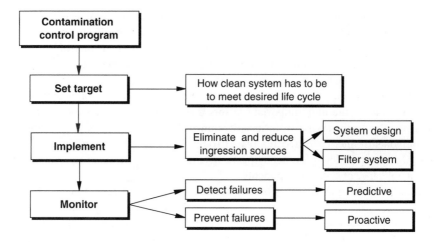

FIGURE 4.1   Contamination control program.

Contamination control is thus concerned with contamination and its effects on a system in design, manufacturing, commissioning, and operation/maintenance phases of system life:

Design: formulating objectives and determining goals, setting specifications for fluid cleanliness in a system, identifying possible root causes of contamination problems.

Manufacturing: using materials that will minimize effects of contamination on wear of components (wear resistance), applying manufacturing techniques that will result in clean components and will inhibit production (wear) or ingression of particles (seals).

Commissioning: using approved test procedures, following correct system flushing procedures, using correct fluids, seals, etc.

Operation and maintenance: following correct operating procedures, implementing monitoring procedures and corrective actions.

Implementation of a contamination control program will fail unless management is fully supportive. Management must produce an environment that will support rethinking of current maintenance practices in order to introduce a "cleanliness ethic" in the plant. To be successful the program must aim to move decision making down to the workplace by giving operators/maintenance staff authority to act in a knowledgable manner. Thus, the personnel must be trained to have knowledge and understanding of the program, techniques, and tools. The program goals must be stated in terms of *what* is required and not *how* it should be done. The program must include proper reporting and assessment procedures. Some factors (constraints) that must be considered are:

- *Money*: investment in equipment, cost of maintenance staff, insurance, cost of spares, equipment loss, production loss
- *Machines*: type of equipment (mobile, stationary), complexity, failure rates, effectiveness of contamination control system
- *Methods*: selection of techniques, applicable monitoring techniques, detection and prognostic methods, sampling procedures, data acquisition — permanent units, data collectors, data processing — knowledge based, computer based, manual
- *Materials*: type of contaminants, types of wear, failure modes
- *Minutes*: expertise acquisition time — learning curve, sampling turn-around time, time to achieve guidelines
- *Manpower*: current experience of staff, technical levels and skills of staff, training required

4.2 Contamination Control Program

Setting up, implementing, and monitoring a contamination control program requires thorough knowledge of the plant that will be included in the program. The reliability of diagnosis and prognosis of the monitoring program will be directly affected by the selection of detection methods; thus, the selection of a proper mix of monitoring methods that will be able to identify changes in machine condition is important to the success of the program. Specifically, setting up a contamination control program comprises the following tasks:

- Identification, by using functional and failure analysis techniques, of failure modes and effects
- Analysis of tribological systems present in selected machines to determine root causes of wear and contamination conditions in each machine
- Identification of root cause and tribological symptoms of machine state that should be monitored and selection of sampling points
- Selection of techniques, guidelines, and procedures to carry out root cause and tribology-based condition monitoring of a machine

A flowchart of setting-up tasks is shown in Figure 4.2. Only when the results of these tasks are known can the decision be made of how the contamination control program should be executed. Experience shows that if setting up of contamination control does not follow the above general procedure the results of the program will not gain predicted benefits.

4.2.1 Failure and Criticality Analysis

Failure analysis techniques are usually employed to detect potential safety problems in newly built or existing systems and application of these techniques is now being extended to analysis of new, still in the design stage, systems. In engineering practice there is a large number, approximately 60, of available techniques for failure analysis. Some of them are (Figure 4.3):

- Failure mode analysis (FMA) — identification of unwanted conditions of the system
- Failure modes and effects analysis (FMEA) — identification of effects of component failure on the system operation and safety
- Hazard analysis (HA) — identification of potential hazards during system operation
- Failure modes, effects, and criticality analysis (FMECA) — identification of effects of component failure on the system operation and safety, probabilities of occurrence and their criticality

The FMEA analysis is concerned with hardware failures and malfunctions, it does not include hazards due to errors caused by human operators, effects of environment, and other operating and hazardous conditions outside the scope of design limits of the components. A functional failure is defined as a component's or system's inability to perform the intended function (e.g., piston speed is low). A structural failure is defined as the component's failure due to changes in material (e.g., fluid viscosity), geometry of component parts (increased orifice size), or tribological action (e.g., increased friction) (Figure 4.4).

The basic objective of FMEA is to identify all possible failure modes, that is, how system/parts can fail (e.g., hydraulic line leakage, valve seized) and deduce the consequences of these faults, that is, what happens if failure occurs (e.g., piston speed is too low, no flow through the valve). Other information sought during analysis is the nature of the failures and whether there are any redundancies, backup components, or sensors that can safeguard components against failures. The FMEA technique supplemented with analysis of criticality of failures becomes FMECA. The basic objective of this analysis is to identify possible failure modes of the components, deduct what are the consequences of these failures, and determine criticality of each failure mode. A methodology of FMEA is shown in Figure 4.5. Guidelines for preparing FMEA/FMECA are the subject of a number of national and international standards [13–16].

Although FMEA/FMECA appears to be a rather straightforward task, its execution requires a clear understanding of system operation, availability of details of system components and their interaction, a

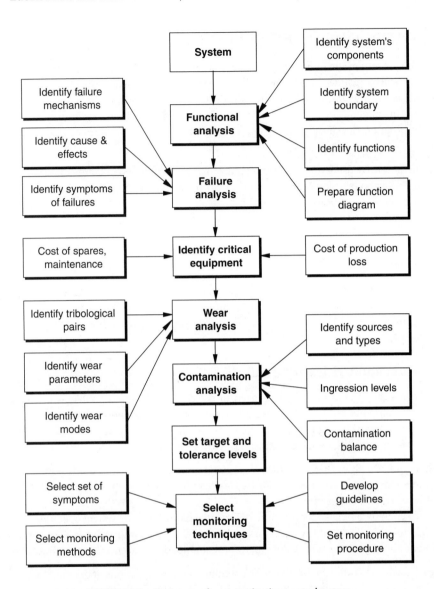

FIGURE 4.2   Setting up of a contamination control system.

knowledge of the functions that the system and its various parts must perform, and a knowledge of interfaces with external systems. As a system may operate in various modes (e.g., emergency mode, start-up mode, test mode) the analysis should be carried out for all modes of system operation.

Failure analysis of complex modern mechanical engineering systems is very time consuming and thus attempts were, and are, made to automate it in application to fluid power systems [17–19]. The current approaches to automated, computer-aided FMEA of hydraulic power and control systems are based on investigation of suites of models (functional, behavioral, structural, and teleological) using qualitative reasoning, fuzzy logic, and other approaches [20–24].

4.2.2   Functions of an Engineering System

The Oxford dictionary definition of a system is "a set or assemblage of things connected, associated, or interdependent so as to form a complex unity." Implied in this definition is the purpose of such a set which, in engineering terms, is the transmission of generalized information that can be in the form of energy,

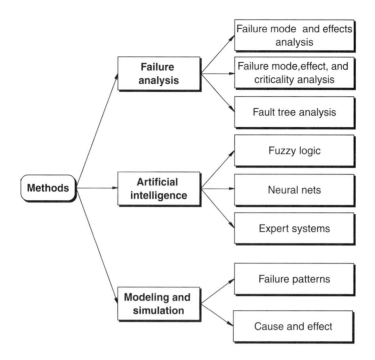

FIGURE 4.3 Failure analysis methods.

mass, or control information. The purpose of a mechanical system is to transform, transmit, and generate mechanical forces and motions, that is, mechanical systems are concerned with energy and work. The action of the system in meeting the system's objective is described in terms of functions that the model (system) must perform to process input information into output information (Figure 4.6).

Whenever an engineering problem is investigated, a designer can either investigate the system by direct experimentation on the real system or carry out the investigation on the basis of some type of a model.

The system can be conceptually separated from its environment by defining a boundary that encloses the system and a number of points that are common with other systems or the environment. These points are information transfer terminals, and the information passing through these points will define the character of interaction with other systems and the environment. After defining the boundary of the system and points of information transfer we are able to investigate the system in isolation from other systems and its environment. Although the selection of a system's boundary is arbitrary, the choice of boundary is influenced by the objective of the investigation.

The real engineering system has a great number of these attributes and the investigation of the system would be difficult if all were considered; thus, a skillful designer will select only these attributes that are relevant to the subject of investigation. To simplify the investigative task some attributes are totally ignored while others are idealized, on the basis of a certain set of criteria and assumptions accepted by the designer. The subset of selected variable attributes of the system that represent information transfer between the environment and the system can be separated into two sets. Those variable attributes that represent information transfer from the environment to the system and that can be controlled or manipulated are designated as inputs to the system. Variable attributes that represent information transfer from the system to the environment and can be measured or observed are designated as outputs. Other attributes that represent the physical or geometric attributes of a system, for example, kinematic viscosity of fluid, or that cannot be or were chosen not to be controlled and manipulated are designated as parameters. Parameters do not have to have constant value; for example, viscosity of fluid will vary with system temperature.

The process of delineation of a system boundary and identification of its attributes is called modeling. The objective of modeling is to construct a system, a model, which is a subset of the real, physical system.

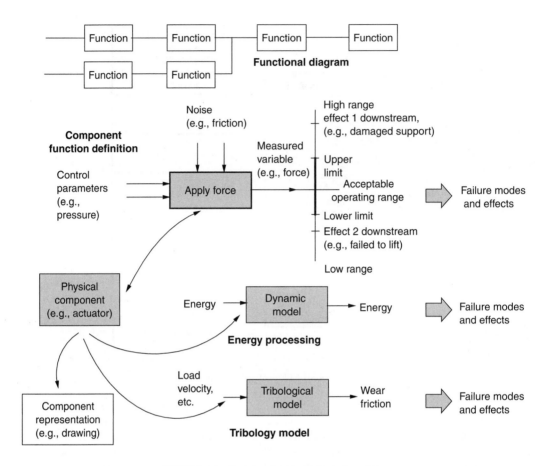

FIGURE 4.4   Functional and tribological models.

Investigation of the model will yield information about functional, behavioral, and structural properties of the real system under consideration. The correct choice of modeling assumptions is the one factor that will usually seriously affect the quality of the system investigation and the magnitude of modeling and simulation tasks. The model of a real system, developed on the basis of such a set of assumptions, should in all important aspects be equivalent to the original real system.

Structural description of a system S defines inner working of the system by showing the interaction between elements (parts) of the system and relevant properties of these elements:

$$S = \{A, P, R\} \tag{4.1}$$

where A is elements of the system, P is relevant properties of the system, and R is interrelation between elements. Such a model describes, for example, a tribological system.

Functional description of a system defines system behavior in terms of input–output relations:

$$O = F(I) \tag{4.2}$$

where F is function relating inputs I to outputs O.

Functional model is used to describe, for example, a mechanical systems where function F represents system transfer function.

In most general cases a system's model is represented as a *blackbox* with technological implementation and structure (topography) of the system unknown. The blackbox power model of a system accepts the inputs, processes them, and transfers the output information to the other systems and environment (Figure 4.7).

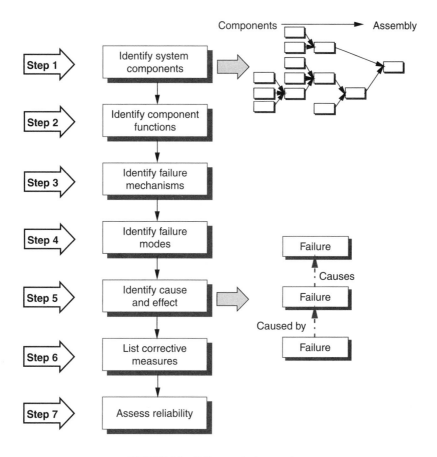

FIGURE 4.5 Failure analysis procedure.

Each technical system consists of a number of subsystems and parts, each performing some function [25]. Thus, we usually may identify a hierarchical structure of the system and identify functions at each hierarchical level (Figure 4.8). Identification of functions and their hierarchy (Figure 4.9) is a rather difficult task but the effort may be greatly reduced by using value analysis techniques like FAST (Function Analysis System Technique) and standards [26,27].

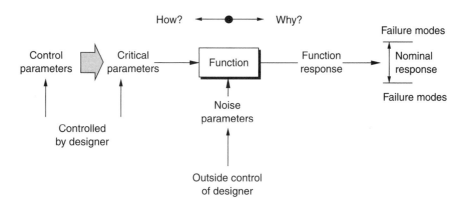

FIGURE 4.6 Function definition.

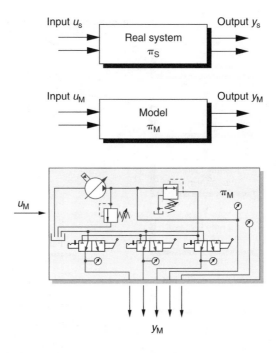

FIGURE 4.7   Modeling process.

In general, most mechanical systems fall into two types: stationary, force transmitting systems (e.g., machine base) and moving parts, power transmitting systems (e.g., hydraulic power unit). Usually a mechanical system is composed of both of these types of systems. When carrying out functional analysis on stationary systems we are concerned with flow paths of forces through the structure, whereas in the case of moving part systems we are concerned with energy flow paths through the system.

Identification of failures is greatly facilitated by the application of Fault Tree Analysis (FTA). The method is based on finding out what are the possible unwanted conditions or failures of the system and finding out what are the operating conditions, component faults and failures that caused these unwanted conditions.

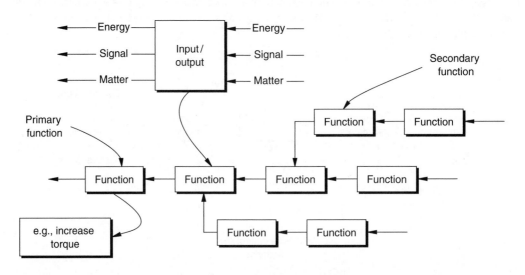

FIGURE 4.8   Functions of an engineering system.

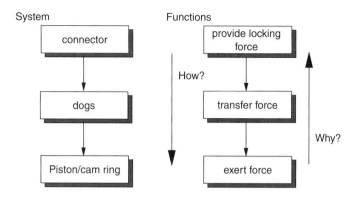

FIGURE 4.9 System-function.

The application of FTA can be backward (we know the condition, or top event, and are trying to find what causes it) or forward chaining (we know the modes of component failures and are trying to find out how these failures affect hierarchically higher components and the system) (Figure 4.10).

The resultant failures can be caused either by a serial combination of preceeding failures in different paths, represented on the diagram as an *and* gate, or by a parallel combination of any preceeding failures, represented by an *or* gate.

FTA technique is well formalized and computer programs are available to automate development and analysis of fault tree diagrams. In the case of a fault tree that represents a system with a built-in redundant path, care must be taken to identify failure modes common across the redundant branches. The occurrence of such faults will reduce the inherent reliability of a redundant system. A typical fault that will have such an effect would be, for example, the failure of a contamination control system that will affect all the components of a hydraulic drive system. Improper maintenance procedure or failure of an indicator or sensor could have a similar effect. In addition, we should also consider the effect of some events, which although they are not directly associated with failure may cause a higher level failure. An incorrectly set pressure relief valve may cause the failure of a pump and therefore a system. Disconnecting a pressure switch may result in the failure of a pump by not switching the system off when the pressure is at the limit.

4.2.3 Types of Mechanical Failures

Failure of a machine or its components is defined as an inability to perform a required function or functions within the constraints imposed by system specification (that is why it is very important to have good equipment specification). We may identify the following types of failures during the life of equipment.

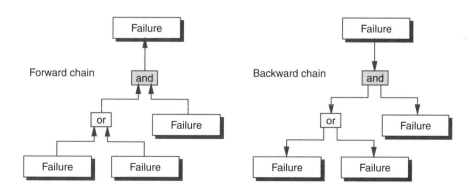

FIGURE 4.10 Fault tree analysis.

4.2.3.1   Incipient Failure

The machine is in incipient failure mode when its performance or condition is gradually degraded and it will eventually fail if no corrective action is taken. This type of failure can often be tolerated, however, its progression must be monitored to prevent occurrence of a catastrophic failure. The decision "to live" with this type of failure will depend on the mechanism of wear. For example, fatigue wear in gears usually progresses in a gradual fashion, thus we may tolerate this type of wear and consider the gears to be in an incipient failure mode. The gears should be monitored to see if there is a change in the rate of wear and if so necessary corrective action should be taken, for example, reduction of the load carried by the gears. On the other hand, detection of scuffing (sliding wear) indicates lubrication problems and as this type of wear progresses very rapidly it may lead very quickly to destruction of the gears and immediate corrective action should be taken.

4.2.3.2   Misuse Failure

The equipment is used beyond the specified limits or is used to perform a function it was not designed to perform. Faults due to wrong maintenance action or operation also fall into this category. For example, replacement of hydraulic fluid with fluid of wrong (e.g., too low) viscosity may lead to excessive temperature in the system and destruction of the pump. An example of wrong operation is closing the inlet valve to a hydraulic pump causing pump failure due to cavitation.

4.2.3.3   Catastrophic Failure

This is a sudden and total failure of the system that is not expected to occur. A typical example of such a failure is jamming of the valve due to contamination.

4.2.4   Phases of Failures

We may identify the following phases of failures:

Wear-in failures: After commissioning of equipment there is a period of time during which various tribological pairs are subjected to large tribological stresses. This often produces excessive wear, which in most cases is not posing a danger to the future operation of the machine. For example, gears during the wear-in (running-in) stage of operation may produce an abnormal amount of fatigue wear particles (pitting). This is caused by abnormal contact stresses between gear teeth caused by asperities and teeth surface irregularities, which are the results of machining process. However, if the gears were correctly designed, the production of wear particles would be significantly reduced after the "high spots" are removed from the gear teeth surfaces. This type of fatigue wear occurring during wear-in period is called "arrested pitting" (under some circumstances the fatigue wear process is not arrested and the gears enter into a "progressive pitting" mode, which inevitably leads to catastrophic failure of the gear set). The wear-in occurs very rapidly. Tests at Monash University have shown that wear-in in gears will occur within a few thousand gear revolutions.

More dangerous failures during the *infancy* period are due to design or manufacturing errors or defects. Very often mistakes are made during assembly, installation, or commissioning stages.

Random failures: Random failures occur in the mid-life of the equipment and are caused by overloads or under-designed components, deficient lubrication, contamination, corrosion, etc. These types of failure are difficult to predict.

Wear-out: Wear-out (degradation) failures are caused by fatigue (both high and low cycle), corrosion, erosion, or by design (planned obsolescence). Failure of a hydraulic system manifested by pump leakage caused by pump wear-out due to a prolonged exposure to contamination is an example of this phase of wear.

The failure rate for each phase of failure may be represented using the "bath tub" curve (see Figure 4.11). In the figure, indices A, B, and C refer respectively to wear-in, random, and wear-out failures.

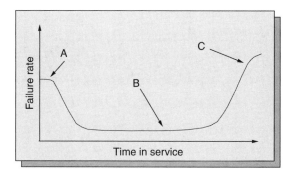

FIGURE 4.11   Bath tub curve.

4.2.5   Causes of Failures

Mechanical failures are most commonly caused by (see Figure 4.12):

Machine (design and manufacturing): lack of knowledge of loads, wrong manufacturing procedures, design deficiencies — tolerances, clearances, fits, strengths, rigidity, manufacturing deficiencies — residual stresses, surface finishes, stress concentrations. Assembly procedures — overstressing, clearances, fits, damage to seals, damage to surfaces, manufacturing defects.

Material: inadequate material properties, wrong material selection criteria, mismatch of material properties with environment (e.g., corrosion), wrong combination of materials, surface treatment, hardness.

Man: inadequate training, inexperience with a new type of system, reliance on computer analysis, incorrect operation of the system, erroneous or omitted design calculations (stress analysis).

Methods: incorrect specification (wrongly specified constraints), lack of quality control, inadequate manufacturing methods, inadequate design methods, inadequate maintenance methods, incorrect operating procedures (service conditions — overloading, overheating, contamination, lubrication), maintenance procedures (contamination, wrong replacements, improper procedures). Incorrect handling, damage during transport (fretting).

Money: emphasis on cost at the expense of quality of design and manufacturing.

Minutes: wrong timing of maintenance activities, cutting corners to meet time schedules, deterioration of material properties over time.

The above list is not exhaustive and other factors may also contribute to failures. In addition to the above factors failure can be caused by effects, that is, qualitative factors, like environmental conditions, radiation, noise. Investigation of failed components shows that most failures can be attributed to:

- Excessive tribological stresses resulting in wear and corrosion. This type of failure is usually gradual and results in changes of geometry of the affected components.
- Excessive elastic deformation.
- Excessive plastic deformation.
- Fracture of component.
- Misuse or abuse.
- Assembly errors.
- Manufacturing defects.
- Improper maintenance.
- Fastener failure.
- Design errors.
- Improper material.
- Improper heat treatments.

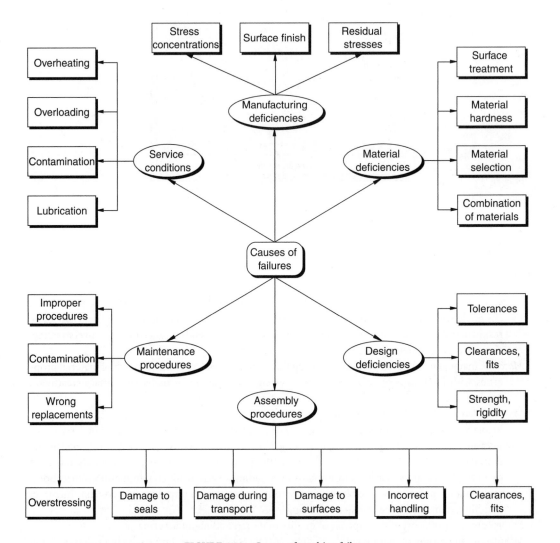

FIGURE 4.12   Causes of machine failures.

- Unforeseen operating conditions.
- Inadequate quality assurance.
- Inadequate environmental protection/control.

To illustrate some causes of failures we may consider rolling element bearings. Bearings are the most common elements in mechanical equipment and should be very long-lasting components. The bearings should have a very low wear rate, and they rarely fail due to material faults. Why then do bearings fail earlier than their design life? Experience shows that most bearings fail by accident (96% of failures according to some literature sources) and approximately 10% of these failures are due to fatigue. A fair number of bearings fail due to lubrication problems, seal problems, and improper mounting. Some of the bearings are already damaged due to mishandling before they are assembled on the machine (e.g., fretting damage during transport). Typical causes of damage to rolling element bearings are shown in Figure 4.13.

4.2.6   Tribological Analysis

The study of wear in mechanical systems is part of a scientific discipline called tribology (*tribo* — to rub in Greek). Wear in mechanical systems is the result of tribological action, and is defined as the progressive

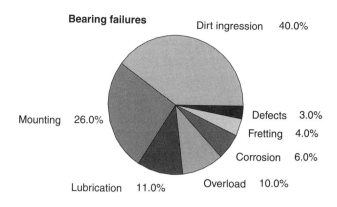

FIGURE 4.13   Causes of bearing failures.

loss of substance from the surface of a solid body due to contact and relative motion with a solid, liquid, or gaseous body. Tribological analysis of a machine leads to identification of types of tribological actions in the machine and therefore identification of possible wear modes and their severity. This information combined with identification of internal and external contamination sources is used to develop sampling and detection procedures of monitoring and diagnostic purposes.

Wear results in producing wear particles and in changes in the material and geometry of the tribologically stressed surface layers of the components forming a tribological pair, that is, two components that are in contact with each other. Normally wear is unwanted; however, in certain circumstances, for example, during running-in, wear may be beneficial.

A typical example of a tribological pair is a pair of gears in which tribological action occurs on the meshing surfaces. Elements of a tribological pair may be in direct contact (e.g., not lubricated gears) or contact between these surfaces may be via interfacing medium (e.g., lubricant) that modifies interaction between the elements of a tribological pair. Examples of tribological systems are listed in Table 4.1.

The body and counterbody are members of a wear couple where wear is of particular importance. The interface medium may have a wear reducing effect (e.g., lubricant) or wear increasing effect (e.g., dust). The environment may also play a major role in the wear process.

The external operating variables that act on the elements of the tribological system form the operating variables of the tribological system. The wear characteristics describe the nature of material loss that occurs through the action of the operating variables.

Tribological action requires both contact between the wearing couple and relative velocity between elements of the couple. Examples of such actions are shown in Figure 4.14. External actions on the element, for example, bending or shear are not considered to be tribological actions.

TABLE 4.1   Examples of Tribological Systems

	Elements of Tribological Systems			
System	Body	Counterbody	Interface	Environment
Guide bush	Guide	Rod	Lubricant	Air
Disk brake	Pad	Disc	—	Air
Gear box	Pinion	Gear	Lubricant	Air
Hydraulic pump	Slipper	Plate	Hydraulic fluid	—
Hydraulic motor	Piston	Cylinder block	Hydraulic fluid	—
Directional control valve	Spool	Valve block	Hydraulic fluid	—
Rolling element bearing	Roller	Race	Lubricant	Air

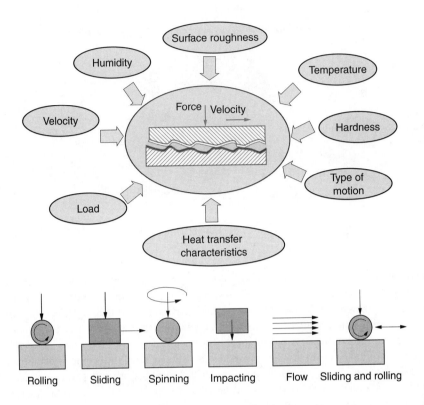

FIGURE 4.14   Tribological actions.

A flow chart of the analysis of the wear process is shown in Figure 4.15. System operating parameters that must be considered in analysis of wear are [28]:

- Normal load over tribological contact area (exception — fluid flow)
- Relative velocity between wear couple (exception — erosion due to fluid flow)
- Thermal equilibrium state — operating temperature
- Duration of wear process — period of application of tribological stress

Identification of wear mechanisms includes investigation of material characteristics, wear processes present, and characteristics of wear/contamination particles (Figure 4.16). The characteristics of debris that are generated during tribological action and the appearance of surface differ for different wear processes.

Sources of wear in mechanical systems are shown in Figure 4.17. Changes in the quantity of debris generated in a system with size of particles and severity of wear are shown in Figure 4.18.

The regime of wear depends on a number of factors that are shown in Figure 4.19. Manufactured surfaces are far from smooth when viewed under a microscope. Typically the surface consists of a multitude of peaks (called asperities) and valleys randomly distributed over the surface. When two surfaces with similar hardness are brought into contact, the number of asperities touching and the contact areas are very small. However, as the normal load increases the contact stresses in the points of contact (asperities) rapidly increase and the plastic flow value of the material and the contact points deform plastically in such a manner that the total contact area is now finite (i.e., increases) as shown in Figure 4.20.

Wear is a complex process involving the removal of material from sliding surfaces. The complexity of the process can be seen in the plethora of defined wear modes. Two modes, abrasive wear and adhesive wear, are probably the best known. Although these two modes are well defined, in practice one seldom finds pure adhesive or abrasive wear but rather both processes are evident on the same worn surface. Other well-recognized forms of wear include fretting, particle erosion, corrosive wear, and fatigue wear.

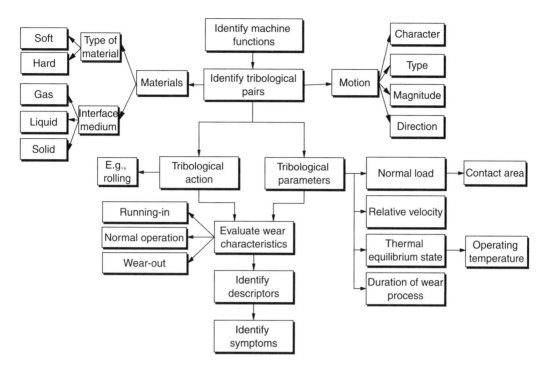

FIGURE 4.15 Tribological analysis of a system.

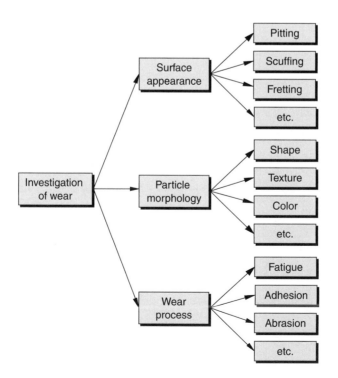

FIGURE 4.16 Investigation of wear.

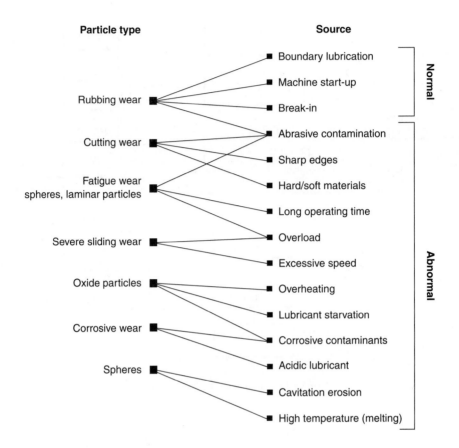

FIGURE 4.17   Sources of wear particles.

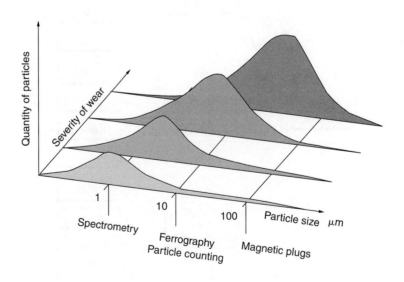

FIGURE 4.18   Progression of wear.

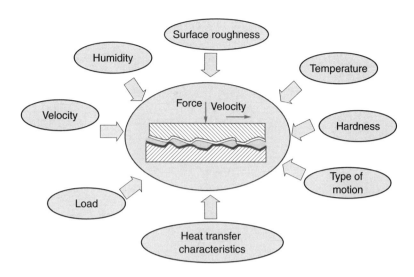

FIGURE 4.19 Tribological parameters.

Wear is a natural occurrence and occurs when any two surfaces rub together. The designer cannot eliminate wear but, by understanding the mechanisms of wear, he or she can try to minimize wear by proper selection of materials, lubrication, and design.

There are a number of wear mechanisms that describe wear; however, the terminology used may be confusing to the beginner (e.g., scoring is called scuffing by some people and abrasion by others). The most

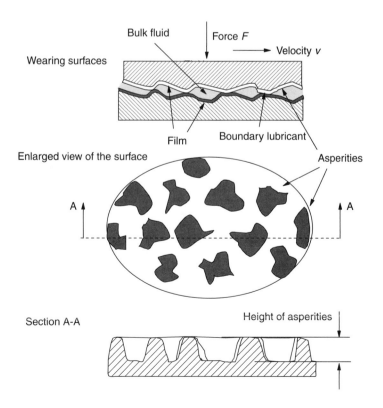

FIGURE 4.20 Contact areas between surfaces (dry friction).

comprehensive classification of modes of wear is that employed by Godfrey [29]. He classifies wear into the following categories:

- Adhesive wear
- Mild adhesive wear
- Severe adhesive wear or scuffing
- Abrasion
- Erosion
- Fatigue
- Delamination
- Corrosive
- Electro-corrosive
- Fretting corrosion
- Cavitation damage
- Electrical discharge
- Polishing

Each of these types of wear categories has its own mechanism and symptoms. In practice, wear usually results from a combination of a number of wear processes, which may be concurrent or follow each other. Before a wear problem can be diagnosed, contributing wear mechanisms must be identified. Previous history of failures (and results of postmortem) may provide useful information in this regard.

4.2.6.1   Adhesive Wear

Adhesive wear is the result of formation and rupture of interfacial adhesive bonds (e.g., "cold welded junctions", "scuffing") (Figure 4.21).

The volume of material removed during adhesive wear is normally high during the initial "running-in" period in which the high asperities are plastically deformed and sheared and some material is removed by micro ploughing of the softer material. This initial high wear rate is usually followed by a marked decrease in the rate at which the material is removed, referred to as a stabilized (equilibrium) wear (see Figure 4.22).

An important characteristic of adhesive wear is the transfer of metal from one surface to the other. Although transfer of material can also occur during abrasive wear, fretting and corrosive wear does not in itself identify any particular form of wear; simultaneous occurrence of adhesion and transfer identifies severe wear conditions resulting from dry sliding. The character of adhesive wear can be influenced by sliding velocity, load, presence of lubricant, material properties, and presence of a gaseous environment.

If the load and speeds are high, the equilibrium condition may not be reached and the wear rate may remain high causing premature failure of the machine component. As the speed and load is varied, the

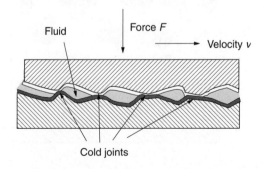

FIGURE 4.21   Mechanism of adhesive wear.

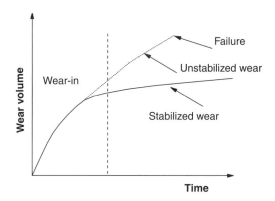

FIGURE 4.22 Running-in process.

wear regime may change from mild adhesive wear (also called continuous wear, rubbing, or normal wear by some authors) to severe adhesive wear in which case the machine will fail catastrophically.

Mild adhesive wear has the following characteristics:

- Repetitive wear on metal surfaces
- Small flakes of metal
- Less than 10 μm in size
- Occurs throughout life of machine

Mild (normal) wear particles are shown in Figure 4.23.

Severe adhesive wear or *scuffing* can be recognized by gross amounts of transferred material and is characterized by:

- Metal to metal contact — no lubricant film
- Visible parallel striations on co-surfaces
- Particle sizes between 5 and 5000 μm (usually 15 to 50 μm)

This type of wear is caused by lubricant loss, overloading, and wrong oil. It typically occurs in places were sliding occurs — tips of gear teeth, pistons, loose bearings — typical particles are shown in Figure 4.24 and Figure 4.25.

Surface films have a marked influence on both friction and adhesive wear. Wear can be reduced by employing low shear strength surface films on a hard base material. This technique allows the load to

FIGURE 4.23 Normal wear.

FIGURE 4.24   Severe sliding wear — 1000× (max. particle sizes 35 μm).

be supported through the soft film by the base material, while any shear takes place within the surface film. Suitable types of film include oxides, chlorides, sulfides, other reaction products, soft materials such as lead, silver, copper, and many other nonmetallic materials. To minimize the bond strength between asperities, the sliding pair should be selected in such a way that the surface film is mutually insoluble in the base metals or form intermediate compounds.

4.2.6.2 Surface Fatigue

Surface fatigue wear is caused by tribological fatigue stress cycles resulting in crack formation in surface regions of contacting surfaces and separation of material ("pitting") (see Figure 4.26).

Fatigue is recognized by surface and subsurface cracks. Surface fatigue occurs when repeated sliding, rolling, or impactive loads are applied to a surface. These loads subject the surface to repeated cyclic stress that initiates cracks near the surface. With time these cracks spread, link up, and form discrete particles that will detach from the surface and can contribute to three-body-wear abrasions. Rapid deterioration of the surface may occur due to the flaking off of further fragments. Edges of the cracks are usually sharp and angular and are distinguished from corrosion pits by the presence of other cracks at the edges of pits and surfaces with "beach" marks or ripples indicating progressive stepwise cracking. Fatigue wear is characterized by:

- Subsurface stressing in gears, rolling element bearing, etc.
- Large particles between 5 μm and 5 mm in size (usually 15 to 50 μm)

FIGURE 4.25   Sliding wear particle — 450× (particle size approx. 60 μm).

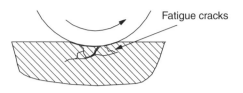

FIGURE 4.26 Fatigue wear — formation of spalls.

Fatigue wear can be observed in rolling element bearings, the pitch line of gear teeth, and some other tribological pairs where high contact, cycling, stresses are present. Surface fatigue or rolling-contact fatigue is particularly important in the operation of ball and roller bearings, and is usually caused by excessive loads. A fatigue type particle generated during failure of a bearing is shown in Figure 4.27.

The fatigue wear particles are often accompanied, especially in the case of rolling element bearings, by spheres (<5 μm diameter, Figure 4.28) and so-called *laminar* particles.

The exact origin of spheres is still in dispute. It has been observed that spherical particles appear, and at constant load on the bearing the quantity of spheres rises exponentially, long before any indication of bearing trouble, characterized by spalling, is detected. Laminar particles are generated by the flattening of fatigue particles during their passage between rolling elements and cages (Figure 4.29).

Gears are subjected to both sliding and rolling motion; thus, both adhesive and fatigue wear is present depending on gears loading speed and position (Figure 4.30). Adhesion wear may occur at the root and tip of gear teeth as these regions of teeth are in sliding motion. Typical gear fatigue particles are shown in Figure 4.31 and severe wear (scuffing) particle in Figure 4.32.

The fatigue wear may occur at the pitch line (pitting). We recognize two types of pitting:

- *Arrested pitting*: Arrested pitting occurs in the initial stages of gear operation when there are large local contact stresses. Usually if the gears are operating within design limits this pitting stops after a while and gears will reach their design life.
- *Progressing pitting*: In progressive pitting mode the damage becomes progressively more severe until the gear fails.

4.2.6.3 Abrasive Wear

Abrasion (cutting) wear is caused by the removal of material by ploughing (micro-cutting process). Abrasion is recognized by the presence of clean furrows (grooves) in the direction of sliding. There are two basic mechanisms of abrasive wear — two-body and three-body (see Figure 4.33).

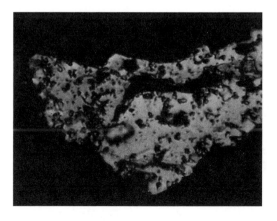

FIGURE 4.27 Rolling fatigue wear particle — 1000× (particle size approx. 60 μm).

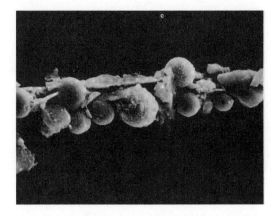

FIGURE 4.28   Spheres.

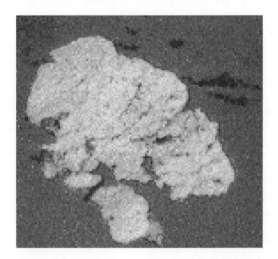

FIGURE 4.29   Laminar particle. SEM 2500× (approx. 40 μm).

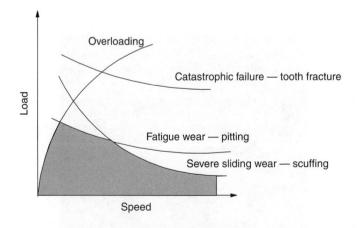

FIGURE 4.30   Wear in gear sets.

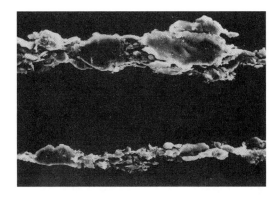

FIGURE 4.31   Gear fatigue particles. SEM 1000× (maximum particle size 25 μm).

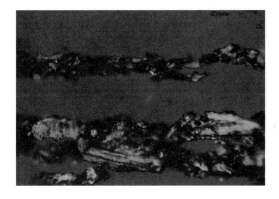

FIGURE 4.32   Gear scuffing (adhesion) particle. SEM 2500× (max. particle size 25 μm).

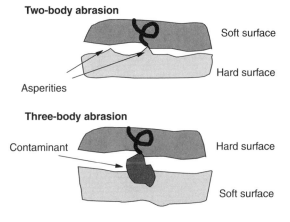

FIGURE 4.33   Mechanisms of abrasive wear.

FIGURE 4.34   Cutting wear.

During the two-body abrasion process, grooves are produced by hard asperities scraping the softer material. The wear can be controlled to some extent by careful selection of relative hardness and surface finish of the sliding pair during design, although other factors such as micro-structure and elastic modulus can have some influence. It is common practice to make one surface considerably softer than the other in order to confine any damage to the component that is easily replaceable. The harder surface is also given a good surface finish to minimize abrasion by protruding asperities.

During the three-body abrasion, hard particles that have been generated due to severe adhesive wear or grit from an external source get embedded in a softer material and provide a cutting tool. The wear can be controlled by eliminating dust or grit from the sliding surfaces by seals and minimizing the abrasion from other wear debris, that is, by reducing fatigue and adhesive wear in the system. A special case of three-body abrasive wear can arise when hard wear particles and external contaminants become partly embedded in the softer counter-face and act like sand paper cutting grooves on the harder counter-face and produce fine swarf referred to as *wire-wooling*. Typical cutting wear particles are shown in Figure 4.34.

Factors that contribute to the appearance of abrasive wear are low-quality machining, contaminants, and misalignment of components of tribological pairs. Abrasive wear is characterized by:

- Abrasive contact between two surfaces
- Two-body wear (metal/metal) — softer metal is removed
- Three-body wear (metal/contaminant/metal) — harder metal is removed by contaminant
- Curls of metal (like lathe swarf) — appearance of particles
- Particle sizes between 5 and 5000 μm (usually 10 to 85 μm)

4.2.6.4   Erosive Wear

Erosive wear is a special case of abrasive wear and is produced by the impingement of sharp particles on a surface, see Figure 4.35. The particles are often transported in a moving liquid or gas and can also occur due to cavitation of pumps at inlet.

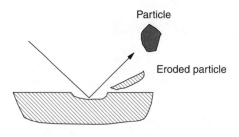

FIGURE 4.35   Erosion process.

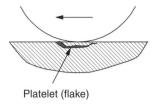

Platelet (flake)

FIGURE 4.36   Delamination process.

The worn surface is characterized by short grooves due to the micro machining action of the abrasive particles. Important factors that affect selection of materials for components, which could be subject to erosion wear by abrasive particles, are: velocity, impact angle, hardness, quantity, density, and surface contour.

At low impact angles harder materials offer the best wear resistance to erosion, whereas at large impact angles the appropriate material depends on the particle velocity. At low impact angles, the abrasive particles remove the material by a cutting process (abrasive wear); hence, hard materials offer the best solution. At large impact angles and low velocities rubbers and resins are used because these materials can absorb the energy of impact much more efficiently than harder materials.

4.2.6.5   Delamination

Delamination wear is the loss of metal due to formation and propagation of subsurface cracks parallel to the surface (see Figure 4.36). Delamination occurs due to the plastic flow of the material in conjunction with sliding causing subsurface cracks to propagate parallel to the surface (see Figure 4.37).

4.2.6.6   Corrosive Wear

Corrosive wear has been defined as a wear process in which the predominant factor affecting chemical or electrochemical reactions is the environment. As chemical reaction rates increase with increase in temperature, corrosive wear tends to accelerate at higher temperatures. The surface of metals subjected to corrosive wear appears as rough pits or depressions that are irregular and discolored.

FIGURE 4.37   Delamination particle (approx. 60 μm).

4.2.6.7   Electro-Corrosive Wear

This type of wear results from stray currents, galvanic action, and streaming potential.

4.2.6.8   Fretting Corrosion

Fretting is a form of surface damage that occurs when loaded surfaces are subjected to small amplitude oscillations. To reduce fretting corrosion the obvious solution is to eliminate oscillations. However, many systems are required to oscillate, in which case it may be possible to increase the magnitude of oscillations or to rotate one surface continually. The other method to reduce fretting corrosion is to reduce the amount of oxygen present at the contacting surfaces by using antioxidants and surface coatings.

4.2.6.9   Cavitation Wear

Cavitation wear produces very rough but clear surfaces that may sparkle under intense light. This type of wear occurs at pump inlets.

4.2.6.10   Electrical Discharge

Electrical discharge across a gas or oil film can cause micro craters or tracks in which there is evidence of molten metal.

4.2.6.11   Polishing Wear

This type of wear produces a smooth mirror finish and can occur in engine bores, piston rings, etc. The basic mechanism is attributed to fine scale abrasion, corrosion, and electrolytic corrosion.

4.2.7   Contamination

Contamination of a fluid is a general term encompassing debris in hydraulic fluid — generated in the system or ingressed from outside. Solid, hard, and sharp particles (metallic, rust, sand) may contribute greatly to component wear; whereas soft, nonmetallic particles (seal residue, fibrous material, paint) will usually have an operational effect due to blockage of clearances and orifices. Even a small quantity of large solid particles will invariably cause catastrophic failure of hydraulic components; on the other hand, a large quantity of small solid particles may cause catastrophic failures due to silting. Thus, the effect of both large and small particles on system integrity must be considered when selecting a filtration system.

4.2.7.1   Built-In

There are basically two types of built-in contaminant, one that is introduced during filling-up with "new fluid" and the other resulting from system construction. "New fluid" is not necessarily "clean fluid." Tests at the Centre for Machine Condition Monitoring, Monash University carried out under the auspices of a Basic Research Program showed that new fluid samples may have contamination levels well in excess of acceptable levels. Thus, filling a system with a new fluid through a suitable filter is of great importance to give a system a "clean" start! Hydraulic system components are manufactured with great care. However, inevitably after manufacturing, assembly of the system, and system flushing a certain amount of contaminants remain in the system. Typical built-in contaminants include burrs, sand, weld splatter, and paints. The efficiency of the removal of built-in contaminant during flushing is dependent on the velocity of the flushing fluid; high velocities are required to dislodge built-in particles. Most manufacturers provide system flushing procedures that should be followed to the letter.

4.2.7.2   Ingested

The environment contributes greatly to system contamination. The contamination enters the system via the fluid reservoir (air breathers and access covers), and any sealing pairs (cylinders seals, pump, and motor seals). It is estimated that approximately 50 to 60% of ingressed contaminants enter via the cylinder seals and it can be expected that the amount of ingression will increase with seal wear. Ingression rates from 0.03 to 0.2 g/h were reported in the literature [30,31].

4.2.7.3   Maintenance Generated

Following improper maintenance procedures may be a major reason for widespread contamination of the system. A "new fluid" added to the system without appropriate precautions is a major source of contaminant. Other activities, like replacing a fitting, may add 6,000 to 60,000 particles >5 μm to the system. Thus, following proper maintenance procedures is extremely important in maintaining the clean condition of the system.

4.2.7.4   Internally Generated

The internally generated contaminants are the result of wear processes (abrasion, adhesion, fatigue), corrosion, cavitation, oxidation, and fluid breakdown. From the operational point of view the rate of wear in hydraulic components cannot be great — otherwise the performance of the components will be affected due to resultant leakage (pumps, valves) or friction. The generation of the contaminant will be accelerated by the contaminant already present in the system, thus the rate of generation of the contaminant is closely aligned with the magnitude and rate of the other sources of contamination (ingested, maintenance-induced, or built-in). The actual amount of wear will depend on the size and the material of the solid contaminants, the ratio of the particle size to the working clearances, the shape of the particles, the lubrication regime (boundary lubrication, hydrodynamic), the pressure, the flow velocity, and the fluid type.

Presence of wear and contamination in hydraulic and lubrication systems is probably the most important factor that affects their longevity and reliability [32–35].

4.2.8   Mechanical Modes of Failure

In the literature there are many different classifications or categorizations of mechanical failure modes. Categorization of failure modes proposed by Collins [36,37] is presented in Table 4.2 and is a basis for describing the common mechanical modes of failure listed in Table 4.3.

TABLE 4.2   Categories of Mechanical Failures

	Category	Subcategory
Manifestation of failure	Elastic deformation	
	Plastic deformation	
	Rupture or fracture	
	Material change metallurgical	Chemical
		Nuclear
Failure inducing agents	Force	Steady
		Transient
		Cyclic
		Random
	Time	Very short
		Short
		Long
	Temperature	Low
		Room
		Elevated
		Steady/transient
		Cyclic
		Random
	Reactive environment	Chemical
		Nuclear
		Human
Failure locations	Body type	
	Surface type	

TABLE 4.3   Mechanical Failure Modes

Category	Subcategory	Category	Subcategory
Elastic deformation	Force induced	Wear	Surface fatigue
	Temperature induced		Deformation
Yielding			Impact
Brinnelling			Fretting
Ductile rupture		Impact	Fracture
Brittle fracture			Deformation
Fatigue	High-cycle		Wear
	Low-cycle	Fretting	Fatigue
	Thermal		Wear
	Corrosion		Corrosion
	Fretting	Creep	
Corrosion	Direct chemical attack	Thermal relaxation	
	Galvanic	Stress rupture	
	Pitting	Thermal shock	
	Intergranular	Galling and seizure	
	Selective leaching	Small spalling	
	Cavitation	Small radiation damage	
	Hydrogen damage	Buckling	
	Biological	Creep buckling	
	Stress	Stress corrosion	
Wear	Adhesive	Corrosion wear	
	Abrasive	Corrosion fatigue	
	Corrosive	Creep and fatigue	

4.2.9   Hydraulic and Lubrication Failures

Proper design of components and systems plays a great part in reliability of hydraulic and lubrication systems. The incorrect design of a contamination control system will inevitably impose severe maintenance problems. In the case of a hydraulic system, looking at Table 4.4 we may notice that leakage, temperature, and contamination control play an important part in counteracting problems in hydraulic systems.

Usually, monitoring of hydraulic systems is limited to monitoring of pressure (and occasionally flowrate) in various parts of the circuit and to visual monitoring of the system for leakages, fluid aeration, water contamination, reservoir temperature, etc. [38]. Pressure measurement allows identification of excessive pressure losses, leakages, etc. Some, more important, techniques are shown in Figure 4.38.

For example, Figure 4.39 shows a diagram of a simple hydraulic system with marked locations of contamination entry points and tribological pairs. Then using this information appropriate monitoring techniques can be selected and criteria for detection of abnormal conditions of the system can be established. Tribological pairs and wear mechanisms in common hydraulic components are shown in Table 4.5.

Visual observation of the state and the behavior of a system (e.g., erratic action of the actuator, loss of motor speed, aerated fluid in the reservoir) were used to detect faults and to identify probable causes using troubleshooting guides widely published by the manufacturers of hydraulic equipment. This type of monitoring is still widely practiced. Although monitoring of pressure and visual observation of the system condition are still valid techniques of condition monitoring, the development of a wide range of sensors for monitoring of vibration, noise, flow, contamination, water content, and other parameters of interest and better understanding of phenomena affecting operation of hydraulic systems (e.g., contamination, aeration, flow forces) led to the development of advanced monitoring techniques that are able to detect early stages of failures. Typical effects of aeration and temperature on dynamic behavior of hydraulic systems are shown in Figure 4.40.

Contamination control in hydraulic and lubricating systems is fundamental to reliability and performance and no longer needs justification. Ideally, the selection and design of a proper contamination

TABLE 4.4 Characteristics of Hydraulic Systems

Characteristics	Result In	Preventive Measure
High power density (ratio of weight/transmitted power)	High operating stresses	Choice of material, design
High speed operation	Noise, vibrations	Proper design of system
High contact stresses (valve plate/cylinder block, gears, etc.)	Fatigue wear	Choice of material, design
High temperature operation	Fluid oxidation, distortion, poor lubrication	Temperature control (heat exchange, reservoir design)
Shock loading	High operating stresses	Pressure control, design
Leakage	Entry of contaminant, increased temperature	Seals, close tolerances, contamination control
Predominance of sliding tribological pairs (pistons, spools, poppets, vanes, etc.)	Damage due to scoring, sliding wear Production of wear debris	Selection of materials, contamination control
High precision of equipment (servovalves, pumps, etc.), small clearances	Very high sensitivity to contamination and wear, silting (valves)	Seals design, contamination control
Low cycle fatigue (high start-up loads)	High stress level	Design, pressure control
Fluid as transmission medium	Sensitivity to temperature, contamination, aeration	Reservoir design, temperature control, aeration control, contamination control
Flow throttling as means of control	Erosion, aeration	Design of components, choice of materials
Pressures/flow pulsation	Noise, vibrations	Noise/vibration control, design
Need to seal pressure volumes	Close tolerances, seals, leakage	Seal design, contamination control
Unskilled operators	Overloading, incorrect operation	Design, pressure control
Poor maintenance practices	Contamination, sealing problems	Contamination control, aeration control
Improper system design procedures	Operational, maintenance, safety problems	Education, training, quality control (design audits)

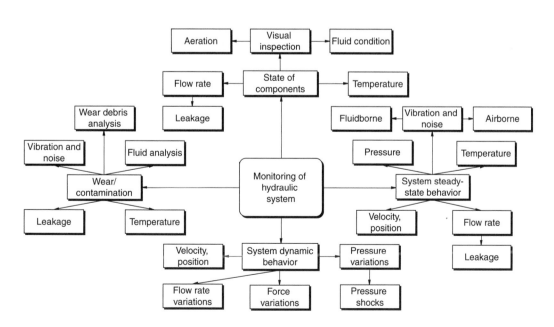

FIGURE 4.38 Monitoring of a hydraulic system.

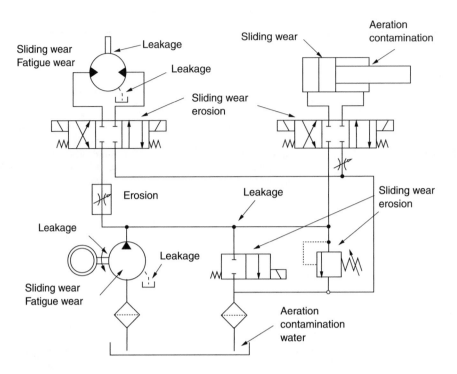

FIGURE 4.39   Wear and contamination in a hydraulic system.

control system should be addressed at the system design stage. However, proper design of a filtration system requires a lot of effort on the part of the designer to find the necessary information about the environment in which the system will operate, the types of components and their tolerances to contamination, user's maintenance procedures, etc. The amount of effort expended on designing an efficient contamination control system usually depends on the complexity and importance of the hydraulic system — for most systems the design task is simply reduced to the selection of appropriate filters on the basis of manufacturers' recommendation.

4.3 Contamination Balance

The technology associated with contamination control is by no means mature. Using available engineering methods, it is neither possible to predict the cleanliness level in a hydraulic or lubricating system during the design phase nor to estimate the remaining life of filters in a system under operation. In contamination control, hydraulic components can be classified into four different groups:

- Components that capture and retain contamination (filters, strainers)
- Components that produce contamination due to wear (pumps, cylinders, valves)
- Components that allow induction of contamination from the system's working environment (breathers, rod seals, rotating shaft seals); manufacturing and assembly residues in the system can be added to this group
- Components that mix the contamination level and allow the contamination level to fluctuate; the reservoir is the main component in this group

Any scientifically based method of design for contamination control must be based on a fundamental relation of equilibrium between particle ingression and the removal of particles. The contamination characteristics of the system are dynamic in nature and dependent on a number of variables and factors:

TABLE 4.5   Hydraulic Systems — Wear Mechanisms

Hydraulic Component	Tribological Pair	Mechanism of Wear
Check valve (poppet)	Poppet/body	Impact, sliding
	Poppet/seat	Flow erosion
Check valve (ball)	Ball/body	Small impact
	Seat	Flow erosion
Vane pumps and motors	Bearing ring	Rolling fatigue
	Vane/rotor	Sliding
	Vane/casing	Sliding
Gear pumps and motors	Rolling element bearing	Rolling fatigue
	Gear pair	Rolling fatigue, sliding
	Gear/thrust plate	Sliding, abrasion
Piston pumps and motors	Rolling element bearings	Fatigue
	Piston/block	Abrasion
	Block/valve plate	Sliding, abrasion
	Piston/slipper	Sliding
	Slipper/swash plate	Sliding
	Block/shaft	Fretting
Directional control valves	Spool/body	Sliding
	Spool	Erosion
Proportional and servo valves	Spool/body	Sliding
	Jet nozzle	Erosion
	Spool	Erosion
Pressure control valves	Spool/body	Sliding
	Seat/poppet	Impact
	Seat and poppet	Flow erosion
	Orifices	Flow erosion
Flow control valves	Spool/body	Sliding
	Seat/poppet	Impact
	Seat and poppet	Flow erosion
	Orifices	Flow erosion
Cylinders	Cylinder rod/gland bush	Sliding
	Seal/cylinder	Sliding
	Swivel bush/pin	Sliding
Accumulators (piston)	Piston/cylinder	Sliding
Rotary actuator (vane)	Bearing	Rolling fatigue
	Vane/body	Sliding
	Vane/rotor	Sliding
Rotary actuator (rack/pinion)	Seal/cylinder	Sliding
	Rolling element bearing	Rolling fatigue
	Gear pair	Rolling fatigue
	Cylinder rod/gland bush	Sliding
	Seal/cylinder	Sliding

- Type of hydraulic system
- Filtration ratios of filters
- Environment
- Expected contaminant ingression rates
- Target cleanliness levels for system components
- Type of filtration
- Duty cycle, fluid loss
- Type of system (e.g., by-pass) and others

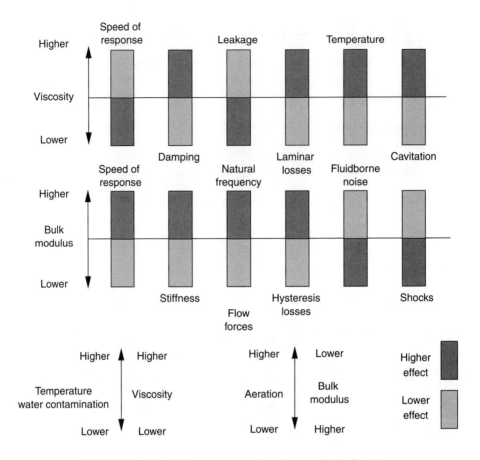

FIGURE 4.40   Effect of temperature and aeration on a hydraulic system.

The basic relationship between particles entering and leaving a system is shown in Figure 4.41 and on this basis three different methods have been developed. The first step in developing such a scientific method was the Oklahoma State University (OSU) model of particles balance in a hydraulic system developed by Fitch [4]. Anderson [42] presented a slightly different approach to determine contamination balance; however, both results are equivalent.

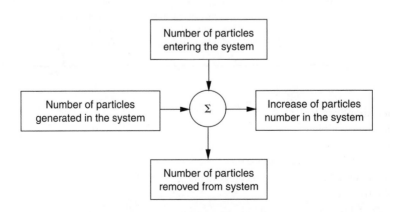

FIGURE 4.41   Contamination balance.

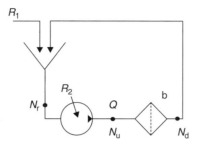

FIGURE 4.42   OSU model.

The development of the *concept of contamination balance*, equating the generation and removal of contamination within the system, provided a mathematical tool to describe some of these interactions and thus led to a better understanding of the contamination characteristics of hydraulic control and lubrication systems.

The importance of considering contamination balance in the system, that is, matching ingression rates to the contamination tolerance levels of components in the system and filter characteristics, is now well recognized. The concept is simple — to maintain the contamination within the allowable contamination level in the system for any amount of contamination that is ingested or generated within a system, an equal amount of contamination must be removed. The concept is, however, difficult to apply in practice especially in the case of multi-branch systems consisting of many hydraulic components, as both ingression rates and contamination tolerance levels, as well as contamination levels in various points of the circuit, are difficult to quantify. The original work at OSU that led to the development of the concept of contamination balance is limited to simple systems.

4.3.1   OSU and Anderson Models

Both models assume a constant rate of wear and contamination. In the OSU model equation for material balance in the reservoir, see Figure 4.42.

$$N_r V = \int (R_1 + R_2)\mathrm{d}t + \int N_d\, Q\mathrm{d}t - \int N_u\, Q\mathrm{d}t + N_o\, V \tag{4.3}$$

where V is volume of fluid, N_o is initial contamination level (number of particles greater than size d per unit volume), N_r is contamination level in reservoir (number of particles greater than size d per unit volume), N_u is contamination level upstream of filter (number of particles greater than size d per unit volume), N_d is contamination level downstream of filter (number of particles greater than size d per unit volume), R_1 is rate of particles ingression (number of particles greater than size d per unit time), R_2 is rate of particles generation by system components (number of particles greater than size d per unit time), and Q is flowrate.

Characteristic size of particle is denoted by d. The ratio of N_d to N_u (for particles size $> \mu$m) is defined as β-rating of the filter.

The contamination level upstream of the filter is expressed by:

$$N_u = \frac{\beta}{\beta - 1}\frac{(R_1 + R_2)}{Q}(1 - e^{-t/\tau}) \tag{4.4}$$

where time constant τ is equal to:

$$\tau = \frac{\beta}{\beta - 1}\frac{V}{Q} \tag{4.5}$$

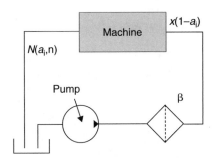

FIGURE 4.43   Anderson's model.

Steady-state contamination level upstream of the filter ($t \longrightarrow \infty$) is then equal to:

$$N_{\text{u}} = \frac{R_1 + R_2}{Q} \frac{\beta}{\beta - 1} \tag{4.6}$$

and downstream of the filter:

$$N_{\text{d}} = \frac{R_1 + R_2}{Q} \frac{1}{\beta - 1} \tag{4.7}$$

These equations assume that β-ratio, Q, V, and particle ingression/generation rates are constant.

A more complete filtration model is provided by Anderson's model. Anderson developed his model to determine equilibrium of particles in a system under the following assumptions (see Figure 4.43):

- During one cycle full volume of oil passes through the machine, that is, if a pump has 100 lpm flowrate and total volume of fluid in the system is 500 l it will take 5 min to complete one cycle (assuming that all oil enters the pump).
- During one cycle machine produces x particles per unit volume.
- The removal efficiency of size i particles, where i refers to characteristic length of the particles, is denoted as a_i, $0 \leq a_i \leq 1$.
- During each cycle $a_i x$ particles are destroyed during passage through system components, lost in the reservoir, or removed by the filter.

After one cycle a number of particles in a fluid volume (particle concentration) is equal to:

$$N(a_i, 1) = x(1 - a_i) \tag{4.8}$$

where $N(a_i, 1)$ is concentration of particles of size i after one cycle. After two cycles concentration is equal to:

$$N(a_i, 2) = x + x \cdot (1 - a_i) \cdot N(a_i, 2) = x + x \cdot (1 - a_i) \tag{4.9}$$

and $a_i [x + x \cdot (1 - a_i)]$ particles are removed. After n cycles particle concentration is:

$$N(a_i, n) = x + x \cdot (1 - a_i) + x \cdot (1 - a_i)^2 + \cdots + x \cdot (1 - a_i)^{n-1} \tag{4.10}$$

and substituting $y = 1 - a_i$

$$N(a_i, n) = x \cdot (1 + y + y^2 + \cdots + y^{n-1}) = x \sum_{r=1}^{r=n} y^{r-1} \tag{4.11}$$

then concentration is equal to:

$$N(a_i, n) = \frac{x \cdot (1 - y^n)}{1 - y} = \frac{x \cdot (1 - (1 - a_i)^n)}{a_i} \qquad (4.12)$$

when $n \implies \infty$, the concentration approaches:

$$N(a_i, \infty) \implies \frac{x}{a_i} \qquad (4.13)$$

We now define quantity α, $\alpha \ll 1$, such that after R cycles

$$\frac{N(a_i, R)}{N(a_i, \infty)} \leq 1 - \alpha \qquad (4.14)$$

and we can calculate a number of cycles R to reach the given percentage of equilibrium concentration $1 - \alpha$:

$$R \geq \frac{\ln \alpha}{\ln(1 - a_i)} \qquad (4.15)$$

As the efficiency of the filter is equal to:

$$a_i = \frac{\beta_i - 1}{\beta_i} \qquad (4.16)$$

where β_i is filtration ratio, we may now express Equations (4.12), (4.13), and (4.15) in terms of β_i:

$$N(a_i, n) = x \frac{\beta_i^n - 1}{\beta_i - 1} \beta_i^{(1-n)} \qquad (4.17)$$

$$N(a_i, \infty) = x \frac{\beta_i}{\beta_i - 1} \qquad (4.18)$$

$$R \geq \frac{\ln \alpha}{\ln(1 - (\beta_i - 1)/\beta_i))} \qquad (4.19)$$

Thus as the efficiency of filtration for small particles is small, a large number of cycles is required before equilibrium is reached.

Both OSU and Anderson models show that the following factors will affect length of time needed by the system to reach particle equilibrium:

- Filtration — number of times a particle of a given size and material, on average, passes through the filter. The better the filter, the shorter the time to equilibrium. Deterioration of filter (due to gradual clogging of the filter, by-passing etc.) will affect particle removal efficiency — especially small particles.
- Fluid cycle rate — expressed in volume per unit time divided by the volume of lubricant in the system. Varies from five times per minute to once per hour or longer. It affects material loss by filtration.
- Location of slow moving oil — the particles may settle or adhere to surfaces, for example, bottom of sump.
- Dispersion quality of lubricant — these additives (detergents) prevent agglomeration of the particles and also discourage their adherence to the surface, thus increasing their life expectancy.
- Breakdown of particles — during repeated passage through wear contacts the large particles will be reduced to smaller sizes.
- Oil loss due to leakage will remove some particles from the system (but it is not the recommended way of controlling contamination!).
- Oxidation and chemical attack will change the size of particles.

The model used by Anderson, similar to the OSU model, has the following deficiencies:

- Assumption of constant wear rate (constant x). This assumption is only applicable to normal wear.
- Assumption of constant a_i ignores removal of particles due to oil losses and deterioration in filter efficiency.
- All wear debris producing components and contamination ingression points were lumped together and represented as a single source of particles. A real system has many points where wear is produced and contamination from the environment is ingressed into the system. The rate of wear generation and contamination ingression varies between components.
- Only one fluid path is considered. Real systems usually have more than one fluid path. The particle concentration in each flow path will depend on the flow rate and the particle generation/ingression rate by the components in the path.
- Changes in contamination condition due to leakages, oil addition, changes in efficiency of filters etc. cannot be included in the model.
- Each model gives concentration of particles in the reservoir. In practice, we are interested to know the exposure of various components to wear and contamination particles.

Nevertheless both OSU and Anderson's models provide better insight into the behavior of contamination control system and provide a basis for sampling guidelines. The models show that large particles will reach equilibrium much sooner than small particles, thus care should be taken to assure that sufficient time has elapsed to allow concentration of particles in the system to reach an equilibrium before the samples are taken. This should be taken into consideration when using monitoring methods that are not able to detect small particles; for example, spectrometric oil analysis (SOAP) (see Figure 4.44).

4.3.2   Dynamic Contamination Control (DCC)

The concept of contamination balance provided sufficient insight to permit recognition and understanding of the controlling parameters of the system and to permit adequate description of the influence of each parameter on the system as a whole. To describe quantitative relationships between variables and parameters describing fluid contamination in hydraulic control systems, a mathematical model in the form of a Dynamic Contamination Control (DCC) was developed at the Centre for Machine Condition Monitoring, Monash University in Australia [39,40]. A DCC allows analysis and optimization of contamination control in a fluid power and lubrication system. The model of the contamination control system is solved using

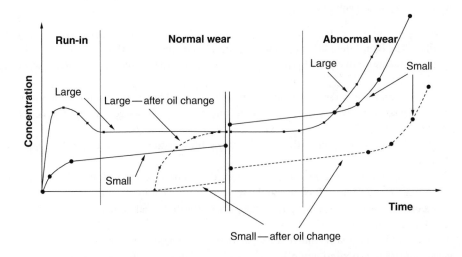

FIGURE 4.44   Equilibrium of small and large particles in a system.

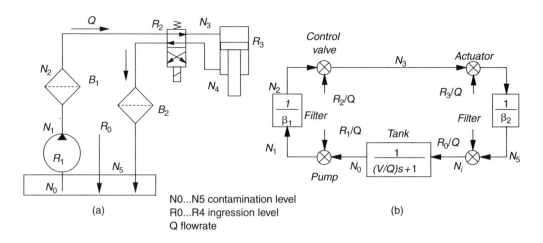

NO...N5 contamination level
R0...R4 ingression level
Q flowrate

(a)   (b)

FIGURE 4.45   DCC Model (Monash University).

digital simulation, the results provide information about the contamination level at various locations in the circuit. The model also allows investigation of the contamination levels at various points in the circuit in response to variation in the size distribution and the level of ingressed or generated contamination. It also allows investigation of the effects of changes in the filtration ratios of filters and their locations in the system. This model has been proven to be accurate through experiments carried out at the Norwegian University of Science and Technology [41].

4.3.2.1   Single Path Model

A single path hydraulic system, shown in Figure 4.45(a), is used as an example to illustrate the DCC model. The system includes a return line and pressure line filters, a hydraulic pump, a reservoir, a directional control valve, and a cylinder.

   The contamination control system is represented by a block diagram in which each block represents the function of a single hydraulic component in the contamination control system. The variable of interest in contamination control is contamination level N_i (number of particles larger than a specified size per unit volume of fluid). Thus, the input and output of the block for an individual hydraulic component are the upstream and downstream contamination levels at the location of the component. The DCC model of a whole hydraulic control system can be obtained by linking the contamination block diagrams of the components of the system. Output of a component block diagram becomes the input of the following component block diagram according to the hydraulic circuit diagram. The DCC model of the system is shown in Figure 4.45(b).

   Some of the variables and parameters in a contamination control system such as flowrate Q, filtration ratios of filters, and particle ingression rates R vary with time. They may also be affected by changing size distribution of contaminant particle, changing duty cycle, filters clogging, or changes in operating environment. Although these variations in values of parameters are not currently modeled, this technique can be readily extended to include these variations in the model when and if the information about interaction between various parameters becomes available. In our example, as the objective of the model is to use it in assessing final levels of contamination in the system when the system contamination levels reach the state of equilibrium, we assume mean flowrate Q, and constant filtration ratios β_1, β_2 and ingression rates $R_0, \ldots, R_3$. The block diagram of the DCC model shows the location in the system where contamination ingress and/or wear debris generation occurs (summing junctions) and the location of filters. The topology of the block diagram is identical to the modeled real system. The contamination level at any point in the circuit can be directly calculated from the closed-loop DCC model. For the system shown in Figure 4.45, the following equation can be written:

$$N_i = \frac{N_0}{\beta_1 \beta_2} + \frac{R_1}{\beta_1 \beta_2 Q} + \frac{R_2}{\beta_2 Q} + \frac{R_3}{\beta_2 Q} + \frac{R_0}{Q} \tag{4.20}$$

where N_i, N_0 are input and reservoir contamination levels (number of particles $> d \, \mu$m per volume), β_1, β_2 are filtration ratios, Q is mean flowrate in the system (volume per time), and $R_0, \ldots, R_3$ are contaminant ingression rates (number of particles $> d \, \mu$m per time).

The dynamic change in contamination level in the reservoir is described by the following equation:

$$N_0 = \frac{N_i}{(V/Q)s + 1} \tag{4.21}$$

where V is reservoir volume and s is Laplace operator.

Substituting Equation (4.20) into Equation (4.21) we obtain:

$$N_0 = \frac{N_i}{(V/Q)s + 1} = \left\{ \frac{N_0}{\beta_1 \beta_2} + \frac{R_1}{\beta_1 \beta_2 Q} + \frac{R_2}{\beta_2 Q} + \frac{R_3}{\beta_2 Q} + \frac{R_0}{Q} \right\} \frac{1}{(V/Q)s + 1} \tag{4.22}$$

and

$$N_0 - \frac{N_0}{\beta_1 \beta_2} \frac{1}{(V/Q)s + 1} = \left\{ \frac{R_1}{\beta_1 \beta_2 Q} + \frac{R_2}{\beta_2 Q} + \frac{R_3}{\beta_2 Q} + \frac{R_0}{Q} \right\} \frac{1}{(V/Q)s + 1} \tag{4.23}$$

then

$$N_0 \frac{\beta_1 \beta_2 [(V/Q)s + 1] - 1}{\beta_1 \beta_2 [(V/Q)s + 1]} = \left\{ \frac{R_1}{\beta_1 \beta_2 Q} + \frac{R_2}{\beta_2 Q} + \frac{R_3}{\beta_2 Q} + \frac{R_0}{Q} \right\} \frac{1}{(V/Q)s + 1} \tag{4.24}$$

and rearranging:

$$N_0 \frac{[(V/Q)s + 1] - 1/\beta_1 \beta_2}{[(V/Q)s + 1]} = \left\{ \frac{R_1}{\beta_1 \beta_2 Q} + \frac{R_2}{\beta_2 Q} + \frac{R_3}{\beta_2 Q} + \frac{R_4}{Q} \right\} \frac{1}{(V/Q)s + 1} \tag{4.25}$$

finally, the contamination level at exit from the reservoir is described by:

$$N_0 = \frac{\{ R_1/\beta_1 \beta_2 Q + R_2/\beta_2 Q + R_3/\beta_2 Q + R_4/Q \}}{[(V/Q)s + 1] - 1/\beta_1 \beta_2} = \frac{N_m}{[(V/Q)s + 1] - 1/\beta_1 \beta_2} \tag{4.26}$$

where

$$N_m = \frac{R_1}{\beta_1 \beta_2 Q} + \frac{R_2}{\beta_2 Q} + \frac{R_3}{\beta_2 Q} + \frac{R_4}{Q} \tag{4.27}$$

is the equivalent contamination input to the system, and the contamination level in the reservoir is equal to (see Figure 4.46):

$$N_0 = \frac{N_m}{[(V/Q)s + 1] - 1/\beta_1 \beta_2} = \frac{N_m}{(V/Q)s + (1 - 1/\beta_1 \beta_2)} \tag{4.28}$$

Filtration ratio (β-ratio) of a filter is defined as the ratio of contamination input (measured as a number of particles above a certain size per fluid volume) to contamination output. The β-ratio is measured by subjecting the filter to a multi-pass test ISO 4572-1981 and may vary from 2 to 70 (β-ratio $= 1$ indicates no filtering action). As good quality filters have β-ratio much greater than 1, then:

$$1 - \frac{1}{\beta_1 \beta_2} > 0 \tag{4.29}$$

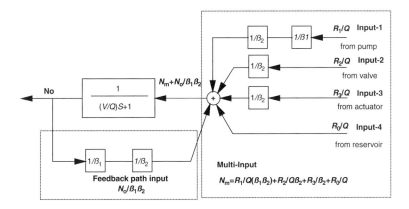

FIGURE 4.46 Reduced model.

In time domain solution (4.28) yields:

$$N_0(t) = \frac{N_m}{1 - F(\beta)} - \frac{N_m}{1 - F(\beta)} e^{-(Q(1-F)/V)t} \qquad (4.30)$$

where:

$$F = \frac{1}{\Pi \beta_i} \qquad i = 1, 2, \ldots, n$$

Time constant of the system is defined by:

$$T = \frac{Q(1 - F)}{V} \qquad (4.31)$$

and time to reach equilibrium is approximately equal to:

$$T_0 \approx 4T = 4\frac{Q(1 - F)}{V} \qquad (4.32)$$

4.3.2.2 Multi-Path Systems

Most hydraulic systems include more than one flow path — the above outline modeling approach can be easily applied to systems having multiple flow paths each having different flowrate and different ingression rates. A DCC model of a multi-branch system that has three flow branches is shown in Figure 4.47. Each branch flow can be treated as a fraction of total flow, thus:

$$Q_i = A_i Q \qquad k = 1, 2, \ldots, n \qquad (4.33)$$

where Q_i is flowrate in ith branch of the circuit and A_i is fraction of total flow in ith branch.

It can be seen that:

$$\sum Q_i = Q$$
$$\sum A = 1 \qquad (4.34)$$

For the contamination system shown in Figure 4.47 that has three parallel paths, we may write:

$$Q_1 + Q_2 + Q_3 = Q \qquad (4.35)$$

To convert the original system to feedback form we express N_4 as follows:

$$N_4 = A_1 N_{13} + A_1 N_{24} + A_1 N_{33} \qquad (4.36)$$

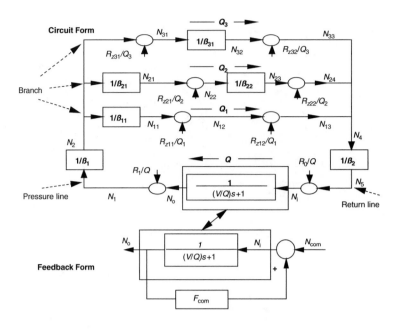

FIGURE 4.47   Multi-branch contamination model.

where:

$$N_{13} = \frac{N_0}{\beta_1 \beta_{11}} + \frac{R_{z11}}{A_1 Q} + \frac{R_{z12}}{A_1 Q} + \frac{R_1}{\beta_1 \beta_{11} Q}$$

$$N_{24} = \left(\frac{N_0}{\beta_1 \beta_{21}} + \frac{R_{z21}}{A_1 Q} \right) \frac{1}{\beta_{22}} + \frac{R_{z22}}{A_2 Q} + \frac{R_1}{\beta_1 \beta_{21} \beta_{31} Q} \qquad (4.37)$$

$$N_{33} = \left(\frac{N_0}{\beta_1} + \frac{R_{z31}}{A_3 Q} \right) \frac{1}{\beta_{31}} + \frac{R_{z32}}{A_3 Q} + \frac{R_1}{\beta_1 \beta_{31} Q}$$

and

$$A_1 = \frac{Q_1}{Q}, \quad A_2 = \frac{Q_2}{Q}, \quad A_3 = \frac{Q_3}{Q}$$

Input contamination level N_i is equal to:

$$N_i = \frac{N_4}{\beta_2} + \frac{R_0}{Q} \qquad (4.38)$$

and substituting the above expressions we obtain the expression for N_i:

$$N_i = \left(\frac{N_0}{\beta_1 \beta_{11}} + \frac{R_{z11}}{A_1 Q} + \frac{R_{z12}}{A_1 Q} + \frac{R_1}{\beta_1 \beta_{11} Q} \right) \frac{A_1}{\beta_2}$$

$$+ \left\{ \left(\frac{N_0}{\beta_1 \beta_{21}} + \frac{R_{z21}}{A_1 Q} \right) \frac{1}{\beta_{22}} + \frac{R_{z22}}{A_2 Q} + \frac{R_1}{\beta_1 \beta_{21} \beta_{31} Q} \right\} \frac{A_1}{\beta_2}$$

$$+ \left\{ \left(\frac{N_0}{\beta_1} + \frac{R_{z31}}{A_3 Q} \right) \frac{1}{\beta_{31}} + \frac{R_{z32}}{A_3 Q} + \frac{R_1}{\beta_1 \beta_{31} Q} \right\} \frac{A_1}{\beta_2} + \frac{R_0}{Q} \qquad (4.39)$$

We can now express N_i in this form:

$$N_i = N_{com} + F_{com} N_0 \tag{4.40}$$

where N_{com} is combined contamination input to the reservoir and F_{com} is combined filter effect.
For our circuit:

$$N_{com} = A_1 \frac{R_1}{\beta_1 \beta_2 \beta_{11}} + A_2 \frac{R_1}{\beta_1 \beta_2 \beta_{21} \beta_{22} Q} + A_3 \frac{R_1}{\beta_1 \beta_2 \beta_{131} Q} + \frac{R_{z11}}{\beta_2 Q}$$

$$+ \frac{R_{z12}}{\beta_2 Q} + \frac{R_{z21}}{\beta_2 \beta_{22} Q} + \frac{R_{z22}}{\beta_2 Q} + \frac{R_{z31}}{\beta_2 \beta_{31} Q} + \frac{R_{z32}}{\beta_2 Q} + \frac{R_0}{Q} \tag{4.41}$$

In general, these equations for a multi-path circuit become:

$$F_{com} = \sum_{i=1}^{i=m} \frac{A_i}{\Pi \beta_i} \tag{4.42}$$

and

$$N_{com} = N_{pr} + N_{br} + N_{rt} = \sum_{i=1}^{k} \frac{R_i}{Q} F_{com} + \sum_{i=1}^{m} \sum_{j=1}^{l} \frac{R_{rij}}{\Pi_n \beta_{mn}} + \frac{R_0}{Q} \tag{4.43}$$

where N_{pr} is contamination level in the pump delivery line, N_{br} is contamination level in each branch flow path, N_{rt} is contamination level caused by contamination ingression in the reservoir, R_{rij} is the jth ($j = 1, 2, \ldots, l$) ingress rate in the ith ($m = 1, 2, \ldots, m$) branch path, n is number of filters in the i-line between ingression source R_{zij} and entry to the reservoir, k is number of ingression sources in pressure line, β_{mn} is filters' β-ratios, m is number of flow branches, and l is number of the ingression sources in the branch.

A time solution for a linear multi-branch system can be obtained using Equation (4.30) and time to reach equilibrium is calculated using Equation (4.32) in which we replace:

$$N_{com} \Rightarrow N_m$$
$$F_{com} \Rightarrow F \tag{4.44}$$

The modeling procedure outlined above is shown in Figure 4.48. Typical packages that can be used to model the contamination system are Matlab's Simulink, ACSL, or Vissim.

4.3.2.3 Simulation

Using the above modeling approaches we may assess contamination levels analytically. However, an analytical approach to this task is rather tedious as in practice we are interested in likely effects of system parameter changes (e.g., β-rating of filters, locations of filters, size of reservoir volume) on contamination levels at particular locations in the system. Availability of simulation packages makes it easy to investigate the system and allows experimentation with different parameters affecting contamination of the system. We may also apply a contamination model in parallel with a dynamic model of the system to investigate the effects of changing contamination characteristics of the system on system operation.

A Vissim simulation block diagram model of the system presented in Figure 4.45 is shown in Figure 4.49. An example of results obtained from digital simulation is shown in Figure 4.50.

Modeling and simulation of contamination in hydraulic systems using the approach outlined in this section allows evaluation of contamination levels in the system and makes it possible to investigate the effects of design changes. This technique could easily be extended to include determination of other characteristics of contamination of a control system, for example, assessing the dirt capacity of the system

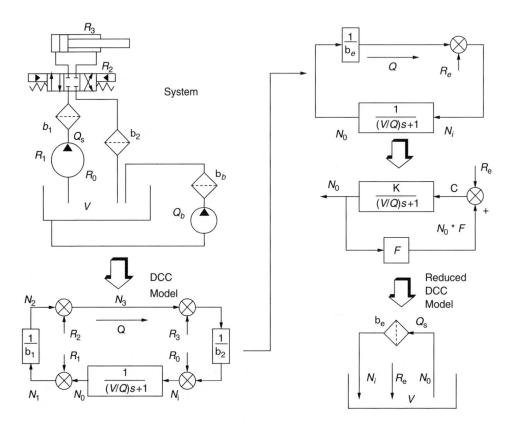

FIGURE 4.48   DCC modeling procedure.

that will provide information about how often a filter must be changed. Also, nonlinear effects, for example, filter rating change with exposure to contaminant could be easily included.

4.4 Monitoring Procedures

A contamination control program is an important component of proactive maintenance and predictive maintenance strategies (see Figure 4.51):

- *Predictive maintenance* strategy is based on monitoring the condition of the system and, when changes in system condition that could lead to failure are detected, corrective action is taken. This strategy is now commonly used in many industries. The monitoring techniques based on the investigation of the wear debris and contaminant carried by a lubricating or hydraulic fluid were recognized long ago as valuable detection and diagnostic tools. Analysis of lubricant/working fluid characteristics provide additional information about the state of a machine, possible wear, and friction regimes and thus in combination with particulate-based analysis enhance both monitoring and diagnostic characteristics of tribology-based condition monitoring methodology. The onset of damage in machinery can be detected by examining the wear and contaminant particles present in the lubricant. The shape, size, size distribution, and number of wear particles present in a sample as well as the morphology and the condition of the particles give many clues as to the state of a machine and the possible incipient damage, and thus help the analyst to judge the likely future performance of the machine. The reliability of diagnosis and prognosis of the condition monitoring program will be directly affected by the selection of detection methods. Selection of a proper mix of monitoring methods that will be able to identify changes in machine condition is thus important.

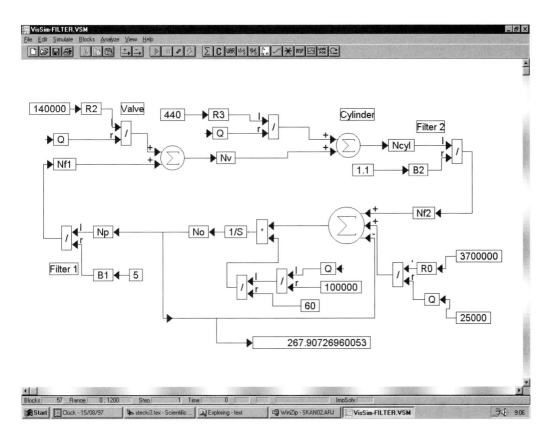

FIGURE 4.49   Block diagram of contamination system shown in Figure 4.46.

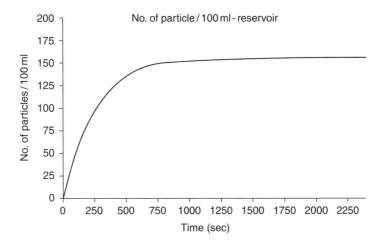

FIGURE 4.50   Result of simulation — level of contamination in the reservoir.

- *Proactive strategy* is based on the monitoring of root causes of failures and taking corrective actions to eliminate or minimize them. Although monitoring procedures closely resemble those used in predictive maintenance, the purpose of monitoring is to provide guidance for eliminating conditions that could lead to failures; Figure 4.52 shows the comparison between proactive and predictive monitoring approaches.

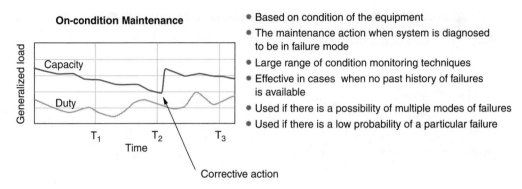

* Based on condition of the equipment
* The maintenance action when system is diagnosed to be in failure mode
* Large range of condition monitoring techniques
* Effective in cases when no past history of failures is available
* Used if there is a possibility of multiple modes of failures
* Used if there is a low probability of a particular failure

* Based on monitoring of system duty and changes in system elements
* Eliminates conditions which could lead to failure of the system
* Does not replace on-condition maintenance
* Includes monitoring of root-causes of the failures

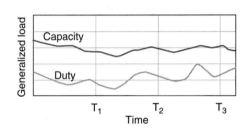

FIGURE 4.51   Comparison of predictive and preemptive maintenance.

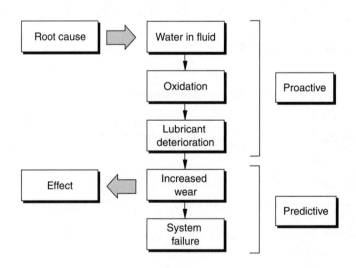

FIGURE 4.52   Relation between proactive and predictive monitoring strategy.

The monitoring procedure is shown in Figure 4.53 and consists of a number of steps:

1. Sampling:
    (a) Obtaining particulate and lubricant/working fluid that are representative of tribological state of the machine
    (b) Obtaining information about presence of conditions that may cause failure, for example, presence of water, high temperature

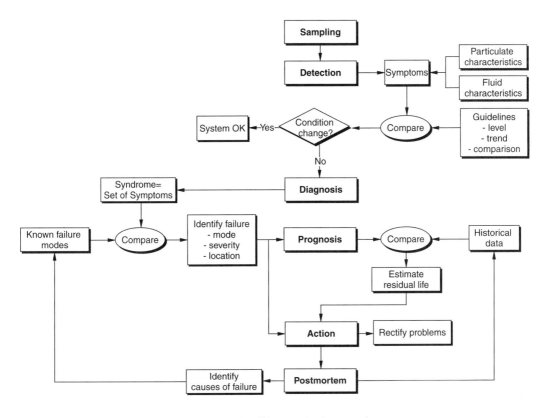

FIGURE 4.53 Condition monitoring procedure.

2. Detection
    (a) Detecting symptoms of an abnormal condition of a machine that indicates an incipient failure of a machine by comparing the results of particulate and fluid analysis with previously determined guidelines
    (b) Detecting changes in properties of particulate and fluid analysis with previously determined guidelines to determine initiation of condition for failure
3. Diagnosis
    (a) Diagnosing, by comparing a set of symptoms with a known syndrome of machine condition, the mode, severity, location, and mechanism of the wear process that may lead to failure
    (b) Diagnosing what causes changes in system conditions, for example, leaking seal resulting in contamination
4. Prognosis, on the basis of historical data and diagnosis, of the future state of the machine and remaining service life before failure
5. Action required to correct and/or rectify identified problems
6. Postmortem
    (a) Investigation of failed components
    (b) Investigation of what caused changes in system condition

The monitoring program must be designed to evaluate fluid properties, fluid contamination, and wear/contamination condition of the system. Proactive maintenance strategy is based on similar steps; however, it is concerned with monitoring root causes, that is, system parameters and variables that indicate presence of conditions, which may lead to development failure symptoms.

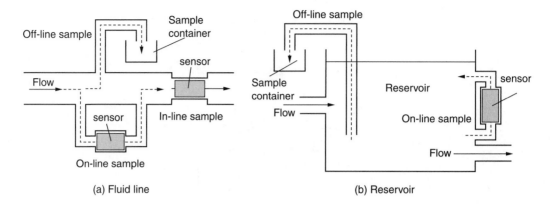

FIGURE 4.54   Fluid sampling methods.

4.4.1 Sampling

Usually sampling is treated implicitly; that is, it is considered to be a part of the detection step. However, the importance of obtaining a sample representing the true condition of the fluid, fluid contamination, and machine wear/contamination warrants treating sampling as a separate step in the condition monitoring procedure. The whole process of condition monitoring is based on the premise that the quality of sampling is acceptable and that samples provide a representative indication of the condition of the machine.

Methodology of sampling varies depending on the type of monitored equipment. There are three basic methods of taking samples of a lubricant or a working fluid (see Figure 4.54):

- *Off-line* (sampling from reservoir or fluid lines, taking grease samples). In this method a sample fluid is removed from the machine for further manual processing.
- *On-line* (sampling from reservoir of fluid lines). In this method a sample of fluid is drawn from a fluid line or reservoir by the test instrument and automatically returned to the reservoir or fluid line after the test. On-line particle counters use this method of sampling.
- *In-line* (fluid lines). In this method a sensor or instrument is placed in a fluid line and provides continuous measurement of parameter/variable of interest. Magnetic plugs use this method of sampling.

Methods of taking a sample from systems that use liquid lubricants will depend on the method of delivery of the lubricant to components that require lubrication. Systems using splash lubrication may be sampled off-line or on-line. All three methods can be used for sampling of liquid lubricants in recirculation lubrication systems and for sampling of working fluids. A sample should be taken at the time when concentration of debris in a system reaches equilibrium for a given set of operational parameters.

When a hydraulic or recirculation lubrication system operates under steady wear and contamination conditions, concentration of the wear and contaminant particles in the fluid (number of particles per unit volume) achieves an equilibrium level for each given set of operational parameters. To achieve such an equilibrium level, since wear debris is continually generated in any operational mechanical system and contamination also always enters the system, the rate of particle removal from the fluid must be at the same rate at which they are generated or ingressed. Each machine has a characteristic operating time necessary to return to its normal equilibrium level. The equilibrium level will be upset when wear or contamination conditions change due to an increased rate of wear or contamination.

Some factors that influence the time required to reach equilibrium level are listed below:

- *Filtration efficiency:* Filtration efficiency is related, in the case of hydraulic filters, to their β-ratings. In general the better the filters (high β-ratings), the shorter the time to particle equilibrium.

- *Particle cycle rate:* Particle cycle rate is equal to the ratio of pump flowrate (volume per unit time) to the volume of lubricant or hydraulic fluid in the system. The higher the cycle rate, the more often the particles are exposed to the filter. The particle cycle rate, that is, the number of times a particle of a given size and material may, on the average, pass through the filter is affected by the size of the reservoir.
- *Dispersive qualities of the fluid:* In systems where the fluid contains detergent additives to prevent agglomeration of the particles and to discourage their adherence to surfaces, particle life expectancy is increased.
- *Locations of slow moving fluid:* If there are locations where the fluid moves slowly, the particles may settle down or adhere to the surfaces. Bottom of sumps, oil tanks, pipe bends, etc. are examples of such locations.
- *Particulate loss:* Known and suspected particle losses are the result of the following factors:
  (a) Placement of filters in the system
  (b) Settling rate of particles in the reservoir, in sharp bends, silting
  (c) Adherence of particles to solid surfaces
  (d) Breakdown of particles during repeated passage through tribological pairs (e.g., gears, bearings)
  (e) Oxidation
  (f) Chemical attack
  (g) Loss of lubricant or working fluid from the system
- *Fluid change intervals:* Systems where the oil change time is of the same order as the time taken to reach equilibrium are difficult to diagnose and are particularly prone to misinterpretation. This is especially so when diagnostic techniques give a warning on achieving certain absolute particle concentration levels. In these cases the particle concentration level should be recorded through the fluid life cycle and compared with a predetermined norm for that system.
- *Fluid addition:* Following the addition of a lubricant or a hydraulic fluid to the system to replenish fluid loss, the concentration of wear debris and contaminant in the fluid will be lowered. Thus, concentration of debris in samples taken after addition of fluid should be adjusted to account for such changes.
- *Location of filter:* In-line filters can profoundly modify the particle size distribution in the machine's fluid. The filter changes the particle population in the following ways:
  (a) A filter lowers the concentration of particles in the fluid.
  (b) The average particle that remains in the fluid of a filtered system is generated more recently than in the case of a system with no filter.
  (c) A filter removes large particles more effectively than the small ones so that the concentration of larger particles is reduced. Thus, to detect the presence of large particles that are indicative of severe wear and/or contamination in the system, samples should be taken from locations in the system upstream of the filter.

4.4.2 Detection

The objective of predictive monitoring is to determine by analysis of a sample whether the condition of the machine has changed sufficiently enough over a period of time to indicate incipient failure of the machine and thus warrant the diagnostic step. After sampling, detection of changes in the condition of a machine is the second, most important step in the condition monitoring procedure. For example, presence of silica in a sample is a symptom of dust contamination, presence of solid particles larger than 10 μm is a symptom of abnormal wear. Each measured or observed change in a characteristic of particulate or lubricant/working fluid and of any tribological variable or parameter can be considered a symptom of machine condition (see Figure 4.55). Usually a set of symptoms is required to reliably identify the condition of the machine. Effectiveness of detection is dependent on the selection of a set of symptoms that is representative of machine distress, availability of detection techniques to detect these symptoms, and availability of guidelines with which the symptoms are compared. In proactive monitoring the objective

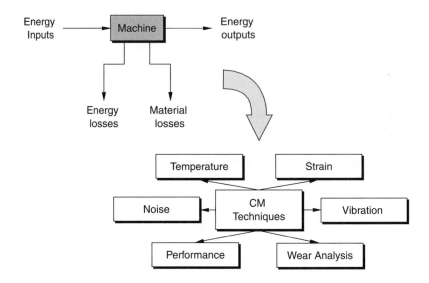

FIGURE 4.55   Condition monitoring symptoms.

is to identify changes in basic parameters of fluid, for example, moisture level, contamination level, and temperature, which are root causes of subsequent changes in the wear regime of the plant. Each change in a measured or observed root cause characteristic of lubricant/working fluid, or a system (e.g., increased temperature) or in contamination level can be considered a symptom of an operational and/or maintenance problem.

4.4.2.1   Symptoms

A set of symptoms reliably describing root causes (proactive) and the state of the machine (predictive) should be determined after a systematic evaluation of the monitored equipment using failure analysis techniques, followed by analysis of tribological systems present in the machine.

The above procedure leads to identification of pertinent characteristics of a system that should be monitored in order to detect any changes in root causes and in the health condition of a machine or a system.

4.4.2.2   Guidelines

Detection of root causes, damage, failure, or malfunction is carried out by comparing results of analysis of characteristics of the particulates or fluid with predetermined criteria (guidelines). The three approaches to detection are shown in Figure 4.56.

Establishing allowable absolute level and trend guidelines requires keeping of historic data on the history of the machine. The levels/trends provided in literature are conservative and may be used only as starting points for setting detection limits.

- Absolute level: An allowable absolute level is used to detect changes in the state of the machine by comparing the results of analysis with the allowable absolute level of system variable or parameter that defines the symptom. For example, the *caution* level may be set at 10 parts per million of Fe (ferrous material) for spectrometric oil analysis.

  Using absolute levels for detection purposes is not very satisfactory as the same machines operating under a different load condition or in a different environment may show different absolute values of symptoms.

  Levels are used when the detection technique provides a single number that indicates a value of some condition monitoring variable. Spectrometric oil analysis (SOA), particle counting, and direct reading ferrograph techniques are examples of detection techniques that provide this type of information.

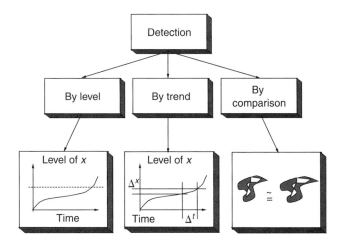

FIGURE 4.56 Detection methods.

- Trend: Trending of analysis data over time and comparing it with an allowable trend offers a better indication of changes in the health condition of the machine and should be used in preference to detection based on levels. For example, for particle counting the *caution* level may be set at 5 particles larger than 10 μm per 10 h of operation.
- Comparison: When a sample yields qualitative data, the detection process is based on comparison of qualitative properties of the sample with a reference model or pattern. Thus, assessment of wear/contaminant debris is carried out by comparing shape, morphology, and color of the debris with reference to particles whose origins are known. For example, presence of parallel striations on the surface of a particle may indicate the sliding mode of wear.

Effective detection of changes in the machine's condition is based on a number of factors listed below:
(a) *Representative samples.* The samples should be representative of the condition of the machine.
(b) *Processing of samples.* The procedure for processing of samples should be clearly laid down and strictly adhered to.
(c) *Reporting and data recording procedures.* The procedures for recording and reporting of data should include relevant information about the machine, samples, and results of analysis.
(d) *Detection techniques.* A suitable mix of detection techniques should be used to detect a relevant set of symptoms.

4.4.2.3 Results of Detection

The result of the detection step is a list of measured/observed symptoms and assessment, based on the results of comparison with guidelines, of their severity. The severity of symptoms is usually classified as low, normal, caution, high, or critical.

4.4.3 Diagnosis

Diagnosis is the task of identifying the current state of the machine by comparing the set of symptoms of the machine state identified in the detection step with a known syndrome that defines a certain state of the machine. For example, a partial set of symptoms to diagnose the condition of a gear box is listed in Table 4.6.

Syndromes of machine state used for diagnosis may be identified by carrying out FMEA, FMECA, or other type of failure analysis. In addition, historical data and other documentary evidence of past maintenance actions should be maintained in order to develop a good data base for future use.

TABLE 4.6   Gear Box — Partial Set of Symptoms

Symptom	Method	Strategy
Abnormal wear morphology of particles	Ferrography, filtergram	Predictive
Solid particles larger than 10 μm	Particle counting	Proactive/predictive
Water contamination	IR spectrometer	Proactive/predictive
Dust contamination, presence of silica	AE spectrometer	Proactive
Lubricant deterioration	Total acid number	Proactive

Effective diagnosis is dependent on:

- Knowledge of syndromes of known conditions of the machine.
- A reliable and complete set of symptoms (identified during the detection step), for example, increased trend of wear debris quantity in successive samples.

If a measured set of symptoms is a subset of a required set of symptoms, then the reliability of diagnosis is diminished and might lead to errors in diagnosis. Diagnosis relying on only one symptom might be unreliable, and often such a diagnosis will be wrong. There are, however, cases when a single symptom is dominating and it is the only one that is monitored.

An even more disastrous situation can, and often does, occur when the wrong condition variable (a symptom) is monitored as then no diagnosis is possible. For example, a gas turbine that was monitored using vibration technique obtained a clean bill of health shortly before the turbine failed catastrophically due to a bearing failure (hydrodynamic bearings were not monitored).

4.4.3.1 Results of Diagnosis

The result of diagnosis is the identification of the mode, severity, location, and mechanism of wear or contamination and changes in chemical/physical properties of the lubricant/working fluid that caused the change in the machine's health condition.

4.4.4 Prognosis

The objective of the condition monitoring program is to be able to predict when the machine will need to be serviced or replaced. The prediction of the residual life of a machine is very dependent on availability of historical data on the past condition of the machine and history of failures of the machine or of similar machines. The most difficult point in prognosis is to decide how long it should be operated before taking corrective action. To be able to make this type of decision the results of diagnosis should be correlated with the following data on the monitored machine or similar machines:

- Records of past failures — records on monitoring, diagnosis, and corrective action taken
- Records of reasons why this failure occurred — postmortem reports
- Records of times lapsed between detection of conditions indicating failures and times when failures actually occurred

In practice it is difficult to develop a data base for prognosis as population of machines that are critical to a given production process may be small; thus, no sound statistical data can be obtained.

4.4.4.1 Approaches to Prognosis

The prediction methods are still in the research stage of development. The usual way to predict residual life is to apply some type of curve fitting into data obtained from detection/diagnosis steps on the assumption that the failure will follow a trend discovered by monitoring. Techniques like Grey Theory, polynomial fitting, and more advanced artificial intelligence techniques (neural net and fuzzy logic) are being applied to provide recognition of the pattern of failures and prediction of future trends in the progression of machine failure.

4.4.4.2 Results of Prognosis

The prognosis step should ideally result in an estimate of how much time is left before the machine will fail in service.

4.4.5 Action

Maintenance action (replacement, repair, etc.) required to bring the machine to an acceptable operational state is dictated by the results of diagnosis and prognosis.

4.4.6 Postmortem

In the event of machine failure, postmortem examination should be carried out to determine the causes of the failure, appearance of failed components, etc. This should be related to the results of detection, diagnosis, and prognosis to provide a historical record for future use. It must be stressed that reliability of diagnosis and prognosis of machine condition monitoring will greatly increase if such examinations are carried out and are properly recorded.

In most companies the postmortem on failed machines is seldom carried out routinely, and even when it is carried out the records are often not complete. When the problem is fixed there is generally another more urgent job than completing the postmortem report.

4.5 Training

Implementation of contamination control requires thorough understanding of the aims of the program, knowledge of a system, and fluid monitoring technology. Contamination control should become an important component of quality improvement process in the majority of industrial, aviation, mining, and metalworking companies that use fluids for power/control. Contamination control should be incorporated in a company quality improvement program which aims to increase reliability, productivity, and profitability of the plant. A successful implementation of a contamination control program is, however, conditional on the availability of clearly defined procedures, well-trained personnel, and monitoring equipment. Training of personnel should include three groups of personnel:

- *Management:* To be successful the program requires involvement and acceptance by all personnel involved in design, manufacturing, and operation/maintenance of the system and this must be fully supported but not imposed by management. Management must fully understand that well-implemented contamination control will be profitable if all essential components of the program are in place. Management must set life-cycle improvement targets that must be achieved using contamination control and then must provide necessary resources — manpower, equipment, training, etc. The life improvement obtained via contamination control should be checked against targets and the program adjusted if necessary. Management must understand that a contamination control program is based on a proactive rather than a reactive approach, that is, it is aimed at avoiding failures rather than at catching failures. Furthermore, management must provide a working environment in which a contamination control program can be successful. This may require structural, functional reorganization of the workplace, change of maintenance strategy, and acceptance of new ideas, new technology, and new attitudes. Management training should include the fundamentals of a contamination control program, impact on other quality improvement programs, profitability assessment, strategy, organization, and resources needed.
- *Supervisory staff:* Engineering staff must be trained to understand the contamination control approach. Implementation of contamination control should not be the responsibility of only the maintenance personnel as staff dealing with purchasing, design, manufacturing, commissioning, and operation can have pronounced influence on the success of the program (e.g., equipment purchased or designed without consideration of how it will fit into the plant's contamination control).

Basic training of engineering staff associated with a contamination control program should include, in addition to basic contamination control training, exposure to the following:

(a) Tribology (lubrication, wear, hydraulic, and lubricating fluids)

(b) Hydraulic power and control systems (principle of operation, troubleshooting)

(c) Functional and failure analysis (FMEA, FMECA, mechanical failures analysis)

(d) Tribology-based monitoring and diagnostics (wear debris analysis, monitoring techniques, sampling, interpretation, etc.)

(e) Condition monitoring (predictive and proactive)

• *Operators and maintenance staff*: Operators and maintenance personnel are best placed to detect any changes in operating regimes of the machines — the most important first step in both predictive and proactive monitoring. Maintenance staff dealing directly with all aspects of contamination control, for example, sampling, filter changing, and servicing of equipment, must be trained in the basics of contamination control so that they can perform their work knowing what they are trying to achieve. They should also be equipped with tools that will facilitate detection without waiting for lab results (e.g., filtergrams, portable contamination comparators) and thus react to changes in a very short time. Training should include instructions on use and capabilities of various monitoring tools, sampling, and sensory detection. Basic understanding of lubrication and tribological processes will be an advantage in the execution of their tasks.

4.6   Conclusions

We must recognize contamination control as a vital factor in operation of high-performance systems. Contamination control is more than just ordering a filter; we must understand various factors that are involved in assuring proper contamination control. A contamination control program integrates our knowledge of contamination control and its effect on quality improvement programs. To be successful, a contamination control program requires management and personnel commitment to quality improvement and understanding that its success is based on clearly defined goals, staff education, selection of the right technology, and an efficient management system.

References

[1] "ISO 16889:1999 Hydraulic Fluid Power Filters — Multi-Pass Method for Evaluating Filtration Performance of a Filter Element", ISO, 1999.

[2] "ISO 4406:1999 Hydraulic Fluid Power — Fluids — Method for Coding the Level of Contamination by Solid Particles", ISO, 1999.

[3] "NAS 1638 Cleanliness Requirements of Parts used in Hydraulic Systems. National Aerospace Standard", NAS, 1964.

[4] E.C. Fitch, *An Encyclopaedia of Fluid Contamination Control for Hydraulic Systems*, Hemisphere Publishing Corp., Stillwater, OK, 1979.

[5] L. Chao, A System Approach to Monitoring and Controlling of Contamination/Wear in Hydraulic Control Systems, PhD thesis, Monash University, Australia, 1995.

[6] L. Chao and B.T. Kuhnell, "The Comparison and Development of On-line Wear & Contamination Monitors", *The Bulletin of Centre for Machine Condition Monitoring*, Vol. 4, No.1, pp. 20.1–20.8, 1992.

[7] C.H. Pek and J.S. Stecki, "On-Line Contamination Monitoring Using Nephelometry", *Proceedings of the 2nd International Machinery Monitoring and Diagnostic Conference*, pp. 121–125, Los Angeles, 1990.

[8] T. Hong and J.C. Fitch, "Model of Fuzzy Logic Expert System for Real-Time Condition Control of a Hydraulic System", *Proceedings of the 3rd International Conference on Machine Condition Monitoring*, Windsor, U.K., 1990.

[9] L. Chao and J.S. Stecki, "Intelligent Expert System for Contamination Control in Hydraulic Control Systems", *Proceedings of 6th International Conference on Industrial and Engineering Applications of Artificial Intelligence and Expert Systems*, pp. 538–543, Edinburgh, June 1993.

[10] J.S. Stecki and L. Chao, "Computer-Aided Techniques in Contamination Control, Design and Performance", *Eighth Bath International Fluid Power Workshop*, pp. 126–140, Bath, UK, Research Studies Press and John Wiley & Sons, 1996.

[11] G.J. Schoenau, J.S. Stecki, and R.T. Burton, "Utilization of Artificial Neural Networks in the Control, Identification and Condition Monitoring of Hydraulic Systems — An Overview", *SAE 2000 Transactions*, Vol. 109, *Journal of Commercial Vehicles*, Section 2, pp. 205–212, 2000.

[12] J.S. Stecki and G. Schoenau, "Application of Simulation and Knowledge Processing in Contamination Control", *SAE 2000 Transactions*, Vol. 109, *Journal of Commercial Vehicles*, Section 2, pp. 331–347, 2000.

[13] "MIL-STD-1629A, The Procedures for Performing a Failure Mode, Effects and Criticality", DOD, 1984.

[14] "SAE J 1739: Potential Failure Mode and Effects Analysis (FMECA) Reference Manual", SAE, 1994.

[15] "ECSS-Q-30-02A Space Product Assurance — Failure Modes, Effects and Criticality Analysis (FMECA)", European Cooperation for Space Standardization, 2001.

[16] "BS5760:1991: Reliability of Systems, Equipment and Components, Part 5, Guide to Failure Modes, Effects and Criticality Analysis (FMEA and FMECA)", BS, 1991.

[17] D.R. Bull, W.J. Crowther, C.R. Burrows, K.A. Edge, et al., "Approaches to Automated FMEA of Hydraulic Systems", *Proceedings of I. Mech. Congress Aerotech95*, Seminar C505/9/099, Birmingham, U.K., 1995.

[18] W.J. Crowther, C.R. Burrows, K.A. Edge, et al., "Automated FMEA for Hydraulic Systems", *Proceedings of 12th Fluid Power Technology Colloquium, IFAS*, Aachen, Germany, 1996.

[19] P.G. Hawkins, R.M. Atkinson, et al., "An Approach to Failure Modes and Effects Analysis Using Multiple Models", *IFMA Proceedings*, V4, Athens, Greece, June 1996.

[20] D.R. Bull, J.S. Stecki, K.A. Edge, and C.R. Burrows, "Failure Modes and Effects Analysis of a Valve-Controlled Hydrostatic Drive", in Fluid Power Engineering: Challenges and Solutions, *Tenth Bath International Fluid Power Workshop*, Research Studies Press Ltd., pp. 131–143, 1997.

[21] C.E. Peleaz and J.B. Bowles, "Using Fuzzy Cognitive Maps as a Model for Failure Modes and Effects Analysis", *Information Sciences*, Vol. 88, pp. 177–199, 1996.

[22] J.S. Stecki, F. Conrad, and B. Oh, "Software Tool for Automated Failure Modes and Effects Analysis (FMEA) of Hydraulic Systems", *Fifth JFPS International Symposium on Fluid Power*, Nara, Japan, 2002.

[23] J.S. Stecki, "Maintenance Aware Design (MAD) Methodology", *The Eighth Scandinavian International Conference on Fluid Power*, SICFP'03, May 7–9, Tampere, Finland, 2003.

[24] J.S. Stecki and B. Oh, "Failure Analysis Using Energy Disturbance and IA Techniques", *The Eighth Scandinavian International Conference on Fluid Power*, SICFP'03, May 7–9, Tampere, Finland, 2003.

[25] G. Pahl and W. Beitz, *Engineering Design: A Systematic Approach*, Springer-Verlag, Berlin, 1988.

[26] L.D. Miles, *Techniques of Value Analysis and Engineering*, McGraw-Hill, New York, 1972.

[27] "ECSS-E-10 Space Engineering — Functional Analysis", European Cooperation for Space Standardization, 1999.

[28] H. Czichos, *Tribology. A System Approach to the Science and Technology of Friction, Lubrication and Wear*, Tribology Series 1, Elsevier, New York, 1978.

[29] D. Godfrey, *Wear Control Handbook*, ASME Publication, 1980.

[30] R.K. Tessman and E.C. Fitch, "Field Contaminant Ingression Rates — How Much?" *Proceedings of Eighth Annual Fluid Power Research Conference*, paper P74-47, October 1974.

[31] L.E. Bensch and W.T. Bonner, "Field System Contaminants — Where, What, How Much", *Proceedings of Seventh Annual Fluid Power Research Conference*, paper P73-CC-1, pp. 187–193, October 1973.

[32] W. Backé, "Wear Sensitivity of Hydraulic Displacement Units to Solid Contaminats", *Ölhydraulik und Pneumatik*, Vol. 33, No. 6, pp. 29–38, 1989.

[33] R.K. Tessman, "Speed is not a Factor in the Contaminant Wear of Hydraulic Pumps", *BFPR J.*, paper 8, pp. 1–11, 1979.

[34] G. Silva, "Wear Generation in Hydraulic Pumps", SAE Technical Paper, 9011679, 1990.

[35] R.K. Tessman, "Component Contaminant Generation", *Proceedings of Eighth Annual Fluid Power Research Conference*, paper P74-46, October 1974.

[36] J. Collins and B. Hagan, "The Failure-Experience Matrix: A Useful Design Tool", *ASME Journal of Engineering for Industry*, 1976.

[37] J. Collins, *Failure of Materials in Mechanical Design: Analysis, Prediction, Prevention*, John Wiley & Sons, New York, 1993.

[38] B.T. Kuhnell and J.S. Stecki, "Monitoring of Hydraulic System", *Proceedings of 2nd International Fluid Power Conference*, pp. 729–734, Hanghzou, China, March 1989.

[39] L. Chao, J.S. Stecki, and P. Dransfield, "Dynamic Contamination Control Model for Fluid Contamination", *Proceedings of International Conference on Fluid Power Control and Robotics*, pp. 342–247, Chengdu, China, 1990.

[40] J.S. Stecki and L. Chao, "Simulation of Contamination Control Systems", *The Bulletin of Centre for Machine Condition Monitoring*, Vol. 2, No. 1, pp. 10.1–10.6, 1990.

[41] J.S. Stecki and M. Grahl-Madsen, "Distribution of Particulate Contamination in Multi-branch Hydraulic System", *The Sixth Scandinavian International Conference on Fluid Power*, SICFP'99, May 26–28, Tampere, Finland, 1999.

[42] D.P. Anderson and R.D. Driver, "Equilibrium Particle Concentration in Engine Oil", *Wear*, Vol. 56, No. 415, 1979.

Appendix

Standards and Drafts of ISO/TC 131/SC 6 Contamination Control and Hydraulic Fluids

- ISO 2941:1974 Hydraulic fluid power — Filter elements — Verification of collapse/burst resistance.
- ISO/CD 2941 Hydraulic fluid power — Filter elements — Verification of collapse/burst pressure rating.
- ISO 2942:1994 Hydraulic fluid power — Filter elements — Verification of fabrication integrity and determination of the first bubble point.
- ISO/CD 2942 Hydraulic fluid power — Filter elements — Verification of fabrication integrity and determination of the first bubble point.
- ISO 2943:1998 Hydraulic fluid power — Filter elements — Verification of material compatibility with fluids.
- ISO 3722:1976 Hydraulic fluid power — Fluid sample containers — Qualifying and controlling cleaning methods.
- ISO 3723:1976 Hydraulic fluid power — Filter elements — Method for end load test.
- ISO/CD 3724 Hydraulic fluid power — Filter elements — Verification of flow fatigue characteristics.
- ISO 3724:1976 Hydraulic fluid power — Filter elements — Verification of flow fatigue characteristics.
- ISO 3938:1986 Hydraulic fluid power — Contamination analysis — Method for reporting analysis data.
- ISO 3968:1981 Hydraulic fluid power — Filters — Evaluation of pressure drop vs. flow characteristics.
- ISO/FDIS 3968 Hydraulic fluid power — Filters — Evaluation of pressure drop vs. flow characteristics.
- ISO 4021:1992 Hydraulic fluid power — Particulate contamination analysis — Extraction of fluid samples from lines of an operating system.

- ISO/AWI 4405 Hydraulic fluid power — Fluid contamination — Determination of particulate contamination by the gravimetric method.
- ISO 4405:1991 Hydraulic fluid power — Fluid contamination — Determination of particulate contamination by the gravimetric method.
- ISO 4406:1999 Hydraulic fluid power — Fluids — Method for coding the level of contamination by solid particles.
- ISO 4407:1991 Hydraulic fluid power — Fluid contamination — Determination of particulate contamination by the counting method using a microscope.
- ISO/DIS 4407 Hydraulic fluid power — Fluid contamination — Determination of particulate contamination by the counting method using an optical microscope.
- ISO/WD 7744 Hydraulic fluid power — Filters — Statement of requirements.
- ISO 7744:1986 Hydraulic fluid power — Filters — Statement of requirements.
- ISO/CD 10949 Hydraulic fluid power — Component cleanliness — Methods of achieving, evaluating, and controlling component cleanliness from manufacture through installation.
- ISO/TR 10949:1996 Hydraulic fluid power — Methods for cleaning and for assessing the cleanliness level of components.
- ISO/CD 11170 Hydraulic fluid power — Filter elements — Procedure for verifying performance characteristics.
- ISO 11170:1995 Hydraulic fluid power — Filter elements — Procedure for verifying performance characteristics.
- ISO 11171:1999 Hydraulic fluid power — Calibration of automatic particle counters for liquids.
- ISO 11500:1997 Hydraulic fluid power — Determination of particulate contamination by automatic counting using the light extinction principle.
- ISO/WD 11500 Hydraulic fluid power — Determination of particulate contamination by automatic counting using the light extinction principle.
- ISO 11500:1997/Cor 1:1998 No title.
- ISO 11943:1999 Hydraulic fluid power — On-line automatic particle-counting systems for liquids — Methods of calibration and validation.
- ISO/CD 16009 Hydraulic fluid power — Filters — Method for determining filtration performance of an off-line filter element.
- ISO/CD TR 16144 Hydraulic fluid power — Calibration of liquid automatic particle counters — Procedures used to certify the standard reference material.
- ISO/TR 16386:1999 Impact of changes in ISO fluid power particle counting — Contamination control and filter test standards.
- ISO/CD TS 16431 Hydraulic fluid power — Assembled systems — Method for verifying cleanliness.
- ISO/CD 16860 Hydraulic fluid power — Filters — Method of test for differential pressure devices.
- ISO/DIS 18413 Hydraulic fluid power — Cleanliness of parts and components — Inspection document and principles related to sample collection, sample analysis, and data reporting.

Further Reading

1. J.S. Stecki and L. Chao., "Selection of Filters for Hydraulic Control Systems", *Proceedings of International Conference on Maintenance Management*, paper 27, Melbourne, September 1991.
2. L. Chao, P. Dransfield, and J.S. Stecki, "PCEX — Expert System for Evaluating Pump Contamination Sensitivity", *Proceedings of 2nd JHPS International Symposium on Fluid Power*, pp. 527–532, Tokyo, September 1993.
3. L. Chao and J.S. Stecki, "Knowledge-Base for Evaluation of Pump Wear Tolerance", *The Bulletin of Centre for Machine Condition Monitoring*, Vol. 5, pp. 130–135, November 1993.

4. L. Chao, P. Dransfield, and J.S. Stecki, "Expert System Aided Filtration System Design", *Supplement Proceedings of 3rd (ICFP) International Conference on Fluid Transmission and Control*, pp. 6–11, Hangzhou, China, September 1993.

5. J.S. Stecki (ed), Workshop on Advances in Hydraulic Control Systems, Monash University, p. 150, Melbourne, 1996.

6. J.S. Stecki (ed), Workshop on Total Contamination Control, Fluid Power Net Publications, p. 177, Melbourne, 1998.

5

Environmental Implications and Sustainability Concepts for Lubricants

Malcolm F. Fox
IETSI
University of Leeds

5.1   The Environmental Drivers

Lubricants are commercial consumer products and are now subject to analyses, as is the case for any other product, regarding their impact on the environment over a new concept, their "life cycle." The effects of producing and using lubricants in society are wide-ranging, which commences from:

- Their formulation, through to
- Their production, from main base oil and additive component parts
- Their beneficial effects during use
- Concluding with the effects of their eventual disposal

The environmental effects of lubricants can be beneficial or negative. But first, it is helpful to examine the concepts and management techniques that are used to assess the importance of these issues. The framework has changed and the issues of environmental impact and resource sustainability for lubricant production, use, and disposal have risen, and are rising further, in the public domain. These issues may not be popular, but cannot be ignored.

Lubricants must be used more effectively with longer lifetimes for increased sustainability. Their production and use is controlled for environmental reasons. Equally their subsequent disposal after use is now strictly controlled and should be recycled. It is no longer acceptable to tip used lubricants onto the ground, into watercourses, or to burn them without specialized furnaces. That is the nature of the environmental impact of the end use of lubricants, their sustainability, and sustainable development.

5.1.1   Concepts of Sustainability and Sustainable Development

Ecological and economic systems are now widely recognized as being interlinked and that unrestrained development and use of resources leads to serious resource depletion, residual pollution, and survival problems. It follows that economic development should meet the test of "sustainability." The concept of "sustainable development" is the key principle of environmental policy formulation and development and was introduced by the 1987 Report of the World Commission on Environment and Development, "Our Common Future[1]." Chaired by the ex-Prime Minister of Norway, Gro Haarlem Brundtland, it is known as the "Brundtland Report" and defined sustainable development as:

development which meets the needs of the current generation without compromising the ability of future generations to meet their own needs.

which contains the concepts of:

- *Intergenerational equity* — we should pass on to the next generation at least an equivalent resource endowment to allow them to meet their needs.
- *Intragenerational equity* — poverty should be eradicated so that differences in living standards are reduced to meet the needs of the majority of people in the world.
- *Carrying capacity* — ecosystems are not stretched beyond their capacity to either renew themselves or absorb wastes.

The concept of sustainable development therefore includes issues of social progress, economic growth, and environmental protection. These concepts may appear to be abstruse and theoretical but are gradually entering legislation at continental and national levels in the form of detailed environmental law and regulations. These, in turn, become technical requirements and standards and the next section demonstrates how environmental policies are transmitted downward into targets and actions, for lubricants as well as for other products. This explains the background for the changes in policies and standards, which increasingly guide the production, use, and disposal of lubricants.

5.1.2   Sustainable Development at Global, Continental, and National Levels

5.1.2.1   At the Global Level

The concept of sustainable development has emerged over the last 30 to 40 years and its significant markers along the way commence in 1972 at a United Nations conference in Stockholm, Sweden:

1972 — *The United Nations Conference on the Human Environment (UNCHE)* [2] in Stockholm, Sweden, considered the need for a common outlook and principles to inspire and guide the world's peoples to preserve and enhance the human environment.

1987 — *The World Commission on Environment and Development* publishes *The Brundtland Report* [1] bringing concepts of sustainable development onto the international agenda. It also provides the most commonly used definition of sustainable development.

1992 — 180 countries met at the *"Earth Summit"* (the *UN Conference on Environment and Development*) in Rio de Janeiro, Brazil [3] to discuss how to achieve sustainable development. The Summit agreed upon the Rio Declaration on Environment and Development, setting out 27 principles supporting sustainable development. It agreed on a plan of action, Agenda 21, and a recommendation that all countries should produce national sustainable development strategies. The 1992 Earth Summit also established the Framework Convention on Climate Change and the Convention on Biological Diversity.

1997 — A special UN conference [4] reviewed implementation of Agenda 21 (as Rio+5), repeating the call for all countries to have sustainable development strategies in place.

2002 — Ten years after the Earth Summit in Rio in 1992, the *2002 Johannesburg Summit* [5] in South Africa, the *World Summit on Sustainable Development*, brought together many participants including heads of state and government, national delegates and leaders from nongovernmental organizations (NGOs), businesses, and other major groups to focus the world's attention and direct action toward meeting difficult challenges. These include improving people's lives and conserving natural resources in a world that is growing in population with ever-increasing demands for food, water, shelter, sanitation, energy, health services, and economic security. These global policies set the pattern for policies at continental and national levels.

5.1.2.2   At the Continental Level

While the world's continents are legally autonomous and the implementation of global commitments may appear to be different due to differences in legal systems, the overall thrusts of environmental legislation are broadly the same. This section examines the environmental actions of the European Union, and how that has devolved down to the national level, as in the United Kingdom.

The European Union (EU), and its European Environment Agency (EEA), is recognized as one of the most important international innovators for environmental issues, standards, and enforcement, comparable to the United States and its Environmental Protection Agency. There are over 600 EU legislative measures that apply to the environment, although some are either revised or have amended previous legislation. The earliest EU legislation for the protection of the environment commenced in 1970 and applied to vehicle emissions. This was the start of a program closely analogous to vehicle emission control programs in North America and Japan and illustrates the force of international environmental action. The First EU Action Program on the Environment was adopted in 1973 by the six original member nations of the EU. Gradual expansion of the EU over subsequent years led to the current (2004) membership of 25 nations. The Single European Act of 1986 amended the original Treaty of Rome of 1957 to explicitly include measures to protect the environment. Since then, EU Environmental Action Programs have been increasingly aimed at proactively protecting the environment:

1993 — "Towards Sustainability," the Fifth Environmental Action Programme of the EU [6] sought to integrate environmental concerns into other policy areas to achieve sustainable development. It recognized that legislation cannot be used to solve all environmental problems and that other mechanisms are needed to achieve sustainable development, including market-based instruments such as environmental management systems, eco-audits, and eco-labeling.

1999 — changes to the Treaty of Rome were made by the Treaty of Amsterdam [7] to give sustainable development a much greater prominence within the EU.

2001 — The EU's Sixth Environmental Action Programme (2001–2010) entitled "Environment 2010: Our Future, Our Choice [8]," takes a wide-ranging approach to these challenges and gives a strategic direction to environmental policy over the next decade as the EU prepared to expand its number of member nation states. The new program identifies four priority areas as:

1. Climate Change
2. Nature and Biodiversity
3. Environment and Health
4. Natural Resources and Waste

The new program set out five approaches to achieve improvements in each of these areas, emphasizing that more effective implementation and innovative solutions were needed. The EU Commission recognized that a wider constituency must be addressed, including businesses, who can only gain from a successful environmental policy. The five key approaches are to:

1. Ensure the implementation of existing environmental legislation
2. Integrate environmental concerns into all relevant policy areas
3. Work closely with consumers to identify solutions
4. Ensure better, more accessible environmental information for citizens
5. Develop a more environmentally conscious attitude toward land use

5.1.2.3 Sustainable Development at the National Level — the United Kingdom

In its response to the EU Directives, the U.K. Government published its strategy for sustainable development in 1999, "A Better Quality of Life" [9], replacing a previous 1994 strategy document "Sustainable Development — the UK Strategy." The 1999 Strategy highlights the need for integrated policies to achieve a better quality of life for current and future generations by ensuring social progress that recognizes:

1. The needs of every person
2. Effective environmental protection
3. The prudent use of natural resources
4. Maintenance of high and stable levels of economic growth and employment

The U.K. Local Government Act 2000 [10] requires local authorities (government, such as councils at different levels) in England and Wales to prepare a community strategy for promoting, or improving, the economic, social, and environmental well-being of their community. This should also contribute to the achievement of sustainable development of the United Kingdom. The concept of sustainable development is set out at some length to emphasize that it is real, it is a reality now, and will become even more important as a decision parameter to include in business policies in the future.

5.1.2.4 Example — The Reduction of Diesel Fuel Sulfur Content in Europe

An example of the impact of the changes driven by environmental policies at the continental level is the sulfur content of diesel fuel. This is now the major source of urban sulfur dioxide from diesel vehicle vehicles, not because its source has dramatically increased but because other sources of sulfur dioxide from coal and heavy oil heating fuels have sharply decreased over the last two decades by replacement with natural gas. The successive Clean Air Acts of the United Kingdom and the increased use of natural gas have ousted higher sulfur liquid and solid fuels from space heating. Diesel fuel itself has seen a dramatic reduction of sulfur content in the United Kingdom over the last 20 years from 0.3% (or 3000 ppm) to a current (2004) level of ~40/45 ppm in the United Kingdom with projected reductions to a maximum of 10 ppm sulfur content by January 2009 [11].

The driver for these reductions in fuel sulfur content arises from international treaty obligations entered into by Western nations. Good as that is for the (urban) environment particularly, it has very serious implications for diesel fuel pump wear. The sulfur compounds at high concentrations in diesel fuel act as inherent lubricants for the injector pump. Their removal can be effectively countered by a program of fuel lubricity additives, which is being successfully implemented in Europe. But there are serious variations across the regions of the United States, with East Coast U.S. diesel fuel being comparable to EU quality whereas for the rest of the U.S. regions, the High Frequency Reciprocating Rig, the HFRR [12] test, shows much higher injector pump wear rates [13].

5.1.2.5   Environmental Management Systems (EMS) and Life Cycle Assessment (LCA)

There are many different analytical tools that businesses can use to evaluate and improve their environmental performance and to then demonstrate those improvements. It is worth examining two of them — Environmental Management Systems (EMS) and Life Cycle Assessment (LCA).

5.1.3   Environmental Management Systems (EMS)

The very visible and tangible environmental impact of some industries in the last few decades and the increasing pool of sophisticated knowledge among both professionals and the public led to increased pressure on organizations to become more "environmentally friendly." These pressures include questions from shareholders at annual meetings and insurers who are increasingly concerned about contingent liabilities. The contingent liabilities from asbestos injury claims in the insurance industry are a stark reminder.

The introduction of an EMS is one of the best approaches to building environmental protection into an organization. It can be effectively applied to a whole organization and used to ensure consistency of action. Just as quality systems were developed during the 1980s to aid manufacturing efficiency, production, and competitiveness, EMS guidelines and standards were developed to enable organizations to become less wasteful, more profitable, and less detrimental to the environment. The procedures for EMS analyses, quality control procedures, and health and safety programs are very similar and can be integrated into an overall program, leading to the aphorism that "a quality production company is a safe company, is an environmentally responsible company which is also a profitable company" and experience shows this to be broadly true. There are several different standards to which an organization's EMS can be certified, the two most important being:

- *EMAS*, the European Union's Eco-Management and Audit Scheme
- *ISO 14001*, the standard produced by the International Organization for Standardization

Another industry-wide EMS standard that is important for the chemical, petrochemical, and polymer industries is "Responsible Care." This system originated in Canada as a specific system for the chemical industry and was introduced to other countries, including the United Kingdom in 1989 by the Chemical Industries Association [14]. "Responsible Care" incorporates health and safety as well as environmental management and its main principles are set out below. The issue is whether this is a real, continuing commitment or a panacea:

Responsible Care

"Responsible Care is a voluntary scheme under which organisations commit to provide a high level of protection for the health and safety of employees and associates, customers, and the public; and for the environment. Organisations make a commitment to sustainable development and continual improvement by promising to adhere to the following principles:

- *Policy*: We will have a Health, Safety and Environmental (HS&E) policy which will reflect our commitment and be an integral part of our overall business policy.
- *Employee involvement*: We recognise that the involvement and commitment of our employees and associates will be essential to the achievement of our objectives. We will adopt communication and training programmes aimed at achieving that involvement and commitment.

- *Experience sharing*: In addition to ensuring our activities meet the relevant statutory obligations, we will share experience with our industry colleagues and seek to learn from and incorporate best practice into our own activities.
- *Legislators and regulators*: We will seek to work in co-operation with legislators and regulators.
- *Process safety*: We will assess and manage the risks associated with our processes.
- *Product stewardship*: We will assess the risks associated with our products, and seek to ensure these risks are properly managed throughout the supply chain through stewardship programmes involving our customers, suppliers and distributors.
- *Resource conservation*: We will work to conserve resources and reduce waste in all our activities.
- *Stakeholder engagement*: We will monitor our HS&E performance and report progress to stakeholders; we will listen to the appropriate communities and engage them in dialogue about our activities and our products.
- *Management systems*: We will maintain documented management systems which are consistent with the principles of Responsible Care and which will be subject to a formal verification procedure.
- *Past, present and future*: Our Responsible Care management systems will address the impact of both current and past activities."

While EMSs are voluntary schemes they are also market-based tools designed to internalize environmental costs. They demonstrate to producers and consumers the need to use natural resources responsibly and minimize pollution and waste.

EMSs introduce market forces in the environmental field by promoting competition between industrial activities on environmental grounds. The assumption is that the market rewards organizations that establish EMSs with greater market share and, consequently, market pressures encourage others to set up their schemes. The overall outcome should be that more organizations become active in environmental management and their environmental performance will improve. EMSs also tend to spread down through the "supply chain" — once an organization has introduced its own EMS, it starts to look at its suppliers and their environmental performance as well. The *advantages* of an EMS include:

- Improved environmental performance
- Improved public relations
- Improved relations with regulators
- Control of environmental legislation requirements
- Formalizing/coordinating existing systems
- Raises environmental awareness within the organization
- Improves relationships with local communities
- Effective, nonbureaucratic documentation system

*competitive advantage
*marketing tool
*cost savings
*cheaper insurance
*pro-active approach

The *disadvantages* of an EMS can include:

- Exposing environmental liabilities (advantageous if forestalling regulatory action), long-term commitment, including resources
- Implementation and certification costs
- Increased workload initially, which then decreases as not being dependent on the fire-fighting approach for tackling environmental emergencies
- Increased training requirements (which occur anyway in an enlightened company)
- Uncontrolled environmental information provided to the public may not always be advantageous as it highlights areas that an organization might not want in the public domain. Once released, there is no control over what the public or media do with it. But there is a general move toward a requirement for transparency of information, including environmental information and companies will have to learn to live with this approach.

EMS standards do not outline the expected environmental performance of an organization, although they do require the relevant legislation and regulations to be adhered to. Instead, compliance is centered

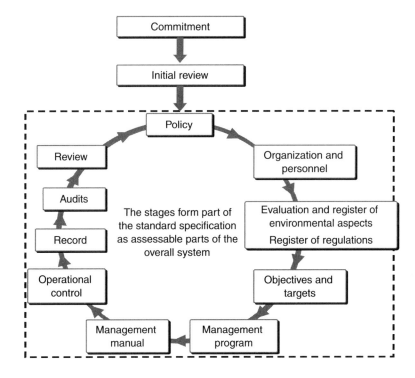

FIGURE 5.1 Establishing an environmental management system (EMS).

on the ability of the organization to meet its own stated objectives. These objectives are expected to change and be modified over time as the organization strives for continual improvement. The basic stages of establishing an EMS are shown in Figure 5.1.

An ISO 14001 analysis requires that an Environmental Statement is published following an initial environmental review and also after the completion of each subsequent audit cycle. It should be designed for the public and written in a concise, comprehensible form. Many companies, especially large companies with high public profiles, are keen to show their environmental credentials. Their insurers increasingly require an EMS to be in place so as to limit and define the company's environmental contingent liabilities, both immediate and long term. There are also *The Valdez Principles*, produced after the oil spill from the Exxon Valdez tanker in Alaska in March 1989, a comprehensive and exacting list of principles that can provide a useful starting point for developing a company environmental policy.

The Valdez Principles

1. *Protection of the Biosphere* — minimize releases of any pollutants that cause environmental damage to air, water, or earth. We will safeguard habitats in rivers, lakes, wetlands, coastal zones, and oceans and avoid contributing to greenhouse effects, depletion of the ozone layer, acid rain, or smog.
2. *Sustainable Use of Natural Resources* — make sustainable use of renewable natural resources like water, soils, and forests. We will conserve non-renewable natural resources through efficient use and careful planning. We will protect wildlife, habitat, open spaces, and wilderness, while preserving biodiversity.
3. *Reduction and Disposal of Waste* — minimize waste, especially hazardous waste, and whenever possible, recycle materials. We will dispose of all waste through safe and responsible methods.
4. *The Wise Use of Energy* — use environmentally safe and sustainable energy to meet our needs. We will invest in improved energy efficiency of products we use or sell.

5. *Risk Reduction* — minimize environmental, health, and safety risks to employees and communities where we operate by employing safe technologies and operating procedures and by being constantly prepared for emergencies.
6. *Marketing of Safe Products and Services* — to sell products that minimize environmental impacts and are safe as consumers commonly use them. We will inform consumers of the environmental impacts of our products or services.
7. *Damage Compensation* — take responsibility for any harm we cause to the environment by making every effort to restore the environment and to compensate those persons who are adversely affected.
8. *Disclosure* — disclosure to our employees and to the public, incidents relating to our operations that cause environmental harm or pose health or safety hazards. We will disclose potential environmental health or safety hazards posed by our operations.
9. *Environmental Directors and Managers* — At least one seat on our Board of Directors is designated for an environmental advocate. We will commit management resources to implement these principles, including the funding of an office of Vice President for Environmental Affairs or an equivalent executive position to monitor and report on our implementation efforts.
10. *Assessment and Annual Audit* — We will conduct and make public an annual self-evaluation of our progress in implementing these principles and in complying with all applicable laws and regulations. We will work toward the timely creation of independent environmental audit procedures to which we will adhere.

5.1.4   Life Cycle Assessment (LCA)

Consumers are increasingly concerned about the environmental impact of products and services. But it is often not straightforward to determine which is the best environmental option because the problem becomes more complicated as it is realized that products, services, and processes have impacts throughout their life cycles, from the "cradle to the grave." Producers are increasingly aware that they must take life cycle responsibility for their products and services. Both producers and consumers need a reliable, credible, and recognized method of assessing environmental impacts for different options and determining which is "best."

Life Cycle Assessment (LCA) is an objective process that is used to evaluate the environmental burdens or impacts associated with a product, process, or activity. The procedure should (i) identify and quantify the inputs/outputs, for example, the energy, the materials used, and the wastes released to the environment, (ii) assess the environmental impacts of the energy and materials used and the wastes released, and (iii) evaluate and implement environmental improvements.

LCA includes the entire life cycle of the product, process, or activity from "cradle to grave" including production, distribution, and consumption. It covers elements such as extraction and processing raw materials, manufacturing, transportation and distribution, use/reuse/maintenance, recycling, and final disposal. LCA is a brilliantly simple concept that is very complex to work through.

Assessing impacts from a life cycle analysis allows the focus to be moved away from "end-of-pipe" remedies for environmental problems toward more effective and efficient remedies that can exist at other points in the life cycle of a product, for example, upstream in the design of the product or process. It also goes beyond assessing individual environmental issues toward an integrated approach that aims to measure overall environmental sustainability.

LCA has been used to define and compare the environmental burdens/impacts for finished products or services and to guide development work. LCA is used by an organization to guide a specific project or set overall strategies. It can provide environmental information to consumers and establish criteria for environmental labeling schemes — the EU Directive on eco-labeling states that criteria should be established using a life cycle approach. Already, companies have shown that they do not wish to have "toxic to ecology" labels on their technical products and have sought alternative nonecotoxic formulations with an equivalent or better performance. LCA is most useful in assessing the environmental impacts of the production of goods and the provision of services, or to compare the impacts of different products, processes, or services. Examples would be, for example, polystyrene vs. china cups, aluminum vs. steel beverage cans, mineral base oils vs. vegetable base stocks.

5.1.4.1   The Functional Unit

It is difficult to make comparisons between unlike products in LCA; therefore, the idea of the functional unit concentrates on the function served by a product. For example, it is unrealistic to directly compare a returnable milk bottle to a carton because the bottle can be reused a number of times. But the more useful functional unit for LCA comparative purposes would be the packaging of, say, 1000 l of milk, which could be by glass bottles, cartons, or other material. Reuse and recycling can then be brought into the overall LCA.

Care must be taken in choosing the functional unit for the LCA because it may alter the outcome of the assessment. There may be thresholds where one product becomes more or less environmentally acceptable than another, for example, milk bottles may need to be re-used a certain number of times before they are more acceptable than cartons.

A Case Study of Life Cycle Analysis

The Netherlands Ministry of Housing, Physical Planning and Environment studied the environmental performance of "crockery" made from polystyrene, paper/cardboard, and porcelain. The functional unit was 3000 uses assuming that the polystyrene was not recycled and the product was used on a "use once/wash once" basis. The basic question was then:

How many times must porcelain crockery be used and washed so that its environmental impact is equal or less than that of disposable polystyrene or paper/cardboard crockery? The answer depended on whether water consumption and waste treatment, energy consumption, air pollution, or landfill wastes are considered in terms of:

— water consumption, where porcelain always does worst, because "disposable" crockery is never washed
— energy consumption, where porcelain is used 640 times before it competes with polystyrene and 294 times before it competes with paper/cardboard crockery
— air pollution, where porcelain needs to be used 1800 times to compete with polystyrene and 48 times to compete with paper/cardboard
— waste, where porcelain needs to be used 126 times before it competes with polystyrene and 99 times to compete with paper/cardboard

Porcelain always overtakes the other products in terms of conserving energy, air, and waste well within the functional unit limit of 3000 uses. But it always has the worst performance in terms of water consumption. The overall problem is then the summation of whether the positive energy, air, and waste impacts for porcelain's use as crockery outweighs its negative result impact for water.

Lubricants do not fit readily into this category because they are not sold by decanted/fed volume, as from large containers, possibly the case in large service center garages, but normally from either sealed containers of 1 l or resealable 5 l. What is relevant is their service replacement interval and their contribution toward energy efficiency by reducing friction.

5.1.4.2   Inputs, Outputs, and System Boundaries

Schematically, any extended industrial system can be represented by Figure 5.2 with the ensemble of operations enclosed within the conceptual box. The conventions are similar to system thermodynamics where the concept box denotes the system boundary separating the system from its surroundings and

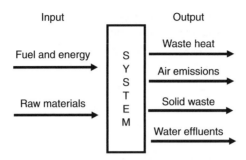

FIGURE 5.2   System inputs and outputs.

defines the system environment. This acts as the source of all inputs to the system and also as the sink for all outputs from the system. The physical description of the system, or inventory, is a quantitative description of all flows of materials and energy across the system boundary either into or out of the system itself, identical to those of conventional thermodynamics. Consequently, relatively few new procedures are needed to manipulate the data. Many arbitrary decisions introduced by some analysts in recent years arise from a lack of appreciation of the close relationships between life cycle analysis and thermodynamics. The system is a physical entity that obeys all physical laws of matter, including laws of mass conservation and thermodynamics. These conditions exclude any nonsense or inconsistencies such as more mass coming out than is put in, or similarly, that more energy is given out than is put in, as in a perpetual motion machine. These laws provide a useful check on the validity of any proposed description. Any violations of the laws invalidates the description.

For extended industrial systems, the data describing the overall performance of the extended system comes from a number of different, separate operations, each taking the output from an upstream operation and processing it into a product for the next operation downstream. Each physical subsystem has the same characteristics as any other system; their function must be specified and obey the standard scientific laws. Figure 5.3 shows the different stages or subsystems throughout the life cycle of a product or operation, the use of materials and energy, emissions to the environment, and the points at which reuse and recycling can occur between subsystems.

An example of *recycling* is the *reuse* of glass bottles, whereas an example of product *remanufacture* is the *recycling* of glass. Material recycling to take the product back to the raw material stage is not applicable here because glass is not converted back to its raw materials of sand and soda. But material recycling can be carried out for ground glass ("cullet"), certain plastics converted into hydrocarbons as raw materials for new plastics, or used lubricants re-refined back into base oils.

> *The further a recycling loop goes back up the flow chart, then in general, the more expensive the recycling process becomes.*

This occurs because each stage of the flowchart adds value to the materials passing through it. Materials recycled back to the beginning of the process will then have to compete with cheaper virgin materials coming forward. Recycled materials will always have to compete with virgin materials on cost, quality for purpose, and availability.

LCAs require input data from a wide range of sources. Unfortunately such data is often inaccurate and incomplete. To address these considerable shortcomings in available information, many industrial organizations have commenced projects that examine and analyze their working practices to give the necessary information for wider dissemination and use. The use of unreliable data cannot be justified and it is essential that any data is reliable and validated for proper analyses. In this way, opposing views of environmental impacts can be examined objectively.

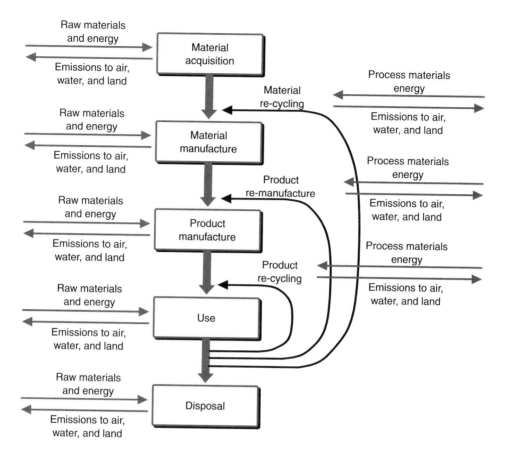

FIGURE 5.3   The life cycle of a product or operation.

5.2 Environmental Implications of Lubricant and Hydraulic Fluids Production

5.2.1 The Size of the Problem

The total global production of lubricant oils, hydraulic fluids, transformer and condenser (capacitor) oils, heat transfer fluids, and metal working coolants is approximately 39 Mt/pa. Of this total mass, the mass of lubricant additives contributes between 1.5 to 2 Mt/pa, and synthetic lubricant base fluids contribute approximately 0.7/0.8 Mt/pa. To put these masses into perspective, "chemicals" have an annual total global production of 400 Mt/pa, which includes additives and synthetic base fluids but not mineral oils. If mineral oils are included and recognized as chemical products, for example, as in the Netherlands, then the lubricants industry is responsible for ∼8.5% of total global chemical production by weight. But the manufacture, at the front end, and eventual disposal at the end of its life cycle of this very large mass of material presents an environmental burden. Therefore, from their reaction activation energies it is not surprising that the environmental aspects of lubricants/hydraulics and their application methods have been of prime importance because of the complexity of local disposal. This continues to be an active area of "research," in reality, mainly development.

5.2.2   Life Cycle Assessment

Proper quantification of the overall environmental impact of lubricants and hydraulics, taken to include the other hydrocarbons described previously, requires a detailed LCA to cover manufacturing, use, and "end of life" fate. A systems approach identifies its boundaries as including petroleum, petrochemical, oleochemical, and engineering industry activities. This is a very complex process due to the broad scope required of the assessment and also because of the particular issues characteristic of the industry and its applications.

An immediate complexity is that lubricants are not produced alone but usually as coproducts in an integrated product network based upon optimized petroleum and oleochemical refining or chemical processing. Consequently, the allocation of resource requirements and environmental impacts to the lubricant/hydraulic elements of these networks must be either exhaustively calculated or considered as necessarily arbitrary calculations. But also, since different lubricant types vary very much in their performance and in the amount required for a particular purpose, a detailed knowledge of the application performance is necessary to define appropriate functional comparisons.

LCA comparisons are made on equivalent outputs, so that a simple comparison of different lubricant types based solely on their resource requirements per mass (kg) or per volume (liter), will give misleading results. Even more so, as lubricants are widely used in many different products and applications, tracing their fate at the end of life can be very difficult because the "end of life" treatment of industrial lubricants used at a single manufacturing site is readily controlled. But tracing the eventual fate of used car engine oils is a more challenging problem because the eventual fate of much lubricant production is not well known. Even when data is, in principle, publicly available it may be regarded as confidential to the product manufacture users and may not be openly published.

Therefore, for LCAs to be publishable generally requires coordination by an independent body that facilitates "pooling" of commercially sensitive information. An example of the extent of this task is given by studies from related chemical industry sectors including the 1995 European Life Cycle Inventory for Detergent Surfactant Production. This study considered energy and resource requirements used to produce seven major surfactant types. There were no assessments of impacts previously described. The overall study took 2 years to prepare and involved 17 technical professionals from 13 companies, a considerable resource input, and the "Eco Profiles" of the European Plastics Industry. A key conclusion from these and other studies was that the technical basis existed to support a general environmental superiority claim either for an individual product type or for various options for sourcing raw materials from petrochemical, agricultural, or oleochemical feedstocks.

For these formidable reasons no comparable comprehensive lubricant LCA has been attempted as yet for lubricant and related fluids. Lubricant and hydraulic fluids have been considered as contributing elements of LCA studies focused on particular applications such as forestry hydraulic equipment applications, municipal cleansing, and domestic refrigerators.

But companies involved in the lubricants and hydraulics industry carry out detailed analyses of their products' environmental impacts in their strategic and internal decision making. It could be argued that the environmental implications of their products have now had to be considered in depth. But the results of these studies are not generally available external to the companies. However, some limited studies have been revealed. Single company studies are limited by access to the company data involved. This restriction can also cause concerns for the independence of the conclusions reached from the commercial interests of that company.

5.2.3   Quantifying Lubricant Chemistry Production

The ideal situation would be to have full "cradle to grave" LCA of lubricant and hydraulic fluid production and use. But this lack of analysis does not imply that no information is available — "Green Chemistry" concepts reinforce and complement those of LCA. This section summarizes some relevant, available information and reviews its implications in terms of Green Chemistry principles to arrive at general qualitative conclusions.

5.2.3.1 Production of Lubricant and Hydraulic Fluid Base Oils — Mass Efficiency and Process Energy Requirements

Any product manufacturing process has distinct steps that can be analyzed and broken down. For lubricant and hydraulic fluids these broadly divide into manufacture of the base fluid, manufacture of the additive, and formulation of the base fluid and additive into a final product. Environmental impacts for each of these production process steps can be assessed. Because each of these process steps involves energy inputs, their efficiencies can be assessed on the basis of each process step efficiency in terms of reaction efficiencies, which is readily measured, and the energy requirements for each process, again readily measured.

For the first point, the typical energy requirement for the petroleum processing of mineral base oil fluid production is estimated as 9 MJ/kg, ~21% of the Energy of Material Resource (EMR), or the total energy of the product, ~42 MJ/kg. Process energy requirements in a modern integrated refinery come from feedstock combustion in several ways. Refineries produce a wide range of products from LPG to asphalt so that all of the input raw material is mainly converted to products or is used for energy production to cause that transformation. The mass efficiency of mineral oil base fluid production is very close to 100% if that part of the feedstock used for energy production is not included in the overall mass and energy balance. Therefore, every part of the crude petroleum input is used in refinery operations; otherwise residual, unusable, and unpleasant materials would accumulate at a fast rate at each refinery.

A considerably smaller proportion of lubricants are based on natural oils. Natural oils have very large and broad market applications in food and agriculture and lubricants are a minor application. To allocate the resource requirements of agriculture to the minor sector of lubricant base fluids is rather arbitrary and, as such, highly sensitive to the assumptions made for the input mix of raw materials, their geographical origin (thus implying transport costs), the related energy inputs of mechanical equipment (tractors, harvesters, and oil extraction machinery), and the energy input into fertilizer production. Taking all of these into account, a typical overall energy requirement of ~3 MJ/kg for vegetable oil production is taken as being typical. It is possible to refine this value further for a particular natural oil but to little numerical effect.

5.2.3.2 The Production of Lubricant and Hydraulic Oil Additives

The wide range of additives used in lubricant formulations, treat rates for individual additive types, and the chemical reactions involved combine to make it difficult to quantify mass efficiencies and process energies for lubricant and hydraulic additive manufacturing. But from the general manufacture of specialized chemicals and polymers it is reasonable to assume that mass efficiencies and process energies will vary considerably between products and in this way affect their manufacturing costs. However, these values and costs are regarded as very commercially sensitive and equally difficult to obtain.

But additives are not unique as chemicals and polymers and there is much published information on similar and analogous compound production of similar scale. This information shows that reaction mass efficiencies for specialist chemicals and polymers are generally 95–98%, the 2–5% losses arising from reactor washouts and filtration losses. These losses reduce if the same product is sequentially made in the same reactor. Therefore, it is readily assumed that lubricant and hydraulic additives are mostly made in a batch reactor. The reaction solvent is usually a mineral oil that is included in the final product, with little further processing of the reaction products. Again, there is very little available information on the process energies for lubricant and hydraulic additives because of the industry's intensely competitive nature. But again by analogy to available values for materials that can be judged as being quite closely related, the estimate for the overall process energy requirements for the manufacture of lubricant and hydraulic additives are close to the EMR of the final products, 35/40 MJ/ kg. Variance of product types can double or halve that value.

Stepping back and reviewing what these values mean, lubricant and hydraulic additives are not produced for their energy content but for the energy or resource that they can save. The insight of an LCA for a friction reducer takes into account the energy that its application will save over its service life. Similar analysis for an antiwear additive must take into account the increased life of a machine given by that additive.

5.2.3.3   Production of Synthetic Lubricant and Hydraulic Base Oils

Very similar analyses to those used previously are used to assess synthetic base fluid production. The benchmark for process mass efficiencies of most synthetic base fluid production is >85%, with lower values for phosphate ester and silicone production by organic compound substitution of a chlorinated reactant such as $POCl_3$ or $SiCl_4$, respectively. These reactions evolve HCl and the reaction efficiency overall depends on whether it is recycled into another reaction or used to make hydrochloric acid. PAOs and PAGs have higher values, produced by addition polymerization that can be configured to give very high yields. Overall, the process energies for manufacturing synthetic fluids are at most comparable to lubricant additives. As these compounds may be produced in a specified state, for example, by varying the polymerization conditions for a PAO, then distillation/fractionation is unnecessary. On this point, the final distillation process energy for mineral base fluids is not needed for some synthetic fluids.

As most synthetic fluids are hydrocarbons, their EMR is very similar to that for the mineral base fluids at ~42 MJ/kg. While it is useful to step back and review what these values are and what they mean, synthetic fluids are not produced for their energy content but for the energy or resource that they can save and their LCA takes that into account.

5.2.3.4   Production of Lubricant and Hydraulic Formulations

Base oil and additives have to be brought together as a formulation for sale and application. This is necessarily a mixing process or liquid blending operation usually in a batch reactor. The mixture must be heated and mixed and if a careful progressive sequence of products is followed, then there should be 100% nominal mass efficiency. From analogous mixing operations, the input energy required for mixing and heating the components during blending is estimated as 3.6 MJ/kg. Longer mixing times may be required for incorporating some types of VII materials because they have the initial physical composition of toffee into the formulation. The length of mixing time must be balanced against the possibility of shear and thermal degradation of the VII.

5.2.3.5   Lubricant and Hydraulic Formulations — Conclusions

Taking the overall composition of a lubricant formulation as being 15% additive and 85% base oil and using the values discussed in the preceding sections on a pro rata basis, then lubricant production has high mass efficiencies, minimum 85% and on average much higher in the 90 to 95%+ region. It has relatively low unit process energies, mainly because of the preponderance of the mineral base oils with their low energy requirements. They contribute to low process energies because of their natural origins and being (a small) part of the very high production volume petroleum industry.

5.3   Environmental Benefits of Lubricant and Hydraulic Formulations

5.3.1   Overview of Lubricant Use

The primary purpose of lubricants is to reduce friction in machines, "machines" defined in the pure engineering sense. It follows that reducing friction in turn reduces the consumption of energy for the same amount of delivered work. Therefore, reduced energy consumption is equal to increased energy efficiency. As almost all energy conversion processes cause some form of environmental pollution, then reduced energy consumption means reduced emissions. This can be a "win-win" situation with considerable returns for improved lubricant and hydraulic fluid performance.

To extend the position of lubricants in the petroleum/petrochemical context, the total lubricant production worldwide as approximately 39 Mt/pa but constitutes only ~1% of worldwide refinery crude petroleum throughput. In comparison, the total production of other nonfuel uses such as solvents, waxes, and bitumen is approximately 1% total petroleum refinery throughput. The main petrochemical feedstocks such as ethylene and propylene are approximately 3% of worldwide refinery crude petroleum throughput, the remaining 95% of crude petroleum refinery throughout is used as fuel in various forms as about

3.5 to 4.0 Bn/Tn/pa. These figures of volumes, masses, etc. are continually changing due to economic activity levels and should be carefully reviewed on the basis of contemporary consumption.

For the developed world, consumption of lubricant and hydraulic fluids is decreasing. Elsewhere in the world, consumption of lubricants is higher per head but decreasing overall. But overall lubricant consumption outside of what is currently regarded as "the developed world" is tending to increase due to increased vehicle ownership and use.

The further context is that crude oil only contributes approximately 40% of world energy consumption, varying in percentage terms from one field of energy to another. Thus, crude oil products contribute heavily to mobile transport but relatively little to fixed energy generation. The other energy sources are coal and gas as fossil fuels, then nuclear energy with renewable energy sources well down the scale at about 1 to 2% but increasing relatively rapidly. The total world energy "system" produces about 22 billion tonnes per annum of carbon dioxide and this emission is now recognized as the major air pollutant contribution by far compared to other nitrogen and sulfur oxide emissions and also solid particulate matter. Carbon dioxide is said to be the main contributor to the "Greenhouse Effect" and its uncontrolled, increasing emissions are associated with global warning.

5.3.1.1   Energy Conservation and Lubricants

After separating out the direct use of fuel to heat premises, plant and processes, most fuel is still used as the energy driving force for machines as transport and power generation. These processes require the relative movement of surfaces in some form of bearing where the lubricants' function is to reduce friction between those surfaces. The lubricant should, overall, reduce the loss of energy into frictional losses and to maximize the conversion of energy into useable work, thereby reducing the energy input. If the "correct," that is, matched for purpose lubricant is used then the efficiency utilization of energy can be improved by up to 10%, more realistically 5%. But improved lubricants properly applied can save even more. To a certain extent this is semantics, for there "is always a better lubricant." But the lubricants have a role to play in overall energy conservation.

5.3.1.2   The Wear and Conservation of Machines by Lubricants

The Jost Report [15] on "Tribology" commissioned by the U.K. Ministry of Technology in the mid-1960s showed that the main benefit of properly applying tribological principles was improved machine life and system reliability. These benefits feed through into the extension of a machine's productive life, thus sustaining material resources and reducing energy consumption. These associated effects conserve nonrenewable fossil fuel resources over the working life of the machine. The report concluded that by effectively applying the existing, 1966, knowledge and techniques in tribology, U.K. industry could make annual savings in operating costs of £515 million at 1966 prices.

5.3.1.3   Working Lifetimes of Lubricants and Hydraulic Fluids

The volume of lubricant and hydraulic fluids used worldwide depends not only upon the number of applications but also on the service life between service life refill of those machines. The determinants are twofold — the prescribed lifetime of the lubricant and hydraulic fluids and also their required replacement. The simple equation is that if the lubricant can be formulated to last longer, then overall demand decreases. Some additional energy is required to modify the base oil, as one example, by more extensive hydrotreatment, but if this doubles the service life of the lubricant then its demand is halved and also its environmental impact pro rata. Further, less used material must be disposed of. It is useful to compare lubricant consumption per major world region using the criterion of M$/GNP.

Western Europe has always used the most highly formulated and sophisticated lubricants for its relatively demanding general automotive performance of high speeds using small engines. The use of lubricants decreased from 0.7 bls/GNP unit in 1990 to 0.6 bls/GNP unit in 1997 and has decreased further to 2004. For the same period, Eastern Europe has decreased from a much greater value of 9.1 down to 6.0 units following its opening up to Western Europe. It shows every sign of moving rapidly toward the Western European average as it increasingly uses more modern lubricant formulations and service intervals. North

America, with traditionally less sophisticated lubricant formulations because of its acceptance of more frequent/shorter lubricant changes, has declined from 1.7 to 1.1 units. This agrees with the general rule that U.S./North American service intervals are usually about half that of Europe.

At the same time, Latin America has declined from 3.1 to 1.6 units, Australia/New Zealand down from 1.7 to 1.1 units, Asia from 1.9 to 1.4 units, Africa from 3.5 to 2.6 units, and overall world consumption from 1.9 to 1.0 units. While these are "broad brush" figures, they are self-consistent. They do not give the complete picture because GNP has increased overall and the values given above need to be translated into total consumption. Over 10 years lubricant consumption has declined from ~40 Mt/pa in 1989 to 37.6 Mt/pa in 1999 in the context of an increased total number of vehicles. This has not been a straight line decrease, fluctuations in demand reflect the worldwide economic climate, as also experienced by the fuel market. There are three major factors that are driving down lubricant demand:

1. A move toward more sophisticated lubricant formulations with longer service lives, particularly the case for Western Europe where 10,000 km service intervals for light vehicles have existed for the last 15 years and now extend well toward 20,000 km for both petrol/gasoline and diesel with some manufacturers offering vehicles with 50,000 km service intervals. Service interval limits are now determined by wear of other components of engine systems, such as spark plug erosion unless advanced designs are used.

2. Heavy diesel vehicle service intervals follow the same trends. A major heavy construction and off-road vehicle manufacturer has a service interval target of 400,000 km for the reasonably near future. The resulting used lubricants may be more heavily contaminated than earlier but their mass is considerably less and their resource requirements are also less.

3. A move toward service refill when the lubricant condition requires it rather than when predicted — "condition monitoring." The actual patterns of use for identical vehicles usually varies between regular long distance, uninterrupted journeys to multiple short journeys with frequent stops, the latter often being more demanding on the condition of the lubricant. Larger engines under demanding operating cycles may consume sufficient lubricant, for example, 6 l/day "top-up" for a total lubricant charge of 180 l, that "top-up" by fresh lubricant is sufficient to maintain a suitably protective lubricant condition in the engine. Increasingly, vehicles have a form of "on-board" engine monitoring that integrates the various types of driving mode experienced by individual vehicles and eventually reaches a predetermined internal value that informs the operator that an engine service is due. This is not a measure of lubricant condition but an integration of the severity of driving modes.

   A feasible proposal, as yet to be offered commercially, envisaged a joint program between a vehicle manufacturer and a national fuel supplier where the manufacturer's vehicles are fitted with transponders ("transmitter/responders") that transmit the current vehicle condition when interrogated as it passes by a fuel service station garage. The information is collated by a central data base and when a service is deemed to be required, a message is sent to the operator/owner. Such a scheme does not require much development because the components are there, transponder technology is well developed in the aerospace industry, medium/larger contemporary vehicles have engine management systems that store performance data, and mobile telephony is very well developed. All that is required is for these systems to be integrated and be shown to be useful. For lubricants, the overall demand would be reduced.

4. By reducing the internal friction coefficients of machines. While automotive manufacturers are improving the overall thermodynamic efficiency of their engines, both petrol and diesel, it is equally recognized that the internal friction of machines can, and should, be reduced. For "machines," the whole automotive power train must be considered, that is, engine, gearbox, transmission, and wheelbearings. Only a small percentage of the available chemical energy from the fuel is transformed into mechanical energy to propel a vehicle (Figure 5.4).

A generous assessment would be 20 to 30% of mechanical energy for movement delivered to the wheels from the available thermal energy. However, the data within Figure 5.4 is for a light vehicle engine

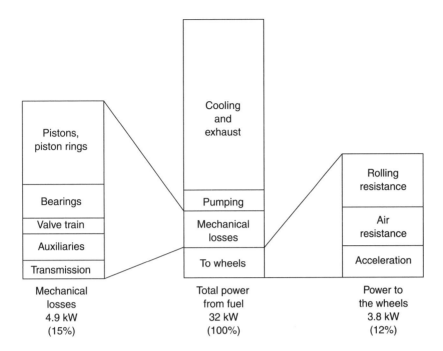

FIGURE 5.4 Fuel energy distribution for a passenger car during an urban cycle. (From B.S. Andersson Company Perspectives in Vehicle Tribology-Volvo-Proc. 17th Leeds-Lyon Symposium on Tribology-Vehicle Tribology, Elsevier, 1991, pp. 503–506.)

undergoing an EU emission test, much more representative of "normal" use than an engine tested at high power level in a cell. During the emission test, only 12% of the available input energy is propelling the vehicle through the driven tires. The rest of the energy is emitted as thermal energy from the radiator/air cooling, in the exhaust, internal energy losses, and a very small amount of unburned fuel, noise, and vibration. Some energy is "parasitic," using part of the engine energy to (increasingly!) drive electrical and air conditioning systems. Some rejected heat is used, advantageously, to heat the vehicle.

The cost implications of these mechanical losses are startling, illustrated by an analysis at current U.K. prices of £0.80/liter:

Mechanical energy losses as a % of fuel energy are 15%

Cost of wasted mechanical energy/liter @ £0.85/liter is £0.148

U.K. total cost/year, 25 M vehicles, 15k miles/year, 6 mi/liter is £9.25Bn

A 10% reduction in mechanical losses by using improved lubricant formulations would save £925M each year from transports costs

This calculation can be repeated for different fuel costs, different vehicle use and currencies, and will still give very significant financial savings.

The obvious objectives are to shift the balance between the amount of heat rejected from the engine and power train and the thermal efficiency of the engine. For engine thermal efficiency applications, simplistic Carnot cycle calculations show that higher thermodynamic efficiencies are achieved by, inter alia, higher cylinder peak operating temperatures. But the higher operating temperatures involved increase the oxidative stress upon the lubricant that must be countered with improved additive technology and base oils with enhanced antioxidancy, such as the extensively hydrotreated mineral base oils and synthetic base oils.

A practical step is to move to lower viscosity base oils in lubricant formulations, provided that the engine design can use them without long-term damage. A major manufacturer has moved from 5W-30 lubricants to 5W-20 for both factory fill and also their proprietary garage brand across almost all of their product range and their associated marques, in Europe and the United States. The environmental benefits are that an estimated 10^5 tonnes per annum of carbon dioxide will not be released into the atmosphere in the United States alone by reducing fuel consumption by $\sim$80 million liters per annum of petrol. In the context of the total volume of fuel consumed by the U.S. automotive industry each year this is not much but every contribution counts.

The problems for the formulator of lower viscosity lubricants are several. For conventional hydrocarbon base oils, lower viscosity is related to lower molecular weight, in turn to higher volatility. But if the engines operate at higher temperatures, then the volatile loss will be much greater. Therefore, base oils of different structures with lower volatilities and higher antioxidancy reserve are required, either as mineral oils extensively modified by further catalytic hydrogenation, or synthetic PAOs or esters. There is a choice between full synthetic ester base oils and extensively hydrotreated iso-paraffin mineral base oils for both reduced viscosity and volatility.

The contribution of these base oils to reducing internal engine friction is that they can only work for hydrodynamic lubrication conditions. But considerable energy is absorbed at boundary lubrication conditions where the hydrodynamic film has broken down and the friction coefficient becomes independent of the lubricant viscosity.

The boundary tribological condition depends upon the nature of the contacting surfaces. These surfaces can be modified by appropriate "surface active" additives. Many compounds have been proposed for this role, some are effective but the most effective are the molybdenum dithiocarbamates, as modified by various peripheral chemical groups, which degrade at the surfaces under severe physical conditions of temperature, pressure, and shear, and deliver molybdenum disulfide-type films onto surfaces with low energy shearable, or lamellar plane surfaces, which reduce friction between the surfaces.

The problem is that the deposition reaction is irreversible and thus uses up the additive. The positive effect of the additive will decrease when its concentration in the lubricant formulation is used up. The degradation products have to move on and some are emitted in the exhaust — molybdenum is not a problem but the sulfur content can be a problem for its deleterious effect on the effectiveness of the catalyst.

Surface active additive technology development has now moved away from metal/sulfur compounds toward derivatives of long chain fatty acid compounds that will give a lamellar effect through the formation of mono-, di-, tri, or multilayer films. The multilayer film effect is analogous to the lamellar structure of the molybdenum thiocarbamates, giving low surface energy structures that reduce friction by minimization of the adhesive forces between the metal surfaces in relative motion. This approach is successfully used in lubricity additives for diesel fuel in reducing injector pump wear.

The friction reduction effects of surface active compounds must be carefully assessed using tribology. It is relatively easy to use a wear measurement device such as a "pin-on-disk" machine, but at high rotation speeds hydrodynamic conditions apply and the determinant is the viscosity of the formulation. However, at low speeds the metal surfaces will contact each other and boundary conditions apply. Under the latter tribological conditions, the friction coefficient can be reduced by up to 20% by the use of surface active additives. However, this reduction is to but one part of the power train. The overall contribution to an increase in fuel efficiency might be of the order of 1 to 2%. To measure a friction reduction on a laboratory machine is promising. The effect must be translated into a long-lasting effect for the service life of the lubricant. The question is not only does the additive "work" by producing an overall friction reduction effect, but also, for how long does it work.

This is a serious current issue for the first decades of the 21st century — there are many compounds that give an initial reduced friction effect. But far fewer give a consistent reduced friction performance over a *longer time scale*, such as the (extended) service life of a lubricant and not lose that activity. Substantial development is undertaken into surface active additive compound systems with enhanced activity for the longer lubricant lifetimes, which are more resistant to oxidation. Satisfying all three is a substantial challenge!

5.4   Lubricant and Hydraulic Fluids as Wastes

5.4.1   Definitions of Waste

The crucial issue in the life cycle analysis of lubricant and hydraulic fluids is how they are treated at the end of their initial useful life. It can be usefully demonstrable to take the used high quality lubricant from a demanding application such as a high speed express train diesel engine and use it in a much lower level of technical demand such as a shunting (switch) diesel locomotive. Then eventually from that application, to take that engine lubricant engine now containing high (~10%) levels of soot and produce a coarse grease for the lubrication of railway points/switches. In reality there is always imbalance between the volumes of lubricant required at each level, in this case with too much waste lubricant produced at the top end by the high performance diesel engines and not enough demand for grease at the bottom end for points/switches.

From this imbalance of demand and supply at various levels, there will always be used lubricant to deal with, as waste or for reclamation/recycling. A cornerstone of waste management legislation is the definition of "waste." But the problem of dealing with waste is that, unfortunately, to different people the term "waste" means different things and this confusion means that there is no simple definition. Whether or not a substance is "waste" is crucial because a waste management licence is required to deposit, recover, or dispose of waste. The answer is determined based on the facts of a particular situation and the pertinent law, irrespective of nation state.

There is no specific legislation for lubricant and hydraulic fluid wastes, if they are to be regarded as waste then they are covered by the general description of waste. The EU legislation provides a community perspective on what is regarded as waste, but national legislation enacts that legislation. The definition of waste in the United Kingdom has changed in recent years to reflect changes in EU definitions and people's attitudes to the recycling and reuse of materials. The U.K. Environmental Protection Act 1990 (EPA 1990) and the Controlled Waste Regulations (1992) made under that Act provide definitions of waste and "controlled" waste for regulatory purposes [16]. Waste defined under these regulations includes effluents, scrap materials, unwanted surplus substances, as well as substances or articles that were broken, worn out, contaminated, or otherwise spoiled. This definition does not deal with materials that are to be recycled or reused nor with the increased public awareness of the problems of waste together with new management techniques such as environmental auditing and life cycle analysis. What is "waste" for one person is increasingly considered a "raw material" for somebody else. The new definition of waste is:

"Waste" means any substance or object in the categories set out in Schedule 2B to this Act (i.e., the EPA 1990)[16] which the holder discards, or intends to, or is required to discard.

Schedule 2B is over-categorized and could be reduced considerably; therefore, some further definitions are necessary:

- The "holder" is the producer of the waste or the person possessing it.
- The "producer" is anyone producing waste, carrying out preprocessing, mixing, or other operations resulting in a change in the nature or composition of the waste.
- A service change lubricant is classified as waste if it has been discarded, disposed of, got rid of by the holder, or it is either intended or required to be so.

Another consideration is whether the substance/object is no longer part of the normal commercial cycle/chain of utility. Some items that may eventually be recycled are treated as waste because they need reprocessing before they can be reused.

5.4.2   Types of Waste

There are various legal definitions of waste in most nation states. In the United Kingdom the EPA 1990 and the Controlled Waste Regulations 1992 made under it divide waste into two categories:

- Controlled Waste — meaning household, industrial, and commercial waste
- Noncontrolled Waste — meaning agricultural, mine, and quarry waste

"Noncontrolled" is a misleading term as agricultural, mine, and quarry wastes are controlled by separate legislation specific to their areas in most nation states. The Waste Management Licensing Regulations 1994 emphasizes "Directive Waste" (referring to the definition of waste in the EU Directive 91/156).

Hazardous and special waste are terms often used for "hazardous wastes," defined in the EU Hazardous Waste Directive (91/689) as referring to wastes that have hazardous characteristics such as corrosive, infectious, or ecotoxic substances. "Hazardous waste" is an EU term. The United Kingdom uses the term "special waste," in general terms defined as waste that is dangerous to life and is controlled by the Special Waste Regulations 1996. The residues from recycling used lubricants are usually classified as "special wastes."

5.4.3 Development of a U.K. National Waste Strategy, "Making Waste Work"

The U.K. Government's Strategy for Sustainable Waste Management in England and Wales, *"Making Waste Work"* December 1995 *(MWW)*, had three key objectives aimed at achieving sustainable development:

- Reduce the amount of waste produced.
- Make the best use of the waste produced.
- Choose waste management practices that minimized the risks of immediate and future environment pollution and harm to human health.

To achieve these objectives, waste management options are ranked as a hierarchy with the most preferred options at the top (Figure 5.5). The recovery of materials is preferred to the recovery of energy.

This follows the same pattern as the EU Principles of Waste Management, which are the prevention of waste as the first priority through reduction and reuse, followed by recovery and then the safe disposal of waste. The U.K. targets for "Making Waste Work" include reduction of controlled waste going to landfill to 60% by 2005, to recover 40% of municipal waste by 2005.

5.4.4 Introduction to the "Waste Strategy 2000"

In May 2000 the U.K. Government published its National Waste Strategy, "Waste Strategy 2000," which had several purposes, a national waste strategy is a requirement of the EU Waste Framework Directive, to help ensure U.K. compliance with other EU legislation such as the Landfill Directive, the Hazardous Waste Directive, and the Packaging Waste Directive, and to help waste management contribute toward

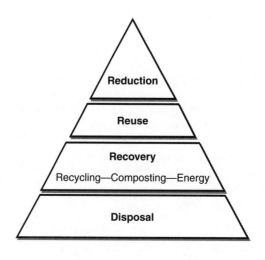

FIGURE 5.5   The waste hierarchy.

sustainable development in Britain. These are intended to deliver the benefits of increasing the diversion of waste away from landfill, encouraging the public to reuse and recycle waste, halting the rise in waste generation, using "the best practicable environmental option" as the basis for waste management decisions, increasing general awareness of the end-of-life impacts of products, and developing markets for recyclable materials. The strategy sets a range of new targets, for achieving what the government considers to be "the appropriate balance between different waste management options" and this will be reviewed by a number of government departments and agencies in 2005, 2010, and 2015, with the mid-point review being undertaken with particular thoroughness.

5.4.5 Waste Reduction

The first step in controlling waste is to reduce the amount of waste produced in the first place. The EU recommends that the following techniques are used and developed for waste prevention:

- Promote "clean" technologies and products
- Reduce the hazardous nature of wastes
- Establish technical standards and rules to limit the presence of dangerous substances in products
- Promote and use recycling schemes
- The appropriate use of economic instruments, eco-balances, eco-audit schemes, life cycle analysis, and actions of consumer information and education as well as the development of the eco-label system

5.4.5.1 Minimizing Resource Use

One of the fundamental tenets of the sustainability concept is to use less material for the same purpose, "less is more," thus "conserving natural resources." General examples include:

- In 1970 an average plastic yogurt pot was 11.8 g, and in 1990 it was 5 g, a 58% decrease.
- A plastic detergent bottle was 300 g in 1970, and 100 g by 1985, a 67% decrease.
- Plastic film for the same application from 180 μm in 1985 to 80 μm now.
- Plastic sacks from 300 to 165 μm, carrier bags from 45 (1990) to 15 μm.
- Average weight of stretch wrap for wrapping pallets now 350 g, replacing 1400 g of shrink wrap used in 1990. Further, use of "soft" adhesives as applied "strips" to hold sacks or containers together could replace plastic wrapping altogether.

5.4.5.2 And Lubricants?

Comparison of the lubricant requirements of a European "sports saloon" of the 1950s, with a contemporary light vehicle, shows that:

- In 1950, the lubricant requirements for a 2.4 l, 100 bhp engine were 9 l to be changed every 1000 mi of a monograde, SAE 40/50 in summer, which was changed to an SAE 30 in winter. Decarbonization of the combustion chamber, of valves and piston crowns was accepted as a necessity at ~25,000 mi. A mechanical rebore of the cylinder block was expected and accepted at 40,000 /50,000 mi, with replacement oversize in-cylinder components. A vehicle lifetime of 100,000 mi was exceptional. The reclamation or recycling of the used lubricant was not contemplated.
- In 2004, the lubricant requirements for an engine of similar output would 4 to 5 l, to last for a service interval change of at least 10,000, more probably 15,000 mi. Decarbonization and reboring of the in-cylinder components is almost unknown. Vehicle lifetimes of over 100,000 mi are common with over 200,000 mi not exceptional. The used lubricant is now expected to be recycled.

The lubricant volume used per 10,000 mi has been reduced by at least 95% but for a much more refined product. After allowing for the small increased energy input to achieve the more refined product, there is still a drastic reduction in resource utilization and energy input.

While reduced internal engine friction lubricants have an important part to play in reducing fuel consumption and extending the useful operating life of engines, the major contributions of lubricants to sustainability is extended operating service intervals to reduce the amount of material used and recycling to produce an acceptable quality base oil for new formulations.

5.4.6 Waste Recovery

After waste minimization has been carried out and the opportunities for reuse explored, the next step is material and energy recovery. The EU Waste Directive recommends that, when it is environmentally sound, preference should be given to the recovery of materials over energy recovery, reflecting the greater effect on the prevention of waste produced by material recovery than by energy recovery. Recovery of materials includes recycling.

5.4.6.1 Material Recycling

It is recognized that material recycling cannot always be done and is sometimes not economically feasible to be done. Used cling film, as an example, needs unraveling, washing, drying, and gentle ironing to be reusable. Or food preparation practices could be changed to reduce or remove the need for the use of cling film.

But where circumstances are suitable, defined broadly as large arisings, clean material and of a single type, then recycling can lead to high quality products and good resource conservation. For lubricant (mainly) and hydraulic formulations, apparently similar fluids can contain different substances, such as different base stocks, base fluids, and additives. The different fluids to be considered for recycling do not have, nor carry, "identifiers" as some plastics have imprinted upon them. An external basis for recognizing the different types of lubricants and differentiating between them is not available.

5.4.6.2 The Collection Problem and the Entropy Model

Before reclamation or recycling, the used lubricant and hydraulic fluids have to be collected to make a viable volume for processing. The collection problem of used lubricant can be compared to a high entropy model seeking to move toward a low entropy solution, which requires an energy input. For after production at a small number of centralized lubricant blending plants, the packaged product is distributed through garages and retailers to individual vehicles, such as the 25M in the United Kingdom. In entropy terms, the system has gone from a low entropy, ordered system to one of high entropy, extensively disordered, system. To reverse this, energy must be expended, which is the energy of collection and part of the overall costing plan.

Used lubricant is collected into garage collection tanks and local authority waste disposal points from which it is tankered to central collection or treatment plants. Tankering itself requires an expenditure of energy, prior collection energies are not usually considered for taking used lubricants to a collection point. There is an increasing "Take Back" expectation of retailing and wholesale organizations, particularly in Western Europe.

5.4.6.3 Lubricants as Waste, Fuels, or for Reclamation/Recycling

There are various processes for reclamation or recycling that have been developed in response to environmental concerns and programs but the extent to which they are used depends on the process and resource economics for a particular intermediate involved. In the final analysis, the recycled feedstock must compete successfully on price and quality with virgin feedstock materials — this is part of the "Waste Paradigm."

Feedstock recycling for lubricants and hydraulic fluids involves separation of the base oil from the used, mixed, materials as the only really useful and recoverable material. The particulates, sooty material and used/unused additives have not been considered for recycling and must be disposed of.

If used lubricants are regarded as a "waste" for disposal, they are then subject to the Special Waste and Signage Regulations of the United Kingdom, or their equivalent in other countries. The additive package in the original virgin oil and its degradation products together with combustion-derived contaminants such as soot, water, and fuel in the used lubricants place them in the "special wastes" category. As such, their producers, transport contractors, and ultimate disposers are subject to administrative measures designed to track their progress "from the cradle to the grave" in respect of their arising, transport, and final disposal, an expensive but necessary process. Therefore, it is advantageous to reclaim or recycle useful components from used lubricants, which are the major and relatively innocuous part, and to deal with the separated minor residues composed of spent additives and contaminants as "special wastes."

5.4.6.4   Reclamation of Lubricants

Some lubricants can be "reclaimed" and then reused, where "reclamation" means low levels of processing such as:

- Straightforward filtration to remove particulates to a standard suitable for the further purpose of the reclaimed oil
- Additive replenishment up to the original specification

Filtering and replenishment is suitable for lubricant and hydraulic fluids that are not used in reciprocating i/c engines, nor exposed to thermal stress and are used in closed systems. The essential distinction is that the lubricants and hydraulic fluids are not exposed to fuel combustion processes and that the direct and indirect products of combustion do not accumulate in the lubricant over time.

An excellent illustrative example of reclamation is the cleaning and replenishment of hydraulic oils in plastic injection molding machines. There are no combustion products and metal wear particles are minimal. The hydraulic oil actuates the clamping of the mold platens under high closure forces, drives the injection screw, and opens the platens at the end of the injection cycle to release the product. The service life of hydraulic oils used in injection molding machines is several years.

Moreover, these oils can be "reclaimed" on site by a mobile filter apparatus. The additives are usually straightforward antiwear and antioxidants whose concentrations can be readily measured and replenished as required to the original specification.

5.4.6.5   Recycling of Lubricants

The composition of used lubricants is that of the fresh lubricant base oil, modified by selective evaporation of the lighter fractions, plus some remaining original additive but usually much more spent additive plus contaminants of heavier fuel fractions, water, soot, and metallic wear particles. Another problem is the variable composition of the base oil component of waste lubricants, for example, Gp. I–III type mineral base oils, plus synthetic polyalphaolefins, PAOs, as Gp. IV and polyol esters as Gp. V.

Recycling of lubricants requires separation processes to produce a "good" quality base oil for reuse in formulations with a minority residue of no use for disposal. Ideally separation processes should have low energy requirements, very good efficiencies of separation of base oil and contaminants to produce an innocuous residue, and a very good quality base oil product. It is very difficult to achieve all of these at the same time in the same process. Reverting to the entropy analogy, to achieve better separation requires a higher level of energy input. The practicable separation processes used to recycle used lubricants are described in another chapter but are recalled briefly as either:

- The *clay/acid treatment process*, which produces ~60% usable base oil product but with a dark color and burnt odor, disadvantageous in quality comparisons to virgin base oil. The clay/acid/additive/ contaminant residue is acid and "tarry," which makes its disposal both very difficult and expensive. Capital investment can be low and small acid/clay treatment plants can be mounted on a trailer. The mobility of such small plants has unfortunately been used by criminals in clandestine operations that produce a low quality, very acid, product. This may be sold as a cheap lubricant or may be blended with lower hydrocarbons of lower taxation levels such as kerosene to produce a "diesel fuel" of appallingly low quality, which is sufficiently acid to corrode fuel systems and engines. This is not helpful to the image of recycled lubricants.

- *The solvent extraction process*, which can produce up to ~75% of usable product with a relatively innocuous residue, using liquid propane. The principle of operation is that "like dissolves like," where the propane as a nonpolar, nonhydroxylic hydrocarbon dissolves the nonpolar hydrocarbons. The usually polar or high molecular weight additives and residues do not dissolve and are separated by filtration. The propane solution of hydrocarbons is then de-pressurized to volatilize the propane leaving a good quality base oil but darker in color than its original and with a burned odor.
- *The vacuum distillation, hydrotreatment, and further distillation processes* give a range of superior quality recycled base oils of significantly less color and odor with a recovery efficiency of up to 95% of that available in the used lubricant. To achieve the higher quality of recycled base oil requires a higher level of input energy to operate the thin film evaporator columns and hydrogenation. While up to 95% of hydrocarbons may be recovered, some of this may be as lower molecular weight fractions that are recycled as fuels. There can be a surprising "carry-over" of trace metals and polymer breakdown products into the base oil, which require further treatment to improve final product quality. A solvent extraction process using dimethylformamide will remove residual sheared polymers and water.

5.4.6.6 Recycled Base Oils and the Waste Paradigm

The Waste Paradigm relates the essential units of materials, from reserves through to wastes and is very instructive for the acceptability of recycled base oil into the product chain. It has the normal form of Figure 5.6.

The crucial insight given by the Waste Paradigm is that if the recycled material is to be an effective contributor to material flows and can compete with fresh raw materials, in this case recycled base oils with virgin base oils, then the recycling process must produce recycled base oil that can compete at the material stage in the scheme above, on the grounds of:

- Quality
- Cost
- Compatibility with additives
- Consistency of supply

The base oil product from the recycling of used lubricants must therefore be of good quality, relative to the original base oil. The recycled product to meet this challenge particularly comes from the vacuum distillation/hydrogenation process that has improved oxidation resistance and enhanced viscosity index, both desirable properties. A typical recycled base oil from this process can be classified as a "better than" Gp. I/"almost as good as" a Gp. II base oil.

5.4.6.7 Residues from Recycling

The residues from the recycling processes are "special wastes" under all nation state regulations and need to be rendered innocuous. The acid treatment/clay/spent additive/soot/filtration processes are extremely difficult to dispose of and are a substantial detriment to the environmental acceptability and economic

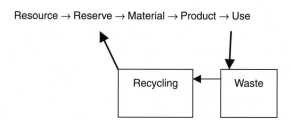

FIGURE 5.6 The Waste Paradigm.

viability of this process. Both the propane liquid/liquid solvent extraction process and the vacuum distillation/hydrogenation/further distillation processes give additive/degraded additive/soot particulate sludges that can be disposed of by incineration, either in special incinerators or in authorized cement kilns. Other separation processes produce residual sludges that are difficult to dispose of and difficult to be acceptably incinerated. It is very expensive to send "special waste" materials to landfill.

5.4.7   Used/Waste Lubricants as Fuels

A convenient route for used/waste lubricant disposal was, until recently, as a supplement for solid and liquid fuels to supplement their calorific values. This was both convenient and economically attractive with a fuel tax rebate of $40 per tonne. The EU Waste Directive removed this subsidy from January 1, 2004 with a complete ban on waste lubricant burning in power (utility) stations from January 1, 2006. After 2005, waste lubricants can only be burned in specially approved incinerators or equally approved cement kilns.

The provisions of the EU Waste Directive in respect of the combustion of raw or partially refined waste lubricants as fuels combine financial and legislative incentives to recycle lubricant components. This cheap, simple, but nonsustainable route to waste lubricant disposal has been closed and recycling must now be seriously considered.

The remaining alternatives of specialized incinerators or cement kiln fuels are controlled by, first, the formidably high costs of specialized incineration and second, the finite number of cement kilns that increasingly use other wastes such as mixed solvent laboratory slops or shredded tire rubber as fuel. Cement kilns are popular disposal sources of "difficult" substances, subject to the EA/HMIP (UK) "authorization" procedure and its essential equivalent in other nation states. The problem is that the capacity of cement kilns to use "wastes" of whatever form or composition is necessarily limited and are not panacea solutions of infinite capacity to the "special waste" disposal problem. A proposal to use "unusual" fuels in any major industrial process in any developed country requires regulatory authorization for that proposal to vary the fuel in that process.

5.4.8   Incineration

Incinerators are designed to destroy the organic components of wastes by combustion. However, most waste has mixed content, whatever its nature, and will therefore contain both combustible organics and noncombustible inorganics. By burning the organic fraction and converting it to carbon dioxide and water vapor, incineration reduces the waste volume and its threat to the environment. Incineration converts waste into an ash residue, known as the product solid and a variety of gases, known as the product gases.

The hazardous nature of these products depends upon the type of waste and the conditions for its incineration. After incineration, the ash volume is considerably less than the original waste, is generally inert, and can be safely landfilled. The heat produced by the incineration process can often be recovered before noxious gases are removed and released to the atmosphere.

5.4.8.1   Waste Incineration Directives

Directive 94/67/EC on the Incineration of Hazardous Waste was adopted in 1994 and implemented by Member States by 1996. This Directive was implemented in the United Kingdom through the existing system of Integrated Pollution Control (IPC) under Part I of the EPA 1990. Its main effect was to introduce more stringent emission standards for operators of hazardous waste incinerators. Many of its provisions also applied to industrial combustion processes using hazardous waste as a fuel, such as cement production.

The new EU Incinerator Directive approved in June 2000 [17] replaces 94/67/EC as well as Directives 89/369/EEC and 89/429/EEC on municipal waste. The new Directive introduces tighter emission and other operating standards for virtually all incinerators, aiming to prevent or minimize harm to the environment media of air, soil, surface water, and ground water from incinerating hazardous waste.

The Directive also applies to co-incineration, occurring at plants that burn hazardous waste as a fuel which are not dedicated hazardous waste incinerators. Authorization may only be granted for a co-incineration process if the hazardous waste burners are located and the waste fed in such a way as to achieve the maximum possible level of incineration. The conditions must specify the minimum and maximum mass flows of hazardous waste and lowest and maximum calorific values of the wastes, the maximum concentrations of pollutants such as PCBs, halogens, sulfur, and toxic metals.

The 2000 Waste Incineration Directive has radically raised emission standards that require high standards of plant and operation. In turn, this has sharply raised the costs of incinerating wastes. In an analogous way to the closing of the waste disposal as fuel option, as described in Section 5.4.7, the provisions of the EU Waste Incineration Directive have effectively closed off incineration of waste lubricants as a route for disposal, emphasizing recycling as the preferred route. Effectively, only the separated sludges from the recycling process can be incinerated.

5.4.9   Waste Disposal to Landfill

Landfill is often regarded as the "last resort" waste management option. The EU recommends that uncontrolled landfill is an operational area that needs special and strong action. An engineered landfill is designed to contain waste and its decomposition products until they present no significant risks to health or the environment. The three main areas of legislative control relating to landfill development are as follows:

- The planning system, which controls the development and use of land in the public interest
- Pollution control legislation, incorporating waste management licensing, and measures for environmental protection from the effects of the landfill, such as protection of existing groundwater resources from landfill leachate
- Regulatory and statutory controls to protect health and safety, and to ensure minimum standards for civil engineering construction

5.4.9.1   The EU Landfill Directive

In April 1999 the EU Directive on the landfilling of waste was adopted [18] and had a significant impact on the European Waste Management industry, principally through the costs of landfill. Its main aims are:

> the harmonisation in all Member States of technical and environmental standards for landfill and to ensure a high level of protection for the environment, in particular of soil and ground water.

Under the Directive, waste is classified according to its characteristics; in parallel, landfill sites are also to be classified as for acceptance of these categories and certain waste types are no longer accepted for landfill. The last includes, among others, liquid wastes that are directly relevant to hydraulic and hydraulic fluid wastes. The Directive includes a requirement that all waste must be treated before being landfilled with the intention of reducing the volume or the hazardous nature of the waste, to facilitate waste handling, and increase recovery.

5.4.9.2   The Landfill Tax

In 1996, a landfill tax was introduced in the United Kingdom to ensure that landfill waste disposal is properly priced to promote greater efficiency in the waste management market and in the economy as a whole. It also applied the "polluter pays principle" so that those producing pollution should pay for its treatment/removal and not use the environment as a free waste disposal resource as well as to promote a more sustainable approach to waste management which produces less waste and reuses or recovers value from more waste.

The tax complements and reinforces the general approach and policies of the U.K. government and also the EU for sustainable waste management. The overall aim of these policies is to increase the proportion of waste managed by using options toward the top of the waste management hierarchy, as in Figure 5.5. The tax allows waste disposal companies to pass the additional costs on to waste producers. In turn this means that waste producers are made aware of the true costs of their activities and so have the incentive to reduce and make better use of the waste they produce. Landfill tax was introduced at £2 per tonne for inert waste and £7 per tonne for active waste. The tax was increased to £10 per tonne for active waste in April 1999 and the Chancellor of the Exchequer announced an annual increase of £1 per year for five years. This charge has acted as a substantial disincentive for the landfill of wastes other than those that cannot be disposed of by other means.

5.4.10   Directions in the Disposal of Used Lubricants

From the examples given, the driving economic force in the EU is to reclaim/recycle used lubricants for their usable materials. The emphasis is upon not using used lubricants as fuels nor to incinerate or landfill them as wastes. This requires adjustments in the lubricant industry, which now has to include the concept of sustainability in its operations and accept recycled base oils into its formulations.

5.5   Environmental Pollution by Used Lubricants

5.5.1   The Polluting Effects of Used Lubricant

When indiscriminately disposed of to land, the transport of used lubricant material depends upon the nature of the subsoil. If the subsoil is clay or clay-rich, then the materials strongly adsorb the lubricant oil mass and will have a very low annual rate of transport, particularly if there is a low rate of hydrological movement. If the subsoil is predominantly sand, then the oil mass is less weakly adsorbed and any transporting force such as gravity or water flow will slowly move the oil deposit through the subsoil. The oil deposit will gradually deplete itself and its immediate vicinity of oxygen and becomes anaerobic, emitting unpleasant odors from the soil.

If the waste lubricating oil is indiscriminately disposed of to water, either directly or by seepage from deposits on land, then surface slicks are formed on the water course that are aesthetically unacceptable. An immiscible surface film on the water course prevents oxygen transfer across the interface between air and water. Very quickly the water course becomes anoxic and odorous. Any waste lubricating deposited onto land will gradually leach original and degraded toxic substances into water and soils over time, dependent upon the nature of the soil. The more volatile components of the waste lubricant evaporate leaving behind the most viscous and heavy chemical components, often as a surface film. The environmental degradation reaction rate of the lubricant becomes diffusion controlled, for example, oxygen diffusion for oxidation becomes diffusion controlled, which will slow down the rate of oxidation.

While hydrocarbons will not readily emulsify, the presence of other compounds from the original additive package or arising from degradation in use will have emulsifying actions to produce sludges, usually unattractive to view.

The environmental implications of the hydrocarbon base oils are that they are difficult to readily dispose of. They have a very high demand of environmental oxygen in the longer term and take a long time to decompose naturally. The hydrocarbon base oils and their intermediate decomposition products will emulsify with "worked" water, such as waves and surf, to produce an unpleasant "chocolate mousse" emulsion at shorelines.

In addition to being unpleasant, the environmental implications of the hydrocarbon base oils are long-lived due to their low reactivity. They cause immense damage to marine life by coating organisms, birds, and animals with a relatively impervious (to oxygen diffusion) oily layer, which also destroys the waterproofing effect of bird's feathers.

5.5.2 The Biodegradability of Lubricants

Biodegradability is increasingly required in developed countries for total loss engine lubricant systems such as chainsaws, outboard motors, recreational vehicles for off-road use, etc. As an example, two-stroke engines for outboards, portable machines, and apparatus are increasingly used; their total loss lubricant systems coat the leaves of vegetation with a fine film of lubricant and also deposit reactive additives into the environment

"Biodegradability" is the ability of a substance to be degraded in the environment by natural events, be they solar radiation, oxidation, or biological degradation. These "events" or "conditions" are clearly very variable; therefore, standard conditions must be established and strictly adhered to. The issues are what those test conditions should be, the extent of degradation, and over what time period the test should be conducted. An additional complication is that the experimental basis on which the biodegradability tests are based were for single substances. However, lubricants are both complex formulations and also variably degraded materials.

There are at least six versions of biodegradability tests [19]. These arise for different conditions, different types of materials, and for different continental regions. As these tests are biological procedures, then variable data can result.

Establishing a reliable biodegradability test program for a company is a very formidable undertaking in terms of resources. A cost-effective approach is to use an established, accredited, independent laboratory with a proven track record in biodegradability measurements once the choice of test and its parameters have been agreed upon.

5.6 The Environmental Future and Lubricants

5.6.1 Annual Statistics and Used Arisings in the United Kingdom

The total mass of new lubricants produced and sold in the United Kingdom is around ~1 Mt/pa, with the automotive and industrial ratios of approximately 3:1. The long-term consumption of all forms of lubricants is slowly diminishing as a result of longer vehicle service intervals and higher lubricant quality.

The waste statistics for lubricants are "unreliable" with high, 50% of mass, "lost" in use by combustion, leaks, and seeps, inappropriate disposal, and "tipping," giving a maximum total U.K. waste lubricant mass of ~500,000 tonnes per annum. This percentage of waste is lower than other developed countries, probably because of the extensive "do-it-yourself" culture in car maintenance in the United Kingdom compared with the "do-it-for-me" culture in countries such as Germany and North America.

5.6.2 Trends in Waste Lubricant Arisings

The global trends in waste lubricant arisings are driven by:

- Longer service intervals, up to 50K km for light vehicles, already leading to reduced lubricant consumption
- Much longer for freight vehicles, 400K km envisaged in the future (moving toward "fill-for-life?")
- Lower sump oil volumes, but
- Higher levels of contamination by soot particulates, additive degradation products, and much higher levels of polynuclear aromatic hydrocarbons (PAHs), some of which are carcinogenic

The main health and safety problems (in addition to the minor problems of fresh lubricant) are PAHs. Increased service intervals sharply increase PAHs at more than 6K (gasoline) miles. This has immediate implications for operators/mechanics dealing with extended use lubricants and "enhanced personal care protection" is required for anyone handling these used lubricants. PAHs are removed from used lubricants by the separation processes described previously, the vacuum distillation/hydrogenation process being the most effective at removing PAHs by far. An additional solvent extraction step can give up to >99.7% PAHs for the vacuum/hydrogenation/solvent extraction process. While this is beneficial for the base oil product quality, PAHs do not disappear and collect in the residue and are part of the disposal problem.

The main disposal route up until 2003/5 is the "soft option" of "clean-up" (to various degrees), then either:

- Use as "cleaner" fuel in power stations/boilers
- Use as "dirtier" fuel in cement production

Separation processes can produce good quality base oils for recycling into new formulations but also necessarily produce toxic residual sludges which are difficult to dispose of, difficult to be acceptably incinerated, or sent to landfill as "special waste," which is increasingly very expensive.

5.6.3   The Environmental Business Economics of Recycling Lubricants

The business economics of recycling used lubricants depend upon various factors such as the internal economics of lubricant production, the external economics of dealing with the environmental costs of used lubricants, the acceptability of recycled materials as lubricants, and the relative cost of virgin base oils.

Recycling used lubricants is currently a marginally viable "business" overall, which is:

- Good for large sources such as bus, freight, and train companies using single formulations of lubricants.
- Not good for recycling collection from individual vehicles using a range of lubricants unless a recycling surcharge/levy is applied. What form that might take is very arguable but a useful analogous example might be the beverage container tax used in many states of the Union; this has the benefit of being fiscally neutral if the user acts in an environmentally responsible way.
- At what level of administration and by which authority should the responsibilities for disposal rest? At present it is the "local authority" if placed within its recycling facilities. Otherwise, the general "duty of care" for "special wastes" applies if the used lubricant is treated as a waste.

The emerging responsibility is moving toward the "vendor" for the "end-of-life" disposal of lubricants, which requires the retailer to take back an equivalent volume of used lubricant when purchasing fresh lubricant, already an emerging issue in the United Kingdom, established in some nation states of the EU. The Packaging Regulations (EA, 95, etc., made under an EU Directive) cover the current packaging of lubricants. From any consideration of lubricant packaging it is clearly evident that it is suitable for purpose, is not overpackaged, and is secure and safe. Therefore, the current packaging of lubricants will probably not be affected by the enhanced requirements of the Packaging Regulations. But in the future, the reuse of packaging through collection and return of containers for refilling will become more important. The quality of the content delivered then becomes an issue.

The EU is moving toward the manufacturer's responsibility for disposal of all contents of vehicle when scrapped, which includes hydraulic fluids and engine lubricants.

5.6.4   Barriers to Acceptance of Recycled Base Oils

Whereas reduced friction lubricant and hydraulic formulations contribute to energy efficiency, their major contribution to sustainability is through extended drain and use periods and their successful recycling to produce good quality base oils for reformulation.

A problem in certain countries is a resistance to the acceptance of recycled products of all forms, regarded as being "inferior" in some way. However, in Western and Central Europe, particularly Germany, recycled materials of proven quality are accepted. Two major manufacturers accept the use of recycled base oils in lubricant formulations for use in their vehicles, subject to low PAH levels in the final formulation. In the final analysis, the vehicle manufacturers carry the responsibility of the vehicle warranty. The barriers to acceptance of oil products containing recycled base oils can be overcome by consumer information and education, to the level of acceptance in Germany.

5.6.5   Relative Costs

Ultimately, if environmental/quality/safety issues can be met, then the relative costs of recycled/virgin base oils are the arbiter. The recent rise in crude oil price, stabilizing at a much higher level, 40$+/bl (early 2005) for light sulfur crude petroleum is due to:

- Continuing unrest in producing countries
- Business/political chaos in Russia
- Industrial action in some countries and political uncertainty in others
- The extraordinary economic development of China and India, drawing in vast amounts of oil products

The net result is that the crude oil production/consumption balance is now so finely balanced that uncertainty raises crude prices further, which quickly passes through into yet further oil product prices. This gives an increased demand for Gp. I oils but there is a global shortage, which is ideal for recycled lubricant base oils to supply.

5.6.6   Conclusions

This chapter has concentrated upon the environmental developments relating to lubricants in the EU and in the United Kingdom as an example of how legislation is framed for effective implementation in individual nation states. But the general direction is the same for all countries, more so for developed countries and less for others. Comparison of environmental law across continents shows how fundamentally similar the underlying purposes of these laws are, even if their procedures of enforcement may appear to differ according to the legal processes of each country.

The message is clear — the unlicensed disposal of used lubricants is an increasingly severe offense and recycling of good quality base oils is now a required operation. This will become an important contributor to the formulation of new lubricant products.

References

[1] 1987—Report of the World Commission on Environment and Development, "Our Common Future," chaired by the ex-prime minister of Norway, GroHaarlem Brundtland.
[2] 1972 — The United Nations Conference on the Human Environment (UNCHE), Stockholm, Sweden.
[3] 1992 — the "Earth Summit" (*the UN Conference on Environment and Development*) in Rio de Janeiro, Brazil.
[4] 1997 — A special UN conference reviewed implementation of Agenda 21 (as Rio+5).
[5] 2002 — Ten years after the Earth Summit in Rio in 1992, the 2002 Johannesburg Summit.
[6] "Towards Sustainability," EU Fifth Environmental Action Programme, 1993.
[7] The EU Treaty of Amsterdam, 1999.
[8] "Environment 2010: Our Future, Our Choice," EU Sixth Environmental Action Programme (2001–2010), 2001.
[9] "A Better Quality of Life," the U.K. Government Strategy for Sustainable Development, 1999.
[10] The U.K. Local Government Act, 2000.
[11] Sulphur Content of U.K. Fuel Oils, *Pollution Handbook 2004*, National Society for Clean Air, U.K.
[12] CEC/93/EF13.
[13] Daimler-Chrysler "HighTech Report," 2/2003, p. 35.
[14] "Responsible Care," the U.K. Chemical Industries Association (http://www.cia.org.uk/industry/care.htm), 1989.
[15] "The Jost Report," a U.K. Government committee report, chaired by Peter Jost, HMSO, Lubrication, Tribology, Education and Research, DES Report, London, U.K., 1966.

[16] Definitions of "waste" and "controlled" waste for regulatory purposes, the U.K. Environmental Protection Act 1990 (EPA 1990) and the Controlled Waste Regulations (1992) made under that Act, *Pollution Handbook 2004*, National Society for Clean Air, U.K.

[17] EU Waste Incineration Directive 2000/76/EC.

[18] EU Council Directive on the Landfilling of Waste, 99/31/EC.

[19] See, for example, S.13.8 by C.I. Betton in *The Chemistry and Technology of Lubricants*, R.M. Mortier and S.T. Orszulik, Eds, Blackie Academic and Professional (VCH Publishers in USA and Canada), 1992.

6

Lubrication Program Development and Scheduling

Mike Johnson
Advanced Machine Reliability Resources (AMRRI)

6.1 Introduction

Effective machine relubrication practices are as critical to the practice of reliability engineering as proper shaft alignment and component balancing. However, while misalignment and imbalance may reveal themselves through the outward symptoms of elevated temperatures, elevated vibration, and loud noises, the symptoms of poor lubrication are often imperceptible.

For many years the industry contended that if one provided enough of the right product at a reasonable frequency that the lubricated components would be sufficiently protected. While this may suffice for some low intensity operating environments, a "best practice" is justified for highly competitive businesses like steel, cement, paper, and automotive production.

A properly devised best practice will incorporate operation and machine specific requirements including:

- Machine criticality and operating environment
- Data collection strategies
- Machine data collection criteria
- Lubricant type, quantity, and frequency requirements
- Contamination control requirements
- Oil analysis requirements
- Activity sequencing
- Planning and scheduling management

This chapter will review the requirements associated with effective machine lubrication practice development.

6.2   Machine Criticality and Operating Environment

Within any given production facility there are variety of machines with a variety of responsibilities. Some of those machines will be built to withstand physical stresses several times greater than the actual production processes will impose. The extent of extra capacity as gauged by the machine designer would be characterized as "service factor." There are differing philosophies as to how much "extra capacity" should be incorporated into a machine design, but it is generally accepted that the greater the "service factor," the longer the machine would be expected to last between required rebuilds. Of course, higher service factors require greater capital investment.

As financial managers became more sophisticated in measuring capital usefulness, engineering departments began to consider whether the extra capital invested would ever be released in the form of higher productivity, and then began to squeeze the "service factors" toward a minimum acceptable level.

As service factors fall, the relative care and attention that should be applied to the machine will necessarily rise in order to sustain equivalent production. The amount of resources dedicated to machine care must be allocated according to that machine's importance to the production process, and as the importance ranking increases the resource allocation should also increase, as illustrated in Graph 6.1. The process for grading the machine's importance to the production environment is called criticality assessment.

An effective criticality assessment will consider various measurable parameters for each machine. Assessment parameters could include several factors, such as:

- Machine function
- Hourly value of machine function
- Machine failure risk to employees
- Machine failure risk to the environment
- Machine failure risk to production quality
- Machine durability (rate of failure, mean time between failures, standard deviation, etc.)
- Speed of failure
- Cost of machine repair
- Operating environment severity

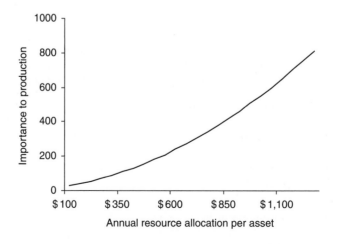

GRAPH 6.1   Criticality ranking and resource allocation.

The operating environment will also influence the degree of detail that the relubrication procedure should entail. Machines operating in a benign environment, such as the environment in an automotive production facility paint shop, will require less attention for contamination control features than would be necessary for equipment operating a steel mill melt shop. Operating environment factors that should influence lubrication plan development include things such as:

- Atmospheric temperature
- Atmospheric humidity
- Atmospheric dust and dirt exposure
- Machine direct exposure to production chemicals, moisture, or solid contaminants
- Machine stability (vibration from external sources)

If a machine has both a high criticality ranking and a high environmental influence factor, then the relubrication practice will need to incorporate measures to adjust frequencies, volumes, and application methods, as well as measures and modification plans to minimize the risk that the environment provides to machine health. A thorough criticality assessment method should incorporate a factor for operating environment.

The site reliability engineer will be able to provide guidance for the amount of detail that each machine requires based on the current criticality standard.

6.3 Data Collection Strategies

Once the list of machines has been established and prioritized from most to least critical, the MLT (machine lubrication technician) may begin to focus on the task of collecting the myriad of details that are required to devise accurate practices. From the outside of the facility it may seem to be a whale-sized task, but with the help of an orderly and systematic approach, the whale can be eaten — one bite at a time.

Each asset must be observed, and the technical details of each characteristic must be recorded as completely as possible. Some of the details may reside in a computerized maintenance management system (CMMS) and may be quickly accessed. It is highly likely that regardless of the state of CMMS development the details required for lubrication practice development will be incomplete. Common details may reside in the CMMS, or in an original equipment manufacturer's (OEM) operations and maintenance records.

For instance, a machine's motor builder, size, type, horsepower ratings, and other similar details may be reflected in these records. Over time, however, motors are repaired, or moved to other service areas, and the original record may not reflect the "as is" state. With that expectation in mind, it is necessary to conduct a physical review of each asset to verify existing information and to supplement with (as built) details. There could be any number of ways to organize the data collection sequence.

Batch process: Environments that are batch oriented sometimes expand in a piecemeal fashion. Figure 6.1 depicts a production setting where the layout location of the production process is not sequential. In this environment one can collect drive train data using location markers in the plant as a sort of geographical reference point.

Continuous process: One approach to organize the data collection exercise is to follow the flow of raw materials through the conversion process. Figure 6.2 shows a sequential view of the major processes for manufacturing Portland cement. Each individual section will contain drive trains operating in tandem or in sequence to the other drive trains. Capturing drive train lube sump details by following the material conversion process also helps with the organization of lube points for a route plan.

Alternatively one may visit each product cell wherever it may exist in the plant layout until all of the drive trains have been reviewed. This process will certainly work, but it does require a higher level of organization to keep the records straight, and will likely require revisiting each machine when it is time to prepare efficient relubrication routes.

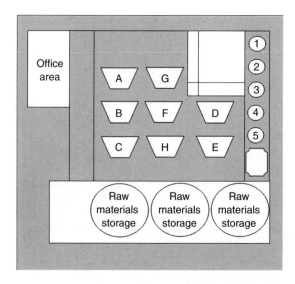

FIGURE 6.1   Batch plant layout diagram.

Planned availability (production demand flow): Where there is a clearly defined risk to personnel if the machine is operating, it may be necessary to either catch the machine when it is idle, or idle the machine at the point that the surveyor is ready to review the compliment of components. An electromechanical robot would fall into this category.

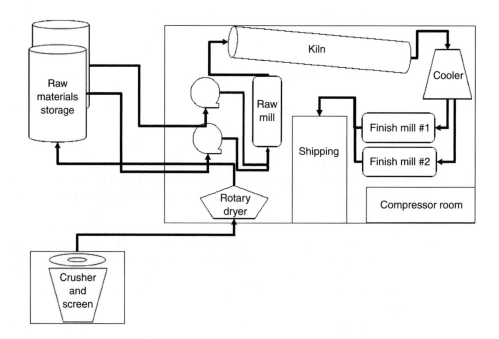

FIGURE 6.2   Portland cement mill major process overview.

6.4 Machine Data Collection Criteria

1. The lubrication survey entails collection of information on every lubricated component on every drive train within the production system. The specific detail that must be collected depends on whether the machine is oil or grease lubricated. Table 6.1 describes the type of component by component detail that is required to build an appropriate practice. The specific detail that must be collected depends on whether the lubricated component requires oil or grease.

2. Oil-based lubrication denotes a fluid sump. The sump may or may not require sophisticated sealing materials to exclude contaminants, fittings to accommodate oil sample collection, and fittings to accommodate contamination control requirements. Two identical machines may have appreciably different relubrication requirements at the time of data collection based on machine criticality, as previously discussed, but that criticality requirement may change. Table 6.2 offers a suggested level of detail for the collection of gearbox sumps. A similar level of system specific details is warranted for bearing, coupling, hydraulic, or circulation sumps.

   Over time increased product demand can influence the criticality assessment, causing the criticality factor to both rise and fall. If the details collected are only sufficiently detailed enough to cover the immediate need, then the process will unnecessarily have to be conducted again at a later

TABLE 6.1 Components that are Common to a Simple Conveyor

Items	Make	Model	No.	Components	Type	Make	Model
Drive train — conveyor A1							
Motor	GE	5KS511SN3260HB	1	2	Bearing	SKF	90BCO3JP3
							75BCO3JP3
Coupling	Falk	1080T	1				
Gear reducer	Falk	2145 Y2B	1	2	Seal	Falk	Type K
					Backstop	Falk	PRT 65
Coupling	Falk	1140T	1				
Head pulley				2	Bearing	SKF	22234CCK/W33
							22234CCK/W34
Snub pulley				2	Bearing	SKF	22230C/W35
Tail pulley				2	Bearing	SKF	22230C/W36
							22230C/W38
Trophing rollers	Cont. Conv	78AH650-66G	192	3	Bearing	Timken	LM 11949
Return roller bearings	Cont. Conv	78AH650-66G	48	2	Bearing	Koyo	UCF208-24

Note: Each individual lubricated component should be accounted for.

TABLE 6.2 Type of Details that Should be Collected Whenever Possible

Component	Make	Model	Details
Gear sump	Falk	2145 Y2B	Type of reductions
			Number of reductions
			Input speed
			Reduction ratio
			Bearing types
			Bearing sizes
			Stated AGMA lubricant grade
			Stated AGMA lubricant type
			Sump temperature
			Oil distribution method (bath or circulation)
			Presence of filtration
			Presence of vent breather
			Number and size of drain port openings

TABLE 6.3   Grease Lubricated Sumps Require Collection of Additional Details

Component	Make	Model	Details
Bearing	SKF	22234CCKW33	Component type
			Component size
			Shaft speed
			Static rating (C/P — for bearings)
			Oil or grease lubricated
			Degree of seal (shielded, sealed, open)
			Seal type
			Relubrication method (manual vs. automatic)
			OEM designated lubricant
			OEM original fill lubricant type if any
			OEM designated lubricant volume
			Operating temperature
			Operating atmosphere — moisture
			Operating atmosphere — temperature
			Operating atmosphere — dust or dirt
			Operating atmosphere — process chemicals

date. Sufficient detail should be collected to enable the development of a "best practice" even if all of the details are not used immediately.

3. Grease-based lubrication denotes a sump that requires systematic refreshment of the lubricant in the sump. The relubrication frequency may be long or may be short, depending on how the machine is operated. As the speed and load increases and the ambient environment becomes more severe the calculated relubrication interval (time between relubrication events) decreases.

   Relubrication of greased components requires a greater dedication to detail than is typical for oil lubricated components. Table 6.3 offers a suggested level of detail that should be captured for bearings that are either oil or grease lubricated. Similar details should be captured for other grease lubricated components.

Table 6.4 proposes general guidelines to follow when collecting information on other types of components. It is not possible in the pages of this chapter to redress data collection parameters for the myriad of types of machines and lubricated components, so the general advice offered is intended to stimulate ideas for a creative pursuit of the maximum level of detail that can be collected in a timely matter.

6.5   Lubricant Type, Quantity, Frequency, Application Method, and Time Stamp Decisions

A wide variety of operational factors will influence the final product selection decision. The factors will be different from industry to industry, and within any given industry from plant to plant, and within the plant, from department to department. Environmental factors and influences should be addressed following review of the fundamental component requirements, assuming that each component was properly sized and constructed from appropriate materials for the intended application.

Product selection will also influence necessary relubrication volume and frequency decisions. Again, there can be many variables. A safe place to begin with any product selection decision process is with the fundamental engineering units as denoted by the lubricated component manufacturer. While there may be misgivings about the quality of some OEM guidance, this is nonetheless the best place to begin this process.

Content in the previous section covers lubricant selection, quantities, and application methods for various lubricated machines and mechanical components. It would be impractical to rehash those details in this short section, but the thought process should be reviewed to see how the decisions will be used in the course of developing an effective machine lubrication practice.

TABLE 6.4   Miscellaneous Factors for Consideration

Component	Make	Model	Details
Sumps			Oil or grease lubricated
			Degree of seal or shielding
			Seal type if known
			Relubrication method (manual vs. automatic)
			OEM designated lubricant
			OEM original fill lubricant type if any
			OEM designated lubricant volume
			Operating temperature
			Operating atmosphere — moisture
			Operating atmosphere — temperature
			Operating atmosphere — dust or dirt
			Operating atmosphere — process chemicals
			Type of lubricated components
			Size of lubricated components
			Rotating speed of lubricated components, if any
Surface coated components			Contact surface length
			Contact surface width
			Surface area
			Surface velocity (speed of surface movement)
			Application method
			Operating temperature
			Operating atmosphere — moisture
			Operating atmosphere — temperature
			Operating atmosphere — dust or dirt
			Operating atmosphere — process chemicals

This section provides very specific advice on plain and element bearings, and only general guidance on where to find the information covering other component types, including element and plain bearings, gearing, couplings, chains, and cables. These individual components may be assembled into a wide variety of operating systems, but the fundamental questions regarding oil film thickness and film type will always lie at the heart of the product selection decision.

Bearings: All machines that have moving parts will have bearings of some type. The bearings may be as simple as flat surfaces mating with flat surfaces, such as the slideway in a machine tool, or may have sophisticated geometries, such as is the case with a ball screw.

Plain and element bearings supporting rotating shafts are most common, and are found in nearly every machine one might imagine.

1. Product type selection for plain and element bearings.

   (a) Plain bearings, also called journal bearings, as shown in Figure 6.3, support and constrain the motion of a rotating shaft. Figure 6.4 shows different styles of plain bearings that may exist for both grease and oil lubricated conditions. The principal forces applied to the shaft are typically radial (perpendicular to the axis of the shaft) but may also be axial (in the same direction as the axis of the shaft).

   Oil film formation is said to be "hydro-dynamic" in nature, and film formation is achieved when the oil accumulates at the contact point between the shaft and the bearing, forcing the shaft to float on the accumulated oil. Since there is no expectation of metallic contact during normal operating conditions, the OEM and equipment supplier would likely recommend R&O (rust and oxidation inhibited) circulating oils for oil bath and circulation systems. Some manufacturers suggest the use of "compounded" oils that are fortified with metal wetting additives to enhance the lubricant's tenacity and surface protection capacity.

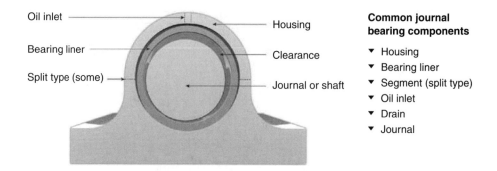

FIGURE 6.3   Common journal bearing components.

Plain bearings are typically oil lubricated, but where slow turning shafts and heavy unit loads create conditions where the pressure wedge may not remain stable, grease may be selected. Guidance provided by Lansdown suggests limits for grease lubrication of plain bearings is based on a limit of 2 m/sec or 400 ft/min of linear shaft speed based on the surface speed of the shaft. This equates to 400 RPM for a 5-cm bearing; less than 50 RPM for an 80-cm bearing. This is to limit churning of the grease that would ultimately lead to grease and bearing failure [1].

Extreme pressure (EP) agents are not typically specified for grease lubrication of plain bearings. Solid film antiwear (AW) additives, such as graphite and molybdenum, are particularly useful for these conditions, and should be considered to help minimize the extent of wear given the higher rate of metal contact. Most greased bearings are once-through applications. As such, there is little risk that the grease will be subject to long-term oxidation stress that oil lubricants will face.

Strong EP and AW chemical additives are not considered to be particularly helpful where soft or "yellow" metals (bronze or babbitt) are in use. A given amount of heat energy must be generated from friction to initiate the reaction between the additive and the metal surface. With steel on bronze, or steel on other soft metals, scoring occurs rapidly enough that the chemical EP and AW additive cannot effectively perform the assigned task. Additionally, these additives tend to generate acids during decomposition that may cause corrosion. The same consideration applies to lubricant selection for oil lubricated bearings.

- Journal bearings support and constrain rotating motion subject to radial loading
- The principle force acts perpendicular to the axis of the shaft
- The journal is the section of the shaft that rests on the bearing
- Sufficient oil viscosity and shaft speed needed to maintain hydrodynamic oil film

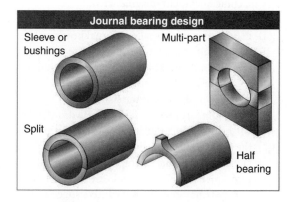

FIGURE 6.4   Different styles of plain bearings. (Noria Corporation. Machinery Lubrication Level II Training Seminar, Slide G2103.)

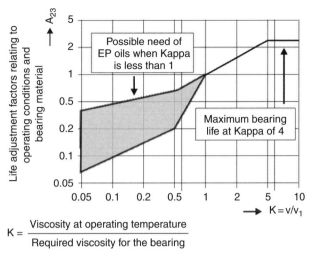

$$K = \frac{\text{Viscosity at operating temperature}}{\text{Required viscosity for the bearing}}$$

Kappa of 1 – 2.5 is desired

GRAPH 6.2 Viscosity ratio range. (Noria Corporation. Machinery Lubrication Level II Training Seminar, Slide 1561. *Source: Lubricants in Operation*, U.J. Moller and U. Boor. 1996. © John Wiley & Sons Limited. Reproduced with permission.)

(b) Product type selection for element bearings is fairly well defined by bearing manufacturers, for both oil and grease selection. The dominant lubrication mode for element bearings is elasto-hydrodynamic (EHD), or "thin film" condition.

EHD conditions provide for complete separation of interfacing surfaces, but with separation ranging from .5 to 3 μm for ball and roller type bearings. Given that the surfaces are intended to be separated, albeit with a very thin layer of oil, the manufacturers suggest the use of R&O inhibited mineral oils for lightly loaded bearings and oils with surface reactive (AW and EP) additives for those applications where normal operation carries a high risk of film collapse.

It is possible to closely estimate (calculate) whether a given bearing in a given set of conditions requires an EP fortified oil or grease through the use of a film thickness ratio called K or *Kappa* factor (K). This is the ratio of actual viscosity at operating temperature divided by the bearing manufacturer's required viscosity at operating temperature. Graph 6.2 shows a viscosity ratio range for which EP additives are highly recommended.

Table 6.5 offers recommendations for oil types based on bearing manufacturer general guidelines for meeting film thickness requirements, and based on practical experience for maintaining lubricant health.

TABLE 6.5 General Guidance for Selection of Lubricant Type for Element Bearings

Proposed Oil type	Element Bearing P/C	Viscosity Kappa	Operating Temperature °C
I, MO	<.15	>4.0	10–80
I, AW, MO	<.15	1–4	10–80
I, AW, S	<.15	1–4	<10[a,b]
EP, MO	>.15	<1.0	10–80
EP, S	>.15	<1.0	>70[a]

Note: [a] Sustained operating temperature.
[b] Risk of startup at or below the low level is high.
I = Inhibited, AW = Anitwear, EP = Extreme pressure, MO = Mineral oil,
S = Synthetic, P = Actual radial load, C = Bearing load rating.

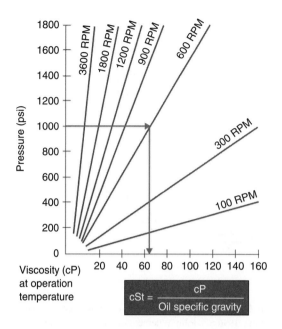

Calculation of journal bearing pressure:

$$\text{Pressure (psi)} = \frac{W}{LD}$$

where:

w = Load on shaft (lbs)
l = Axial length of bearing
d = Diameter of shaft (inches)

Calculation of thrust bearing pressure:

$$\text{Pressure (psi)} = \frac{0.4W}{LD}$$

where:

w = Thrust load (lbs)
l = Width of bearing ring (inches)
d = Average pad diameter (inches)

$$cSt = \frac{cP}{\text{Oil specific gravity}}$$

GRAPH 6.3 Relationship between shaft speed, oil viscosity, and shaft loading. (Noria Corporation. ML II, Slide 1563. *Source: Tribology Handbook*, 2nd ed. Neale, Michael, Page C7.2. Reproduced wih permission.)

2. Viscosity selection for plain and element bearings. Correct viscosity selection is the single most important factor in correct lubricant selection and program design. For all types of bearings and lubricated mechanical components, for both oil and grease lubricated applications, the machine's actual operating temperature will dictate which viscosity grade of the selected product will be necessary.

(a) For plain bearing applications, Graph 6.3 shows the relationship between shaft speed, oil viscosity, and shaft loading (PSI) that is useful to determine the minimum oil viscosity at operating temperature. Once the viscosity target is identified there are two approaches that the practitioner could use to determine the correct starting point viscosity selection as measured at 40°C [2].

   Table 6.6 shows an alternate view of the same relationship [3].

   Observing the following steps, the practitioner may use Graph 6.4 to determine the viscosity starting point (which viscosity at 40°C to select) [4].

   Step 1: Identify the required viscosity point on the vehicle axis and draw a line from left to right across the graph.

   Step 2: Identify the operating temperature on the horizontal axis and draw a line from bottom to top across the graph.

   Step 3: Find the point where the two lines cross.

   Step 4: Identify the next higher viscosity point as the minimum viscosity starting point for the select application.

   This chart is designated for paraffinic mineral oils since it reflects viscosities with VI values at 100. It would be simple to create another chart reflecting the viscosities for either higher or lower VI values.

(b) Viscosity selection for element bearings. There are sophisticated equations that may be used to calculate the actual film thickness for a given set of design parameters. These equations are not easily adapted to actual conditions, and as such bearing manufacturers have provided short-cuts that enable accurate selection with easy-to-follow guidance.

TABLE 6.6 Viscosity Extract from ISO Viscosity Classification[a] (ISO 3448) for Sliding Bearings at a Working Temperature Range of 15–60°C

Rotation Speed Range (min^{-1})	Bearing Pressure[b]		
	Light (<7 bar) (<70 N/cm^2)	Medium (7–17 bar) (70–170 N/cm^2)	Heavy (>17 bar) (>170N/cm^2)
5000–10,000	ISO VG: 10		
2000–5000	ISO VG: 15		
1000–2000	ISO VG: 22	ISO VG: 32,48	
500–1000	ISO VG: 32,46	ISO VG: 68,100	
300–500	ISO VG: 68,100	ISO VG: 100,150	
100–300	ISO VG: 100,150	ISO VG: 150,220	ISO VG: 320,460
50–100	ISO VG: 150,220	ISO VG: 220,320	ISO VG: 460,680
<50	ISO VG: 220,320	ISO VG: 320,460	ISO VG: 460,680

Note: These recommendations are for circulating lubrication; thicker oils are needed for loss lubrication.

[a] ISO VG: kinematic viscosity in mm^2/sec at 40°C.

[b] Bearing pressure $p = F_N/b \cdot d$.

Source: Lubricants in Operation, U.J. Moller and U. Boor. 1996. © John Wiley & Sons Limited. Reproduced with permission.

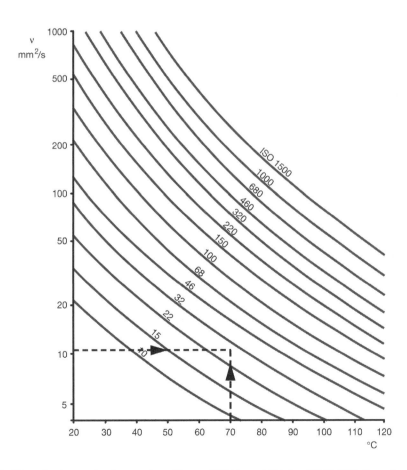

GRAPH 6.4 Viscosity at operating temperature. (*Source*: SKF General Catalog 5000F. 2003. Reproduced with permission.)

TABLE 6.7   Element Bearing Viscosity Selection Criteria

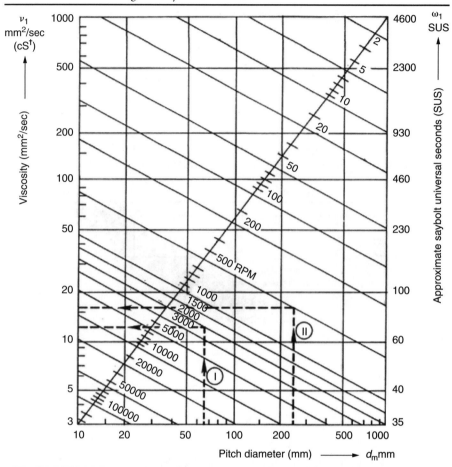

This is a three-step process, as follows:

Step 1. Use Formula 6.1 to estimate bearing Pitch Diameter, d_m,

$$d_m = (QD + ID)/2 \tag{6.1}$$

where QD is bearing outer diameter and ID is bearing bore.

Step 2. Determine shaft rotation speed.

Step 3. Using Table 6.7, locate the intersection between the shaft speed and a line intersection with the calculated d_m value.

Step 4. Draw a line to the Y axis to read the required viscosity in centistokes (mm²/sec).

Step 5. Determine the correct viscosity at 40°C in the same manner as noted above.

The viscosity estimated in the process noted above is the minimum acceptable viscosity for the given conditions to maintain a "fat" oil film in an element bearing. Bearing manufacturers propose that any time the film ratio falls below this level then wear resistant and seizure resistant additives be incorporated to protect surfaces [5].

It is advisable to provide a cushion when making the final oil selection given that change in mechanical, environmental, or production conditions may cause shock loading. The degree of cushion selected should be determined by the type and operation of the equipment, by the general speed of the equipment, and by consideration of other lubricants that may already be available in the work area.

TABLE 6.8 Maximum Bearing Speed nDm Factors

Bearing Type	Oil Lubricated	Grease Lubricated
Radial ball bearings	500,000	340,000
Cylindrical roller bearings	500,000	300,000
Spherical roller bearings	290,000	145,000
Thrust-ball and roller bearings	280,000	140,000

Source: *Lubrication Fundamentals*, Wills George. 1980.© Mobil Oil Corporation. Reproduced with permission.

A manufacturer's general purpose (GP) greases commonly have viscosities ranging between 100 and 220 cSt, depending on these factors. Keep in mind that thicker oils, and thicker grease consistencies, tend to churn, generate heat, and consume energy.

(c) Special consideration for grease lubricated bearings. There are speed limits to the effective grease relubrication of all element bearings. The factor used to gauge the decision to use either grease or oil is called the nDm, which is shown in Formula 6.2.

$$nDm = \frac{(ID + OD) \times N}{2} \tag{6.2}$$

where *ID* is bearing bore, *OD* is bearing outside diameter, and *N* is shaft speed.

Large bearing sizes and high shaft speeds create high nDm values. As the nDm value increases the extent of churning and thickener degradation increase relative to bearing element type. Since each grease will respond differently to the "working" effect that the element produces, and since some greases will soften and some will harden, it is impossible to determine, short of conducting laboratory tests for each grease type, just how well any given grease will withstand work shear and consistency changes.

Bearing element type has an influence on grease life, as shown in Table 6.8. As the calculated nDm value approaches the limits proposed by Table 6.8, the practitioner will need to increase the relubrication interval and be particularly cautious to follow rigorously defined feed rates to prevent churning and lubricant destruction [6].

3. Lubricant supply volume for plain and element bearings. Following selection of the type of lubricant at the appropriate viscometric range, the next exercise is to select the correct volume of lubricant for replacement. As this section deals with developing relubrication practices, the focus will be on replenishment rather than design criteria for sump capacities and distribution system flows.

(a) Oil replenishment volumes are entirely dependent on oil loss through operation, leakage, contamination, or normally scheduled "bleed and feed." Replenishment volumes are dictated by sump capacity requirements. Bearing oil sumps have some mechanism for indicating the normal sump capacity. In most cases there is an external indicator such as an oil bowl, level indicator or gauge, or dipstick.

The lube technician should specify the nature and location of the indicator, and unit of measure if it is not evident from physical observation, but since oil loss is machine dependent, and since machines are not designed to relieve a given quantity of oil through normal use, it is not possible to project before the fact the quantity of oil to be used in replacement.

(b) Grease replenishment volumes are an altogether different story. Grease lubricated applications are nearly always expected to be "continuous loss" systems. As such, planned, consistent, systematic replenishment is necessary to protect the grease in the machine and the machine itself.

• Grease replacement volume for plain bearings is based on the required resupply per hour of operation.

TABLE 6.9 k_g Factor Based on Shaft Speed

Shaft Speed rev/min	k_g
Up to 100	0.1
101–250	0.2
251–500	0.4
501–1000	1

Formula 6.3 used in conjunction with Table 6.9 may be used to calculate the grease replacement volume, as follows [7,8]:

$$Q_g = k_g \times C_d \times \pi \times d \times b \tag{6.3}$$

where Q_g is grease volume per hour, k_g is rotation speed factor (Table 6.9), C_d is diametrical clearance, $\pi = 3.14$, d is shaft diameter, and b is bearing width.

Formula 6.3 should be considered as a reasonable quantitative starting point, but operating conditions should influence actual feed rates. The service factors noted in Table 6.10 could also be used to adjust the time (1 h) for the calculated quantity by multiplying the time value in hours by the actual operating condition factors [9].

This volume should be uniformly distributed during the course of the final time cycle, or to the extent that program management can allow. If the calculated quantity was 6 g/h, then 1 g per each 10 min would be better than 6 g per each 60 min. Continuous feed is desirable.

- Grease replacement volumes for element bearings. Estimating replacement volumes for element bearings is somewhat simpler to do. It should be restated again that the calculated estimates are based on sound design engineering principles, and must be subject to adjustment for "as built" design and operating conditions.

TABLE 6.10 Grease Relubrication Interval Factors for Changing Environmental Conditions

Condition	Average Operating Range	Correction Factor
Temperature	Housing below 150°F	1.0
F_t	150 to 175°F	0.5
	175 to 200°F	0.2
	Above 200°F	0.1
Contamination	Light, nonabrasive dust	1.0
F_c	Heavy, nonabrasive dust	0.7
	Light, abrasive dust	0.4
	Heavy, abrasive dust	0.2
Moisture	Humidity mostly below 80%	1.0
F_m	Humidity between 80 and 90%	0.7
	Occasional condensation	0.4
	Occasional water on housing	0.1
Vibration	Less than 0.2 ips velocity, peak	1.0
F_v	0.2 to 0.4 ips	0.6
	Above 0.4 (see note)	0.3
Position	Horizontal bore centerline	1.0
F_p	45° bore centerline	0.5
	Vertical centerline	0.3
Bearing Design	Ball bearings	10
F_d	Cylindrical and needle roller bearings	5.0
	Tapered and spherical roller bearings	1.0

Note: ips = in./sec, 0.2 in./sec = 0.5 mm/sec.

An SKF formula, Formula 6.4, is a simple and convenient starting point for estimating grease replenishment quantities for element bearings [10].

$$Q_g = D \times B \times 0.005 \qquad (6.4)$$

where Q_g is grease quantity in grams per interval, D is bearing outside diameter (mm), and B is bearing width (mm) (0.114 may be used as the multiplier with English units to provide output Q in ounces.)

Where actual bearing dimensions are not known, a close proximity to the actual suggested value could be estimated by using housing dimensions and factoring again by 0.33. This does not provide exactly the same value for all bearing types given bearing element and construction differences, but it is generally close.

It is suggested that some comparisons should be done per bearing type per bearing manufacturer before proceeding uniformly with this factoring method.

Lubriquip Technical Bulletin #20115 provides formulas and concise direction for determining lubrication volume requirements for bearings and various other types of lubricated components [11].

- Relubrication frequency. The basis for frequency for grease relubricated plain bearings is shown in Formula 6.3, which provides recommended nominal volume per hour of operation.

Formula 6.5 provides a baseline that incorporates operating factors from Table 6.10 for vibration, solid and moisture contaminants, heat, bearing type, and shaft axis, each of which can have an negative impact on the lifecycle and effectiveness of the lubricant in use [12].

$$T = K \times \left[\left(\frac{14,000,000}{n \times (d^{0.5})} \right) - 4 \times d \right] \qquad (6.5)$$

where T is time until next relubrication (h), K is product of all correction factors $F_t \times F_c \times F_m \times F_v \times F_p \times F_d$ (see Table 6.10), n is speed (rpm), and d is bore diameter (mm).

SKF, Timken, and FAG Lubrication guideline publications provide alternate quantitative approaches that are also valid, and could be considered as a strong reference starting point [13–15].

- Lubricant application. There are various methods that may be selected to add the required quantity to the designated component. The high volume and short frequency applications will benefit most from some form of automatic lubrication supply. Semiautomatic (single point automatic lubricator), such as seen in Figure 6.5, or fully automatic (multi-point automatic lubrication system), such as seen in Figure 6.6, provide feed conditions that best meet reliability goals, particularly for high volume and short interval requirements. Most bearings, both element and plain type, can be covered to meet nominal reliability requirements with manual (grease gun) relubrication methods.

4. Time stamp. A nominal value should be assigned to each practice type based on some type of time study. A detailed time study would consist of breaking the whole of the process completely down into individual movement, assigning a typical value for the movement from a standardized table, and then agglomerating the scores. This is useful for highly repetitive tasks. Machine relubrication tasks are repetitive in the sense that the same activities are repeated, but the scale of the repeated activities makes this type of reference prone to errors. A more generalized time assessment can provide sufficient detail to support labor balancing and allocation by department.

(a) Activities that are conducted as stand alone work orders, such as flushing a hydraulic system, should reflect the amount of time required to prepare for the activity, travel to the asset or group of assets, perform the task, travel back, and store materials for reuse at a later date.

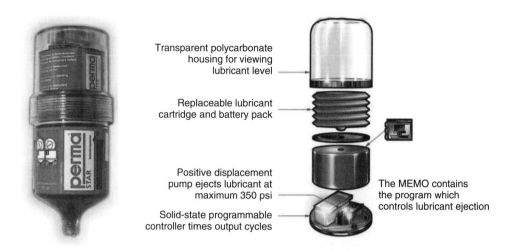

Transparent polycarbonate
housing for viewing
lubricant level

Replaceable lubricant
cartridge and battery pack

Positive displacement
pump ejects lubricant at
maximum 350 psi

The MEMO contains
the program which
controls lubricant ejection

Solid-state programmable
controller times output cycles

FIGURE 6.5   Cross-section of a programmable positive displacement single point lubricator. *Source*: Getting the Most from Single-Point Lubricators, Luis F. Rizo, Elfeo Inc. Reproduced with permission.

(b) Activities intended to be done in sequence with other similar activities, such as bearing relubrication, or topping reservoirs should have an allocation for preparation, travel to and from, and storing gear, but the allocation would be distributed into a series of individual tasks.

Table 6.11 offers guidance for estimated fixed and variable time per procedure type. The fixed value suggests that any time this type of activity is scheduled that this amount of time should be allocated, regardless of the number of the type of activity. This could be considered a one-time charge any time that particular activity is scheduled.

The "Time Per Task" value is a variable for each activity to be scheduled in a sequence in addition to the one time charge as noted above.

If the route required 2 bearing relubrication (greasing) and 1 reservoir change, then the time tally for the route would be:

$$\text{Bearing time allotment} = (1 \times 15 \text{ min}) + (2 \times 3 \text{ min}) = 21 \text{ min}$$

$$\text{Small reservoir change allotment} = (1 \times 15 \text{ min}) + (1 \times 40 \text{ min}) = 55 \text{ min}$$

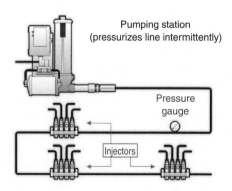

Pumping station
(pressurizes line intermittently)

Pressure
gauge

Injectors

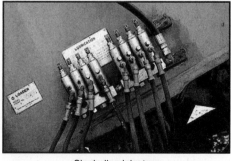

Single-line injectors

FIGURE 6.6   Overview of a single line series automatic lubrication system. *Source*: Automated Lubrication-Benefit & Design Options, Wayne Mitchell/Machinery Lubrications Magazine. 2001. © Lincoln Corporation. Reproduced with permission.

TABLE 6.11 Incremental Time Measurements for Common Lubrication

Lubrication Task Time Estimates		
Task	Fixed Time (min)	Time per Task (min)
Top-up and inspection	15	7.5
Sample (vacuum w/minimess)	15	6
General bearing regrease	15	3
Motor regrease	15	15
Coupling regrease	15	60
Small to medium sump drain and fill	15	40
Small to medium sump drain and fill (w/ filter cart)	15	20
Chain lubrication (aerosol)	15	5
Open gear lubrication (aerosol)	15	10

Once the routes are rolled together these time stamps will play an important role in work balancing to assure coverage.

Following the example for bearings, lubricant selection by type (R&O, AW, EP, Synthetic or Mineral Oil), viscosity (ISO Grade) and/or stiffness (NLGI Grade), and application rate, gear relubrication must receive a similar treatment in the process of designating type, thickness, quantity, and application method in order to develop a precise machine relubrication procedure.

The following sections provide reference points that can be used to find the type of information that will be necessary to develop the requisite level of detail, but will not try to elaborate on the specific details themselves.

Gears: 1. Enclosed Gearing. Excellent advice for selection of lubricant by base oil and additive type, and viscosity is provided by the American Gear Manufacturers Association (AGMA) Technical Bulletins ANSI AGMA 9005 E02 and AGMA 925-A03. These documents provide background and detailed knowledge for viscosity selection under various operating conditions.

2. Open Gearing. AGMA Technical Bulletin ANSI AGMA 9005 E02 provides background and detailed knowledge for product selection, product performance measurement methods, and volumetric application recommendations for open gear applications [18].

Couplings: 1. Excellent advice is available from most coupling manufacturers, including product selection, viscosity, product volume, and application methods per type of coupling. Again, AGMA provides very specific detail on coupling lubrication through Technical Bulletin ANSI/AGMA 9001 B97[19, 20].

Chains: 1. Helpful chain lubrication information may be found at the American Chain Association document "Identification, Installation, Lubrication and Maintenance of Power Transmission Roller Chains in ANSI B29.1 and ANSI B29.3" and "Fundamentals of Chain Lubrication" [21, 22].

Hydraulic Systems: 1. Hydraulic systems are composed of several subsystems that each require specialized consideration for appropriate product selection. The subsystems include the pump, valves, motors, cylinders and other working components, piping, and sump.

Each of these system components may have specialized lubrication requirements that may compete with other systems components based on the design, function, environment, and operation of the hydraulic system. Since the subsystems must function from within the broader context of the whole, some sacrifices may have to be made.

Product type selections can be challenging, but typically resolve to the degree of wear resistance protection dictated by the main system pump, which typically requires AW additives, and viscosity conditions dictated by the pump and working components. Additional attention must be provided for fluid filterability properties in the event that the control system operates with proportional servo control mechanisms.

Particular concern must be given to the dynamic temperature range, particularly if the system operates intermittently and outdoors.

There are a variety of useful, detailed technical resources available in either print or via Web-based electronic print, from a simple overview of system requirements to a sophisticated component-by-component and step-by-step method for evaluating fluid selection [23–25].

6.6 Contamination Control Requirements

Contamination control may be the single largest area for improvement and financial reward in the field of maintenance and lubrication program management. Contamination control interests and technical contribution derive from interests in both the field of hydraulics, the early area of concern for fluid contamination, and tribology.

Hydraulic system contaminants are one of the, if not the single most, biggest contributors to system failure. This position has been widely understood by industry experts for many years. Recently, with the advent of "proactive" maintenance focus, a similar level of interest for contamination control has developed for circulating and noncirculating sumps for other types of mechanical components.

The lubricant supplier strictly speaking has not had a voice in the discussion about fluid contamination control since this specialty has been the domain of hydraulics system component and hydraulic filter component manufacturers. It should nonetheless be part of the fundamental design of an effective machine lubrication care specification.

Abundant evidence exists from leading component manufacturers that shows the relationship between contamination levels and lubricated component failures. Graph 6.5 illustrates the relationship between bearing life and contamination control. Accordingly, the lubrication specialist should be involved in decisions for setting basic contamination control requirements for all systems, even those traditionally not considered candidates for advanced contamination control [26].

Aspects of contamination control improvement considerations fall into two categories: contaminant exclusion and contaminant removal. Actions associated with either will be driven by the reliability ranking of the asset. A high machine reliability priority will justify more extensive development and machine improvement for contamination control.

Contaminant exclusion: The three primary avenues through which contaminants enter tribological systems, whether oil circulation or grease sumps, are breather vent ports, seals and through lubricant replenishment.

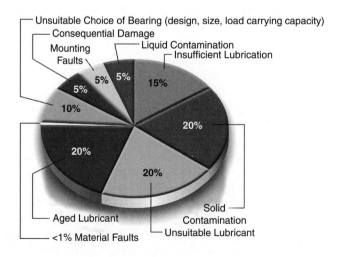

GRAPH 6.5 Bearing failures by failure category. Note: These projected causes of bearing failures are approximations. Failure analysis frequently reveals multiple causes of failure. (From publication No. WL82 102/2ED 12/97. FAG Bearings Corporation. "Rolling Bearing Damage: Recognition of Damage and Bearing Inspection" Printed by: Weppert Gmbh & Company, Originally from Antriebstechnik 18 (1979) No. 3, 71–74.)

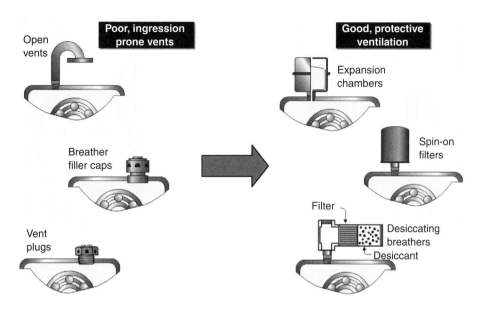

FIGURE 6.7 A single line parallel system, including the controller, reservoir, pump, feed line, and individual injectors.

1. Vent Ports. Oil lubricated circulating and noncirculating sumps must have vent ports to function properly. The vents are designed to allow easy passage of air across a sump opening while minimizing risk of gross, bulk contamination (large particles, process chemicals, rain drops, etc.).

   Most vent ports have threaded openings for easy installation or removal of threaded attachments. Figure 6.7 illustrates a traditional vent port configuration designed for use on a hydraulic or oil circulation reservoir. These types of vents allow for significant ingression of various airborne contaminants, including moisture, and should be upgraded to types shown in the chart. The tighter the contaminant control target (cleanliness target) the higher the quality of the replacement configuration, leading up to Beta designated breather filter elements for the highest priority systems.

2. Shaft and cylinder seal points. Another high ingression point that should be reviewed and modified is the shaft seal or hydraulic piston seal point. Figure 6.8 represents a typical lip seal (top left) configuration contrasted against a typical bearing isolator (bottom right). Lip seals are more useful for containing lubricants than excluding contaminants and only then until the seal condition is degraded due to surface contact with the shaft or three body abrasive wear from contamination.

   Hydraulic cylinder seal and wipers are equipped with similar quality of materials that degrade rapidly once put into service. It should be noted that there are many designs and a great disparity in the quality of materials that may be found on OEM equipped hydraulic system components, based on the quality philosophy of the manufacturer. Performance for these components may be improved through the use of cylinder isolators as shown in Figure 6.9.

   Lubricant handling is another area where minor improvements may provide large dividends. These issues are part of effective lubrication program design, but should be addressed as an item separate from machine relubrication methods development.

3. Passive shields can be installed between lubricated components and sumps and high heat sources as a means for controlling radiant heat transfer. Additionally, passive shields may be used as weather brakes to minimize the effects of direct sunlight, cold, precipitation, or airborne solid debris.

Contamination removal: Filtration has been addressed separately in this text, but should be identified again as a key aspect of precision lubrication care. Selection of filter media follows establishment of contamination control targets and alarm settings, and is an important aspect of the condition control loop.

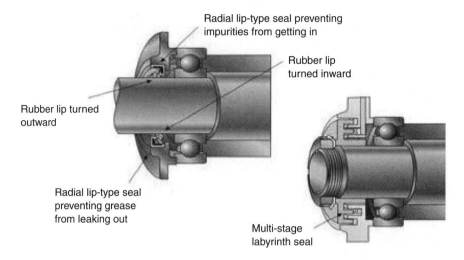

FIGURE 6.8   Comparison of two shaft seals. On the left is a shaft "lip" seal and on the right is a shaft multi-stage labyrinth seal.

Once the target control values are established the lubrication technician should review each system for the existence of moisture, air, heat, and solid contaminant removal systems, and add these according to the level of machine risk.

1. *Thermal control.* Fluid flow around heat exchangers should be reviewed to assure that the area of immediate exposure is not in a dead flow zone and to assure that the heat rate per surface area is low enough to prevent burning of the fluid. Coolers should be reviewed for correct sizing and/or coolant flow conditions. Shell and tube heat exchangers should also be checked periodically for seal and sidewall integrity.

2. *Moisture control.* Reservoirs should be reviewed for evidence of condensation on the tops or sides of the lube oil tanks, and for evidence of moisture accumulation at the bottom of tanks. Super-saturation in the headspace can be controlled through force draft air flow across the head space for large tanks, or through the use of desiccant breathers for small systems.

FIGURE 6.9   Hydraulic cylinder seal options, including the rod boot and the traditional shaft seal and wiper band.

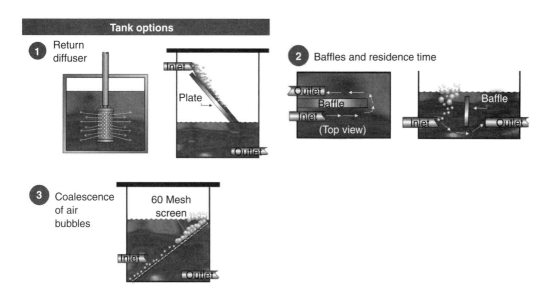

FIGURE 6.10   Options for controlling aeration in hydraulic tanks and other lubricating reservoirs.

3. *Air entrainment control.* Air entrainment, or air saturation, is a problem primarily in systems that have high flows, low fluid dwell times, and where the fluid plunges into the reservoir at the return as suggested in Figure 6.10. These are all issues of system design, and should be addressed as systematic improvement. Symptoms of air entrainment should be noted for consideration during fluid selection and permanent system modifications. Vacuum distillation is useful for a de-gassing system with high entraining levels and may be considered as a useful intermediate step.

4. *Solid contamination control.* Where systems have in-place continuous filtration, the lubrication technician should evaluate the capability of the existing system and either accept or make recommendations for improvements. Where the system, or sump, does not have continuous filtration the technician must make a decision to recommend no change, the addition of side-stream intermittent filtration, or the addition of hard piped continuous filtration. The decision will be driven by the rate of ingression and the system requirements for cleanliness.

   (a) A common condition exists where hydraulic and circulation system designers working independently or alongside the OEM will install filtration capacity that meets minimum component manufacturer operating requirements, but does not meet the owner's reliability requirements. These scenarios are very common.

   The lubrication technician must evaluate each system pressure and return side flow circuits for the presence of full flow element and for the quality of those elements. As noted in the chapter on contamination control, the filter industry relies heavily on statistical quality testing to measure the capture efficiency of individual element designs. Each element should have a beta (β) efficiency rating for the sake of element quality comparison.

   The lubrication technician must verify that the element is β rated at a performance level that meets the expected OEM ISO cleanliness code. Although the β rating does not necessarily guarantee performance (due to the many variables that may exist at any given time for a hydraulic circuit), it is certainly a useful criteria to begin a quality comparison.

   (b) Many sumps exist with cooling or heating systems but no filtration system. If the criticality of these systems is high enough, these systems should be considered for the addition of a low pressure, constant flow filtration circuit either following or in front of the thermal control circuit. Again, any additions should be sized to meet cleanliness requirements with rated (tested) filter designs.

Offline filtration, repairs, and equipment rebuild flushing and
flushing during equipment commissioning

FIGURE 6.11   Example of a portable filter cart, or "kidney loop" filter system used to filter a reservoir during operations.

    (c) Finally, many sumps exist that lack any thermal control or filtration circuits. Bearing sumps and gear sumps commonly do not have these features. Often these sumps have proven their respective toughness through sustained years of production without failure. For these "untreated" sumps, the lubrication technician and reliability engineer should evaluate the system for long-term reliability, and for future production requirements, and decide as to whether the addition of hard piped filter system should be installed or whether the system may benefit sufficiently from the use of intermittent filtration through a side stream portable filter system, as depicted by Figure 6.11.

        A system modification plan should be developed for either the use of fluid quick disconnects or the installation of a continuous flow system, and in both cases the element quality should be evaluated for its ability to meet defined program objectives.

    (d) Where the components are capable of withstanding appreciable contaminant induced abrasive wear, and system reliability is good relative to the production goals, then a clear option is to take no action.

6.7 Oil Analysis Requirements

Oil analysis has been used to help equipment operators monitor oil health, machine wear, and contamination levels. Traditionally oil analysis was conducted by the lubricant supplier to help guide the customer on lubricant change cycles. While this is a worthwhile endeavor itself, there is appreciable value that may also be derived by using oil analysis as a primary control loop for management of the lubricant *insitu*.

    The lubrication practitioner should incorporate oil analysis into the set of machine specific lubrication practices, and integrate these results into the long term plan for *insitu* sump and lubricant management. There are four key areas that the practitioner must address, as follows:

1. Asset selection
2. Test slate selection
3. Alarm types and limits
4. Analysis frequency decisions

Practical guidance for each of these parameters is addressed in the section on used lubricant analysis. A brief review will be provided on each criterion strictly as a means for establishing limits for lubrication practice development.

Asset selection: Oil analysis can be a power tool when applied selectively to assets that provide high production value or cost control value. Early in the exercise the practitioner was directed to review the site's machine criticality assessment to reach an understanding of which of the assets should be viewed more closely, or should receive a disproportionate share of attention for improvements or upgrades.

    Lubricant analysis is a tool that can provide a high return on investment when used properly, providing returns on the order of five to one and higher [27]. Nonetheless, the relative effectiveness of every dollar spent should be considered fully before recommending an oil analysis program.

TABLE 6.12 Gear Lubricants Typical Oil Analysis Test Slate

Test	Enclosed Gears	Open Gears Spray/Splash Lubricated
Viscosity	Routine	Routine
Atomic emission spectroscopy (AES)	Routine	Routine
Water	Exception (3a)	Exception (3a)
Crackle or FTIR		
Karl Fischer		
Particle count	Routine	Not required
Large wear debris[a]	Recommended	Recommended
Rotrode filter spectroscopy (RDE)[a]		
X-ray fluorescence (XRF)		
DR ferrography		
Analytical ferrography	Exception (4, 5)	Exception (4, 5)
FTIR oxidation	Routine	Routine
Acid number (AN)	Exception (7)	Exception (7)

[a] Either RDE, DRF, or XRF should be considered routine to look for large wear particles greater than 10 μm in size.

Assets with high reliability criticality scores that function in aggressive environments (moisture, atmospheric debris, chemical saturation, ambient temperature extremes), that function within operational extremes (heat, cold, constant or intermittent load, isolation), and that have traditionally low MTBF (mean time between failure) scores nearly always warrant the application of an oil analysis routine, regardless of the sump size or preexisting oil sump change intervals. As the criticality score declines, environmental and operating conditions moderate, and overall reliability stabilizes, the relative economic benefit of each dollar applied to oil analysis may be replaced by comparatively higher benefit from other condition assessment methods.

Test slate selection: Following decisions regarding which assets to apply systematic oil analysis, the practitioner next would determine the types of tests to conduct for each asset. The final selection of tests is referred to as a test slate. The test slate will comprise test methods for primary and secondary tests, as was described in the chapter on oil analysis. A gear lubricant test slate is presented in Table 6.12 [28].

Table 6.13 depicts various primary and secondary tests that may be selected depending on the types of measurements that are desired, and the relative strength of the test type for providing useful detail [29].

The most critical assets will receive a broader test slate than the less critical, and will likely receive an additional layer of superficial on-site screens (viscosity, moisture, particle count) on a very frequent basis in support of a more detailed laboratory analysis conducted on a less frequent basis.

Each test type may have a variety of test procedures that could, or should, be considered in arriving at the final definition of test slate methods. The select oil analysis laboratory will likely have constructive feedback to offer on the final choice of ASTM test methods to assign to each parameter in the test slate. These decisions are explored in greater detail in the chapter on oil analysis.

Alarm types and limits: Following the selection of the assets to trend, and the types of tests to conduct, the technician would assign score limits for normal, alert, and alarm status that would be applied to each test. Absolute and statistical mathematical methods are used to calculate alert and alarm score limits. A pass-fail evaluation is generally applied to on-site screens, but depending on the sophistication of the screen, statistical and absolute alert and alarm limits could be applied.

Table 6.14 provides an overview of a system of general industrial and automotive alert and alarm limits.

Analysis frequency decisions: Oil analysis frequency selection, like other aspects of oil analysis program decisions, is driven by machine operational factors. Intervals are shorter where criticality is high and the potential for contaminant level changes is high, particularly as it pertains to solid and liquid contaminants.

TABLE 6.13 Generic Test Slate for Various Machine Types and Production Environments

Methods and Descriptions for Routine Lubricant Analysis	Procedure Standard	Gearboxes	Hydraulic Systems	Circulating Sumps	Non-Circulating Sumps	Gas Turbines	Gasoline Engines	Locomotives and Commercial Diesel Engines	Transmissions	Transformers
1. Elemental Analysis										
a. Wear Metals (Fe, Cu, Pb, Sn, Cr, Al, Ti)	D 4951	R	R	R	R	R	R	R	R	–
b. Contaminants (Si, Na, K)	ASTM D 5185	R	R	R	R	R	R	R	R	–
c. Additives (B, Ba, Ca, Mg, P, S, Ph)	D 6595	R	R[3]	R[3]	R[3]	R[3]	R	R	R	E
2. Kinematic Viscosity										
a. Vis @ 40°C	ASTM D445	R	R	R	R	R	X[1]	X[1]	R	–
b. Vis @ 100°C		X[1]	X[1]	X[1]	X[1]	X[1]	R	R	X[1]	O
3. Moisture Analysis										
a. Crackle Test - Screen	none	R	R	R	R	R	R	R	R	–
b. Karl Fischer Moisture	ASTM D1744 or D6304	E[6]{3a.}	E[6]{3a.}	E[6]{3a.}	E[6]{3a.}	E[6]{3a.}	E[6]{3a.}	E[6]{3a.}	E[6]{3a.}	R
4. FTIR (Fourier Transform Spectroscopy)										
a. Nitration, Sulfation, Oxidation	ASTM E168	R	R	R	R	R	R	R	R	R
b. Soot Concentration		–	–	–	–	–	R	R	–	–
c. Additive Health		R	R	R	R	R	R	R	R	R
d. Other Degradation Byproducts		–	–	–	–	R	E	E	–	–
5. ISO Particle Count	ISO 4406-99, 11500	R	R	R	R	R	E	E	R	R
6. Acid Number (AN)	ASTM D664	E{4a}	R	R	E{4a}	R	–	E	R	R
7. Base Number (BN)	ASTM D4739	–	–	–	–	–	E[4a]	R	–	–
8. Flash Point	ASTM D93	–	–	–	–	–	E[2]{2a.}	E[2]{2a.}	–	–
9. Glycol	ASTM D2982 ASTM D4291	–	–	–	–	–	R	R	–	–
10. Ferrous Density	none	R	E[3,4]{1a}	E[3,4]{1a}	E[3,4]{1a}	E[3,4]{1a}	E[3]{1a}	R	E[3,4]{1a}	–
11. Analytical Ferrography	none	E[5]{10}	E[5]{10}	E[5]{10}	E[5]{10}	–	E[5]{10}	E[5]{10}	E[5]{10}	–
12. Oxidation Stability Performance										
a. RPVOT (Rotary Pressure Vessel Ox. Test)	ASTM D2272					E[7]{4a,c,6}				
b. LSV (Linear Sweep Voltammetry)	ASTM D6971	E[7]{4a,4c}	E[7]{4a,4c}	E[7]{4a,4c}						
c. MCP (MicroPatch Calorimetry)	ASTM WKI3070									
13. Dissolved Gas Analysis (DGA)	ASTM D3612	–	–	–	–	–	–	–	–	R
14. Interfacial Tension		–	–	–	–	–	–	–	–	R
15. Power Factor	ASTM D877	–	–	–	–	–	–	–	–	R
16. Dielectric Strength	ASTM D1866	–	–	–	–	–	–	–	–	R

P = Performance Measurement Test; E = Exception Based Test; R = Routine Test; { } = Exception test is based on a positive of the number in parentheses.

(1) Used to verify multiviscosity oil properties; (2) used to measure for the presence of fuel contamination; (3) conducted following a sharp increase in ferrous wear debris; (4) conducted following a sharp increase in solid particle concentrations; (5) conducted following a rise in PQ and/or DR ferrography; (6) conducted following pass/fail for bulk water; and (7) conducted to assess chemical stability of large reservoirs.

TABLE 6.14   Generic Test Slate Alarm Limits and Alarm Types

Parameter	Units	Critical	Caution	Normal
Chemical index	Index	$+2\sigma$	$+1\sigma$	$< +1\sigma$
Contamination index	Index	$+2\sigma$	$+1\sigma$	$< +1\sigma$
Ferrous index	Index	$+2\sigma$	$+1\sigma$	$< +1\sigma$
DV at 40°C	%	±15	±10	Baseline
FW index	Stat	$+2\sigma$	$+1\sigma$	$< +1\sigma$
ICP — additive	%	±50	±25	Baseline
ICP — wear debris	Stat	$+2\sigma$	$+1\sigma$	$< +1\sigma$
FTIR — O_x, Ni, S	%	±5	±25	Baseline
FTIR — additives	%	±50	±25	Baseline
Fe density (DR, FW, PQ)	Stat	$+2\sigma$	$+1\sigma$	Average
Acid number	Absolute	1.0 >base	0.2 >Inflection	Inflection point
Analytical ferro.	Qualitative	NA	NA	—

Source: M. Johnson and M. Spurlock. Noria Corporation. 2004.

Table 6.15 provides a basis for making systematic sample interval decisions for most common industrial and automotive machine types. The technician should strive to be as consistent as possible when making adjustments from machine to machine to align work schedules.

6.8   Activity Sequencing

Relubrication scheduling programs that include large numbers of assets can become unwieldy unless the whole of the program is divided into manageable parts. How those parts are best managed will vary depending on the size of the production site, the tools available for moving materials between discrete jobs, the timing and logistics associated with relubrication intervals, and the scope of the activity to be conducted. A scheduling approach that may constitute the most efficient method at one site may be a poor choice at another site.

Relubrication activities are fundamentally divided into tasks that are operating condition dependent and tasks that can be conducted without concern for the operating state.

Downtime dependent tasks: Tasks that must be conducted when the machine is in a nonoperating state must be uniquely defined accordingly during the survey process. These tasks will be scheduled around other "downtime" activities for this asset as timing allows. The technician will bundle the identified materials and tools required for the job much like that of repair work order activities.

Scheduling is complicated by the fact that the required task, which might otherwise be done based on a carefully selected schedule, will have to be done around nonscheduled or emergency availability. An example of this is a cooling tower drive sample collection activity. While ideally the machine might be modified to facilitate relubrication regardless of the operating state, until the necessary modifications are made the work must be conducted when the technician can physically access the sump.

Runtime dependent tasks: It is preferable to conduct routine relubrication activities on machines during a running state. This may be considered the default state. If the task may be conducted in either state, then the technician should note that the procedure may be conducted in either state. The majority of activities will be conducted while the machine is in operation.

An ultrasonic assisted bearing relubrication activity would fall into this category. It would be impossible to conduct this type of activity, which is preferable for electric motor relubrication, if the machine is not operating. The oil sample collected from a positive pressure supply line must of course be conducted during a run state. An oil sample from a static line, or a fixed sample port on a reservoir, may be conducted in either state, even though it is preferable to collect oil samples from operating machines.

For machines designated to be lubricated during a running state, the practitioner must next decide whether to schedule activities based on a "whole" machine practice, or based on the nature of the practice to be conducted. Again, the type of machines to be covered will help influence these decisions.

TABLE 6.15   Systematic Method to Select a Sample Frequency Interval

Sample Frequency Generator

1. Select "Best Fit" Default Frequency

Bearings	500 hrs	Gearing, Low Speed	1000 hrs
Chillers	500 hrs	Hydraulics, Aviation	150 hrs
Compressors	500 hrs	Hydraulics, Industrial	700 hrs
Differentials	300 hrs	Hydraulics, Mobile	250 hrs
Engine, Aviation Recip	50 hrs	Transmissions	300 hrs
Engine, Diesel	150 hrs	Turbines, Aviation	100 hrs
Final Drives	300 hrs	Turbines, Gas	500 hrs
Gearing, Aviation	150 hrs	Turbine, Steam	500 hrs
Gearing, High Speed Industrial	300 hrs		

Write Default Here
Default
Hrs

2. Score Application Adjustment Factors

Economic Penalty of Failure - Circle Factor

Very High				Normal				Low
0.1	0.25	0.5	0.75	1.0	1.25	1.5	1.75	2.0

Consider downtime costs, repair costs, and general business interruption penalty.

Fluid Environment Severity - Circle Factor

Very High				Normal				Low
0.1	0.25	0.5	0.75	1.0	1.25	1.5	1.75	2.0

Consider pressures, load, temperature, speed, contaminants in oil, and duty cycle.

Machine Age - Circle Factor

Infant				Middle Age				Old Age	
0.1	0.5	1.0	1.5	2.0	2.0	2.0	1.5	1.0	0.5	0.1

Infant machines are those going through break-in and have operated for less than 1% of expected machine life. Old age machines are those showing symptoms of distress.

Oil Age - Circle Factor

Infant				Middle Age				Old Age	
0.1	2.0	2.0	2.0	2.0	1.5	1.0	0.5	0.25	0.1

Infant oils are those that have just been changed and are less than 10% into expected life. Old age oils are showing trends that suggest additive depletion, the onset of oxidation, or high levels of contamination.

Place Lowest
Circled Factor Here
Adjustment Factor

Target Tightness - Circle Factor

Tight				Normal				Loose
0.1	0.25	0.5	0.75	1.0	1.25	1.5	1.75	2.0

Oil properties that trend extremely close to targets and limits are "tight". Oils that typically trend well within targets and limits are loose. For instance, an oil with a cleanliness target of 13/11 and trends around 13/11 is tight.

3. Sample Frequency = Default x Adjustment Factor

Sample Frequency
Hrs

 Copyright © Noria Corp.

1. *Whole machine scheduling.* Where the machine might be scheduled out of production for a period of time to receive routine care, including relubrication, the sum of the tasks to be conducted may be scheduled to be done all at the same time.

    Industrial vehicle inspection and relubrication activities are conducted based on miles or hours of operation, depending on the type of vehicle. A mine haul truck would likely have schedules based on hours of operation. If the truck has a diesel engine, the schedule might be based on 250 to 500 h operation intervals, which would coincide with suggested oil change intervals. With each passing unit of hours, additional tasks would be added, leading up to comprehensive depot level overhauls. During these planned events the practitioner would prepare to cover all the required lubrication related tasks at one time.

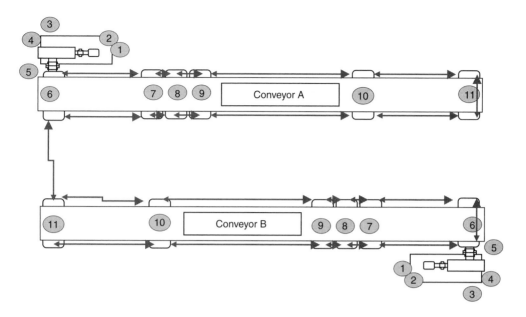

FIGURE 6.12 Top-view of a typical drive train relubrication sequence.

2. *Practice type scheduling.* Contrast this schedule method to a site utilizing conveyors to transport materials rather than haul trucks. Earlier in the process each practice type was defined, the specific lubricant type was defined, and the frequency was defined, all based on sound engineering principles. Once the data has been collected and tabulated, the data can be sorted and ordered in an efficient work flow based on the products, type of activities, and the frequencies that have been defined as appropriate.

For a conveyor drive train the practitioner faces about seven fundamental types of tasks to provide relubrication coverage on a conveyor, including motor lubrication, coupling lubrication, bearing lubrication, routine level check, sampling, filtration, and oil change.

In this example, the belt pulley bearing lubrication activities could be grouped into a route. They are more similar than dissimilar. They will likely require a single type of product for a large number of components. When possible the activities should be bundled, and then set into a linear sequence, and this sequence would be set into a route, or a series of routes, to minimize the number of products carried and the number of steps walked to fulfill the required task.

Figure 6.12 shows a top view of two conveyors sitting adjacent in a material transfer process. Each number depicts an identification of a lubricated component in the drive train, and the arrows depict the preferred sequence to conduct the collective tasks. It would be possible to lubricate the conveyors based on a sequence from A1 through the order of numbered components, but this sequence would create unnecessary steps and consume valuable labor resources.

Similarly, all common component tasks should be delineated such that common functions with common materials on common frequencies may be grouped together and executed using the least amount of time. Examples of routes by common "ingredients" could include:

- 1 month grease routes, grease Z
- 1 month grease routes, grease Y
- 1 week gear drive level check and oil top-up route, gear oil 220
- 1 week hydraulic level check and top-up route, hydraulic oil AW68
- 6 month ultrasonic motor relubrication route, grease W
- 3 month sample route

6.9   Planning and Scheduling Management

Finally, once the routes have been set, walked down, and confirmed, the lubrication technician is ready to begin scheduling the necessary work. Key concerns with lubrication planning and scheduling are completeness, work balancing, and data management.

Design completeness: The lubrication program is not complete until every lubricated component is accounted for, regardless of its relative criticality factor. A cursory check of each asset in the completed database is a minimum requirement. A detailed check should be in order for the high criticality items.

Work balancing: The time stamps for each activity can now be rolled together to reflect the total estimated time for a series of activities. These "route" time stamps can then be used for effective labor allocation and balancing within a work cell or for the entire facility.

Work schedule and work order generation: Modern industrial complexes use computerized maintenance management systems (CMMS) to track work activities by asset type where possible. The CMMS must provide sufficient data (hierarchal) structure so that the individual component can be grouped or tracked. This is often not possible without significant reordering of the CMMS structure, which is not likely to happen. There are several stand alone lubrication scheduling programs that could be used to manage the work process (open, issue, verify, close routes) to the detail level following scheduling orders from the CMMS programs. This does create added complexity to data logging and tracking, and the decision to run two programs will have to be considered on a case-by-case basis.

Work verification: An important final step in assuring effective lubrication program management is verification and tracking of the activities. A lubrication program lead or supervisor should periodically spot check to verify completeness of all scheduled activities, and follow up on notes provided by the lubrication technicians on the printed work orders.

Where a collaborative process exists the skilled lubrication technician becomes the eyes and ears of mechanical supervisors, and can provide direction that will help eliminate "sunk" costs associated with poor equipment configuration and maintenance hostile machine designs.

6.10   Conclusion

The lubrication engineer or technician has many responsibilities that must be covered during the course of devising a machine care program that may be considered best practice. Traditionally, the lubrication engineer has focused the majority of his or her effort on matching the operating requirements of the machine with the appropriate lubricant viscosity, the appropriate grease stiffness, and the appropriate lubricant lifecycle qualities relative to the actual operating characteristics.

Recently additional focus has been applied to developing a comprehensive volume, frequency, contamination control, oil analysis test slate, and thorough but superficial inspection criteria, all designed to promote improved reliability, production quality, and long-term utility. The lubrication engineer must begin the exercise with broadly applicable quantitative tools, and then refine all recommendations based on experience and product capabilities knowledge.

A properly devised "best practice" will incorporate operation and machine specific requirements including:

- Machine criticality and operating environment
- Data collection strategies
- Machine data collection criteria
- Lubricant type, quantity, and frequency requirements
- Contamination control requirements
- Oil analysis requirements

- Activity sequencing
- Planning and scheduling management

Competitive advantage is not limited to acquiring quality production machines or quality maintenance materials. Knowledge will play a vital role for the lubrication program manager and the designated lubrication technicians in achieving maximum value from each dollar invested in the plant tribology program. Leadership and project management skills would also serve the modern lubrication technician. Business planning, skills development, and project management are becoming increasingly vital as well as more and more sophisticated production machines, operating at higher loads and higher speeds, are replacing older, less reliable and less productive assets.

References

[1] Lansdown, A.R. *Lubrication and Lubricant Selection: A Practical Guide.* Antony Rowe Ltd., Chippenham, Wiltshire, 1996.
[2] Neale, Michael. *Tribology Handbook*, 2nd ed., p. C7.2.
[3] U.J. Moller and U. Boor, *Lubricants in Operation*, John Wiley & Sons, New York, 1996, p. 111.
[4] U.J. Moller and U. Boor, *Lubricants in Operation*, John Wiley & Sons, New York, 1996, p. 116.
[5] SKF Corporation Bearing Maintenance and Installation Guide, Page 29. February 1992.
[6] Booser, Bloch — (Secondary Reference: MLII, Slide #1557.)
[7] Neale, Michael, *Tribology Handbook*, 2nd ed. p. A7.5.
[8] Neal, Michael, *Tribology Handbook*, 2nd ed. p. A7.5.
[9] Luegner, Tex, *Practical Handbook of Machinery Lubrication*, 2nd ed.
[10] SKF Corporation Bearing Maintenance and Installation Guide, p. 29. February 1992.
[11] Lubriquip Technical Bulletin #20115 (http://www.lubriquip.com/pdf/20115.pdf).
[12] Luegner, Tex, *Practical Handbook of Machinery Lubrication*, 2nd ed.
[13] FAG Roller Bearing Lubrication Guideline WL81115E. http://www.fag-industrial-services.com/gen/download/1/15/40/37/FAG_Rolling_Bearing_Lubrication_WL81115E.pdf.
[14] Timken Bearing Company. http://www.timken.com/industries/torrington/catalog/pdf/general/form640.pdf.
[15] SKF Bearing Company. http://mapro.skf.com.
[16] Perma Corporation Perma Star Single Point Lubricator. (http://www.permausa.com/sv.htm)
[17] Noria Corporation Machinery Lubrication Seminar ML I (Lincoln Industrial Corporation. http://www.lincolnindustrial.com).
[18] http://www.agma.org/site_search_results.asp (Select "publications," then "ANSI-AGMA Publication Standards," then search for "9005-E02"). The same Web reference provides links to the AGMA 925 standard as well.
[19] Technical Bulletin ANSI/AGMA 9001 B97. http://www.agma.org/site_search_results.asp (Select "publications," then "ANSI-AGMA Publication Standards," then search for "9001 B97").
[20] Falk Corporation. Installation and Maintenance of Double and Single Engagement Gear Couplings Technical Bulletin 458-110, http://www.falkcorp.com/dist-info/main-listofpublications.asp.
[21] http://www.americanchainassn.org/ACAPubs.htm.
[22] Wright, John L., "Fundamentals of Chain Lubrication." *Machinery Lubrication Magazine.* http://www.machinerylubrication.com/article_detail.asp?articleid=316&relatedbookgroup= Lubrication.
[23] http://www.machinerylubrication.com/article_detail.asp?articleid=277&relatedbookgroup= Hydraulics.
[24] http://www.maintenanceresources.com/Bookstore/FluidPower/HandbookHydFluid.htm.
[25] http://www.insidersecretstohydraulics.com/index2.html.

[26] ML I. Slide # 1218a. Ref.: SKF, Pioneer, RP, Idcon, others.

[27] Evans, John and Wearcheck, S.A. *How to Calculate the Effect of Oil Analysis on the Bottom Line.* *Practicing Oil Analysis*, July 2004.

[28] Walsh, Dan, "National Tribology Services. Gear Lube Test Slate Selection." *Practicing Oil Analysis Magazine.* March 2002.

[29] Barnes, Mark, Noria Corporation, "Reality Check: Time to Retool Your Test Slate." *Practicing Oil Analysis* Magazine. January 2003.

7

Lubricant Storage, Handling, and Dispensing

Mark Barnes
Noria Corporation

7.1 Introduction

The impact that inappropriate storage of new lubricants can have on equipment reliability and longevity cannot be overstated. Put simply, an inability to control the quality of a lubricant while in storage, or allowing contamination ingress due to poor storage practices can result in a diminished life expectancy for all equipment, from the smallest grease lubricated bearings to the largest steam or gas turbine system.

Likewise, effective handling and dispensing of lubricant in the plant has an equally vital role to play on the impact that in-service lubricants can have on component life. Exposing lubricants to environmental contaminants such as water and particles during basic lubrication procedures such as oil top-offs or regreasing can compromise both the effectiveness of the lubricant itself and insure that those contaminants are introduced into the oil wetted path, resulting in many commonly encountered failure mechanisms such as abrasion, erosion, surface fatigue, and corrosion.

In this chapter, we explore those factors that must be considered in order to insure that new lubricants retain their ability to provide effective lubrication from the initial receipt of a lubricant, to putting that lubricant into service.

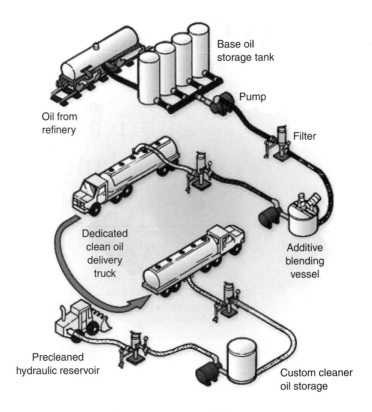

FIGURE 7.1   Typical process for lubricant formulation and bulk transfer.

7.2   Receiving New Lubricants

7.2.1   Lubricant Manufacturing and Delivery

The steps involved in the formulation and ultimate delivery of a lubricant to any industrial facility are illustrated in Figure 7.1. In general, lubricating fluids are manufactured by taking appropriately selected base oils and blending with appropriate additives before delivering to the end user in bulk (as illustrated in the figure) by tote, drum, or individual containers such as plastic quart containers commonly used for passenger vehicle oils. Contamination and base oil/additive degradation can occur at any step of this process.

While the lubricant manufacturer is charged with insuring that the selection of base oils and additive package meets the required performance characteristics of the lubricant in question, the manufacturer is also responsible for insuring an equally important characteristic — that the formulated lubricant is delivered to the end user without levels of contamination that might compromise the integrity of the fluid before it even goes into service. Even before delivery to the end user, significant levels of contaminants, including heat, moisture, and particles can cause premature lubricant degradation in storage, at the manufacturing facility, at the blend plant, or at the lubricant distribution warehouse.

Upon delivery to the end-user, the degree to which these contaminants should be excluded will be critically dependent on application. For example, a highly critical hydraulic application, with very tight tolerances and high pressures will obviously require a significantly higher degree of contamination exclusion than a high viscosity open gear oil, to be used on a noncritical application. Nevertheless, whatever the application, it is still incumbent upon the manufacturer to provide some degree of contamination exclusion.

The onus for new fluid cleanliness cannot however be put squarely on the shoulders of the lubricant supplier. Even if the supplier delivers lubricants below target levels of particle or moisture contamination, oftentimes poor practices at the plant site can compromise the efforts of the lubricant supplier to insure product integrity. Likewise, the cost to control contaminants throughout the manufacturing and delivery process must also be considered. While filtering oil to achieve a desired fluid cleanliness level is always possible, the cost to achieve this level of cleanliness, only to be compromised by poor practices once onsite may render this approach less than optimal. As an example, the typical cost associated with delivering a hydraulic fluid to a cleanliness rating of ISO 16/14/11 (per ISO 4406:99 [1]) is typically around $0.25 per gallon over and above the cost of the lubricant itself.

In order to balance cost vs. benefit, it is generally good practice to set two levels of fluid cleanliness, one for new oil deliveries and one for in-service oils. Typically the new oil cleanliness requirements should be set 1–2 ISO range codes higher than the in-service targets as a compromise between cost and acceptable new fluid cleanliness levels.

Water is perhaps a more harmful contaminant for stored lubricants than particles. While it is true that certain solid contaminants can cause adsorptive additive depletion along with promoting catalytic degradation at higher temperatures, the effects of water washing of additives, along with hydrolytic effects make water a far bigger threat for stored lubricants. It is generally recommended to set maximum permissible water levels below the saturation point of the fluid at all ambient temperatures to which the fluid will be exposed during manufacture, storage, transportation, and dispensing. For some lightly additized oils such as turbine oils, this may mean less than 100 ppm or 0.01% (v/v).

7.2.2 Lubricant Packaging

New lubricant deliveries can be separated into two basic categories — bulk or prepackaged. While bulk lubricant deliveries are typically provided via either tanker truck or rail car, often thousands of gallons at a time, prepackaged lubricants are supplied in a host of containers from large 250 to 300 gallon totes, to barrels (either steel or plastic), or in single use containers such as the plastic quart cans commonly used to sell passenger vehicle engine oils and other fluids.

The choice between bulk vs. prepackaged delivery will to a large extent be governed by cost and usage patterns. For example, while the cost of bulk delivery may range between $0.20 and $0.35/gallon, the associated cost of tote or barrel delivery may range anywhere from $0.35 to 0.55/gallon, depending on the volume of oil used per year. Beyond cost, other deciding factors include cleanliness requirements, storage stability, environmental concerns, and achieving appropriate inventory stock rotation as indicated below. Table 7.1 illustrates the relative advantages and disadvantages of bulk lubricants vs. prepackaged products.

7.2.3 Storage Stability and Inventory Control

The life of any lubricant is strongly influenced by the ambient environmental conditions under which it is stored. Stored outside where the potential for moisture to enter the oil via rain, snow, or humidity, or where high or low summer or winter temperatures are experienced can reduce the life expectancy of any lubricant to a matter of months. By contrast, highly stable oils such as turbine oils, stored in climate

TABLE 7.1 Comparison of Bulk vs. Pre-Packaged Lubricants

	Bulk	Tote	Barrel	One-Shot Container
Risk of supplier cross contamination	High	Medium	Low	Low
Risk of contamination ingress during handling	Medium	Medium	Medium	Low
Storage stability	Medium	Medium	Medium	Medium
Safety risk of handling	Low	Medium	Medium	Low
Environmental spill risk	High	Medium	Medium	low
Distributor inventory aging risk	Medium	Medium	High	High
Handling cost	Low	Medium	Medium	High

TABLE 7.2   Factors Affecting Storage Life of Lubricants

Variable	Increase Storage Life	Shorten Storage Life
Base oil	Highly refined mineral oils, synthetic hydrocarbons, and inert synthetics like silicon-based oils	Lower-grade mineral oils and inorganic esters
Additives	R&O additives	EP additives
Thickener	No	Yes
Storage temperature	Low	High
Temperature variability	Low	High
Container	Plastic containers or liners	Metal drums, especially poorly conditioned ones
Humidity	Low	High
Agitation	Low	High
Outdoor storage	No	Yes

controlled conditions can be expected to last literarily for years. While little research has been conducted to determine the absolute effects of poor storage on lubricating oils and greases, it is generally acknowledged that extreme temperatures, particles, and moisture all contribute to premature oil degradation.

Most at risk are more highly additized oils such as engine oils, transmission fluids, and tractor oils. These oils contain such high concentrations of additives that any slight change in temperature — particularly low winter temperatures — can cause additives to drop out of suspension. Table 7.2 provides a general description of factors that shorten the life of stored lubricants.

Water is perhaps the most deleterious of all contaminants. Certain additives such as zinc-based antiwear additives and organo-phosphates are prone to a hydrolytic reaction rendering these additives ineffective. While this effect is unlikely to occur at lower temperatures, storing oils in highly humid environments above 110° F can lead to significant degradation of lubricating oils.

Other additives are also affected by the presence of water, due to the effects of water washing. Water washing involves the dissolution of the additives preferentially in water compared to the oil. Under these circumstances, these additives, which include organometallic additives such as calcium and magnesium-based detergents and borate-type extreme pressure (EP) additives can be stripped from the new lubricant.

Some additives in new formulations are not completely dissolved in the oil. When the oil reaches service temperatures these additives may finally dissolve, a process known as "bedding in." Other additives by design will never dissolve. For example, some gear oils may be formulated with solid additive suspensions such as graphite, molybdenum disulfide, or borates. These oils should not be stored for prolonged periods because the solid additives are prone to settling and should always be agitated before being put into service.

Even oils that are not prone to chemical degradation or settling can also be compromised by prolonged exposure to heat or water. Stored in steel drums or tanks, the formation of rust can cause rust scale to form, resulting in particle contamination in the oil. Under extreme levels of water contamination, other effects including bacterial growth can also occur, resulting in premature oil degradation and reduced lubricant life once put into service.

In general, long-term storage at moderate temperatures and low humidity has little effect on most premium lubricating oils, hydraulic fluids, and process oils. However, some products may deteriorate and become unsuitable for use if stored longer than three months to a year from the date of manufacture. Table 7.3 provides general guidelines for the maximum amount of time a lubricant should be stored to avoid performance degradation under normal conditions (clean and dry) and moderate temperatures (60°F to 80°F). If a product exceeds its maximum recommended storage time, it should be sampled and tested to confirm it is fit-for-purpose, using the appropriate performance tests outlined below.

Even if an oil is stored appropriately and is deemed to be relatively "storage stable" based on the preceding information, it makes sense to control inventory levels such that excessive quantities are being stored for prolonged periods of time. Likewise, for those products that are constantly being consumed and

TABLE 7.3   Typically Recommended Storage Life for Lubricants

Product	Recommended Maximum Storage Time, Months
Lithium, lithium complex, and polyurea greases	12
Calcium complex grease	6
Motor oils, gear oils	6
Fluids or lubricants with solid additives	3
Turbine oils, hydraulic fluids, R&O oils	18
Emulsion-type hydraulic fluids	6
Soluble oils	6
Custom blended soluble oils	3

replenished, a first-in-first-out inventory control procedure should be adopted so that the oldest stored product is used first.

Concurrent with this, all stored lubricants should be labeled with salient information such as date of manufacture, date of receipt, and a "use-by" date (based on Table 7.3), as well as product information including manufacturer, brand and grade, and inventory control levels (so-called max/mins).

7.3   Storing Lubricants

7.3.1   Bulk Tanks

Bulk tanks offer perhaps the most convenient storage of large quantities of lubricants. Properly designed and maintained, and used with appropriate transfer containers, there is no reason why bulk tanks cannot provide the same degree of fluid handling precision and cleanliness that prepackaged containers provide. Several American Petroleum Institute (API) guidelines exist that relate to the design, fabrication, and commissioning of bulk storage tanks [2,3].

Tanks can be constructed from a number of materials, including stainless steel, mild steel plate, and anodized aluminum. While stainless steel and anodized aluminum are typically higher in price, they typically offer significantly lower maintenance cost over the long term. Mild steel tanks, though the least expensive, are prone to rust and corrosion. As such, it is advisable to treat the inside with a rust preventative or other corrosion prevention coating such as plastic or epoxy.

All storage vessels need appropriate vents or breathers to allow air to enter or exit the tank during oil filling/draining. While it is important that these breathers have a high air flow capacity, it is equally important that they offer a fine enough particle removal size rating to maintain appropriate levels of fluid cleanliness. Quality tank and vent breathers typically offer in excess of 98% particle capture efficiency at 3 μm. Where bulk tanks will be housed outside, or in highly humid environment, any vent or breather should also contain a desiccating material such as silica gel to help remove any trace of water vapor from the air as it enters the tank.

All tanks need to be installed with environmental considerations in mind. In the United States, these guidelines can be found under EPA guidelines CFR280 [4]. Perhaps the most fundamental environmental requirement is the need to have sufficient spill containment to house the complete volume of lubricant that can be stored in the tank. Common practice is to line the bottom of any spill containment with an absorbent such as sand (Figure 7.2).

In order to maintain appropriate levels of fluid cleanliness both inside the tank and during usage, oil should be filtered into and out of the tank. The filter should be rated to maintain appropriate levels of fluid cleanliness based on the application to which the fluid is to be supplied. Tanks should be configured with quick connect to allow for filling and dispensing without exposing the lubricant to the outside environment and any transfer pipes or hoses capped immediately after use.

In addition to contamination control measures, the tank should have an appropriate oil level gauge, a drain to allow for periodic flushing, inspection hatches to allow for entry and cleaning (*note*: confined space entry rules apply here), and an appropriate oil sample valve. For very deep tanks, particularly when

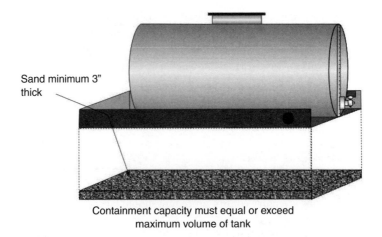

Sand minimum 3" thick

Containment capacity must equal or exceed
maximum volume of tank

FIGURE 7.2   Spill containment recommendations for bulk storage.

lubricants will be stored for extended periods, a series of sample valves at different heights can also be useful to test for signs of additive stratification. Detailed information on best practice for bulk lubricant storage and handling can be found in References 4 and 5.

7.3.2   Totes

Intermediate bulk storage containers (Figure 7.3), often referred to as "totes," offer many of the advantages of barrels, but with the convenience of receiving large shipments direct from the manufacturer. These containers typically hold 250 to 300 gallons of lubricants and are typically fabricated from stainless steel, plastic, or disposable cardboard. Many tote bins come with liners insuring appropriate contamination control. The use of liners also negates the need for secondary external spill containment since the vessel itself provides adequate protection.

Many of the comments regarding bulk storage tanks such as the use of filters to dispense lubricants, appropriately rated vents and breathers, and level gauges apply to tote bins, which are often connected via hard piping, serving almost as a semipermanent bulk tank.

7.3.3   Barrels

Barrel storage is perhaps the most commonly used lubricant storage method. They are available in either steel or plastic, typically holding 55 gal or 205 l. Because of the weight of the barrel and fluid, a typical barrel may weigh in excess of 450 lb, making safe handling and lifting a priority.

Steel drums are the most commonly used. With care, they are reusable, convenient, and inexpensive though they are prone to impact damage (e.g., from fork-lifts or dropping) and internal corrosion. Several manufacturers supply lubricants in plastic barrels. These are typically cleaner when new, do not rust, and are somewhat lighter than steel. Because they are less rigid than steel, plastic drums can be difficult to handle when full. Caution should also be exercised when pumping oil from plastic barrels since the flowing fluid can cause static charges to build-up, resulting in potential fire and explosive conditions. Grounding straps should always be used when dispensing fluids via a transfer pump from a plastic drum.

Storage of drums and barrels outside should be avoided at all costs. The main reason for this is due to the effects of change in ambient temperature and humidity. Even when sealed, barrels have a tendency to breathe (Figure 7.4). During the daytime, heating of the oil and air in the head space of the barrel cause air inside the barrel to escape, while at night when the temperature drops air is drawn back into the barrel. This "breathing" effect causes airborne contaminants, particularly moisture, to be drawn inside the barrel, causing premature lubricant contamination and potential degradation.

FIGURE 7.3   Example of a tote storage tank.

Where outdoor storage cannot be avoided, barrels should be either covered with plastic drum covers in a custom-made free-standing enclosure such as that shown in Figure 7.5, or on their side, with the bungs at 3 o'clock and 9 o'clock so that the air in the head space is at the top of the barrel, and air is unable to escape (Figure 7.6).

7.3.4   One-Shot Containers

Small so-called one-shot containers such as those used for passenger vehicle engine oils and other fluids offer the greatest flexibility in terms of being able to take a sealed, new lubricant to the point of application.

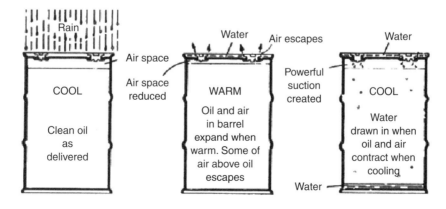

FIGURE 7.4   Breathing of barrels.

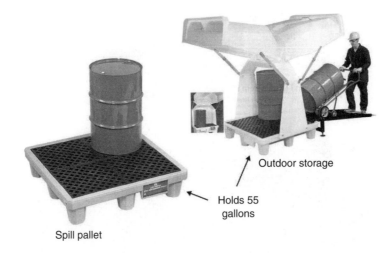

FIGURE 7.5   Free-standing storage enclosure for 55 gallon drums.

This insures that the opportunities for contamination ingression are limited. Likewise, the typically new oil cleanliness levels are much better with these containers while the potential for cross contamination is minimal. While the volume of lubricants used for many industrial applications preclude the use of such containers, wherever feasible, one-shot containers are perhaps the best option.

7.3.5   Grease Storage

Greases come supplied in drums, pails, and small disposable cartridges. Just like lubricating oil, storage stability, product integrity, and cleanliness are the most important considerations.

Beside contamination and environmental stresses outlined earlier, one other major factor to consider with greases is separation of the base oil from the grease thickener while in storage. Under normal conditions (average temperature and humidity) most grease will shed some oil. Whenever this is observed on top of a pail or barrel, it is common and good practice to stir this oil back into the grease. Excessive separation should be considered a cause for concern and should be thoroughly invested with specific ASTM performance tests.

Contamination is a particular concern with greases. Unlike lubricating oils where contaminants can be readily removed via filtration, once a grease is contaminated, it is a virtual certainty that these contaminants will be introduced into equipment if the grease is used. This can result in premature equipment failure due to abrasion and fatigue.

Similarly, cross contamination of different greases through using transfer tools or grease guns for two different products should be avoided. Many different grease thickeners are considered to be incompatible

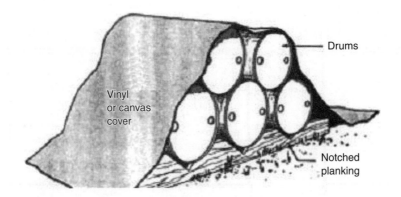

FIGURE 7.6   Temporary storage of barrels. (Note the location of the bungs at 3 and 9 o'clock.)

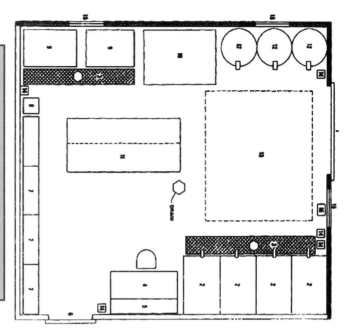

Oil house layout. Some of the features of this layout are as follows: 1 and 6. self-closing doors wide enough for passage of lift trucks or other material handling equipment. 2. Drum racks. 3. Grating and drain. 4. Desk. 5. Filing and record racks. 7. Individual lockers or storage cabinets. 8. Waste disposal container. 9. Solvent cleaning tanks. 10. Purification equipment, soluble oil mixing equipment, or other special equipment. 11. Cabinets and racks for equipment, supplies and small containers. 12. Grease drums with pumps. 13. Parking area for oil wagons, etc. 14. Fire extinguishers. 15. Ventilators. 16. Container of sawdust or other absorbent.

FIGURE 7.7   Typical layout of a lubricant storage room.

and can lead to excessive softening of thickening once the cross contaminated grease is put into service. Just like lubricating oils, greases should be stored with pertinent information such as date of manufacture, date of receipt, "used-by" date (based on Table 7.3), and product data including manufacturer and brand.

7.4   Lubricant Storeroom Design

Ideally, all lubricants should be stored inside in a climate (temperature and humidity) controlled environment. In addition to climate control, the design and layout of the store room should be geared for efficient and ergonomic work flow, with first-in-first-out principles in mind. Figure 7.7 shows an ideal layout for a lubricant storage room, while Figure 7.8 shows the main features to be considered when designing an ideal storage room.

Any lubricant storage room should be designed with enough working space, as well as storage space for as many commonly used lubricants as necessary. In addition, the room should be designed to provide for spill contamination and to comply with any local, state, or federal environmental regulations. Safety considerations, such as explosion-proof lighting and a fire suppression system should also be factored into the design.

In some cases, the provision of small self-contained storage tanks is warranted (Figure 7.9). In this case, new lubricants are pumped from a 55-gal drum into the storage tank. In this case, the oil should be filtered both into the storage tank and from the tank during dispensing using an appropriately rated filter. To avoid cross contamination, dedicated transfer lines, pumps, and filters should be used for each lubricant type and grade, while each tank should have an appropriate desiccant and particle removing breather, either dedicated to each tank (preferred) or gang mounted as shown in Figure 7.9.

7.5   Dispensing Lubricants

The final step in insuring the integrity of new lubricants before they go into service is to insure that the lubricant quality and cleanliness is maintained during dispensing and application. In general, it is good practice to dispense lubricants using a filter cart dedicated to the type and grade of lubricant in use.

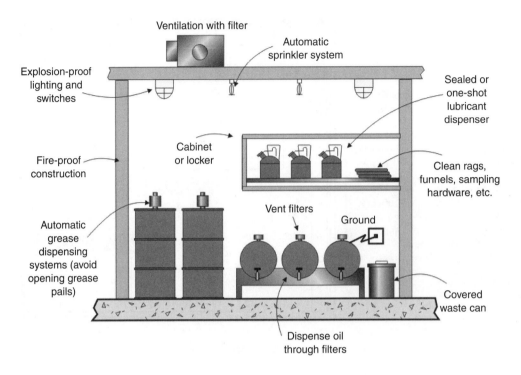

FIGURE 7.8   Key components of a lubricant storage room.

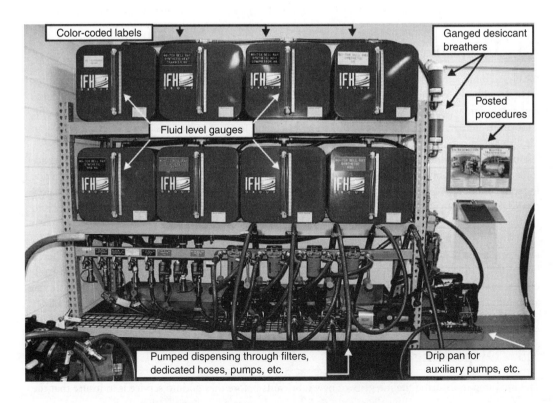

FIGURE 7.9   Ideal configuration of bulk storage containers.

FIGURE 7.10 Skid mounted drum carrier and integral filter transfer cart.

The filter cart should be equipped with a suitable particle removing filter in order to achieve and exceed the desired ISO solid contamination code for any specific application. A variety of appropriate transfer containers and skid mounted systems such as those shown in Figure 7.10 are commercially available, which allow for dispensing fluid using a filter transfer cart.

In some applications such as large volume circulating systems (e.g., turbine), it makes more sense to dispense directly from a bulk storage tank. Under these circumstances, provision should be made to hard pipe to supply lines, again including an appropriately rated filter to insure fluid cleanliness. In either case, as stated earlier, the bulk tank or drum should be equipped with an appropriate desiccant and particle removing breather to insure the cleanliness and integrity of the bulk stored fluid, as air enters the storage vessel and the fluid is dispensed.

For applications where smaller oil volumes are required, for example, for equipment top-offs, or where accessibility is an issue, dispensing from a barrel or from bulk may not be feasible. Under these circumstances, top-off or intermediate transfer containers will undoubtedly need to be used.

Historically, many transfer containers in use are made from galvanized steel. While simple and inexpensive, the use of galvanized containers and other dispensing equipment such as funnels should be avoided since some lubricants, most notably those containing zinc-based additives, have a tendency to react with galvanizing. Instead, rigid plastic containers, such as those shown in Figure 7.11, which include O-ring seals on both the lid and fill spout should be used.

Inevitably, these top-off containers need to be filled either from a bulk dispensing station such as that shown in Figure 7.9 or directly from the barrel. When filling directly from barrels, the barrel oil should be dispensed either using a filter cart or, for lower viscosity fluids, using gravity through an appropriate filter as shown in Figure 7.12. Again, continued cleanliness of the oil in the barrel is insured by the use of an appropriate vent breather.

When using top-off containers and other transfer equipment, preventing cross-contamination is important in insuring equipment reliability. It is advisable to designate filter carts, top-off containers, funnels, and other transfer equipment to one class and grade of lubricant. To prevent accidental cross contamination, it is recommended to label each storage container or barrel with a specific color- and shape-coded tag

FIGURE 7.11   Example of high quality oil top-off containers.

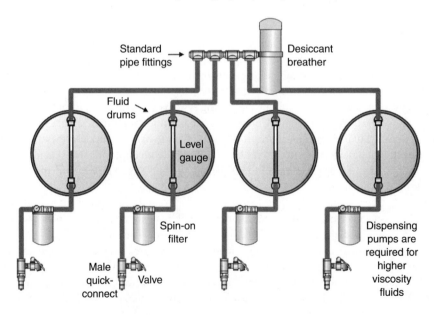

FIGURE 7.12   Bulk barrel dispensing station.

to avoid accidental mixing as shown in Figure 7.13. In this case, the color is used as the primary indicator of fluid type, while the shape can be useful for individuals who may be color blind.

7.6   Insuring Product Integrity

While lubricant manufacturers and distributors typically make every effort to insure new lubricants meet appropriate performance criteria, mistakes can and do happen. Likewise, most distributors are unable or

FIGURE 7.13　Example of color and shape coded transfer equipment.

unwilling to offer any degree of certainty as to levels of contamination that may be present in new lubricant deliveries, particularly with respect to particles and moisture.

Because of this, it is advisable particularly for critical applications to conduct certain quality control (QC) tests to insure that both the lubricant performance properties and cleanliness have not been compromised during manufacture, or due to distributor storage and handling practices. Selecting which QC tests should be performed should be based on product type and the anticipated performance properties of significance. Table 7.4, though not an exhaustive list of all possible tests, provides a list of general guidelines of what tests should be performed.

For bulk deliveries, it is both feasible and generally advisable to perform these quality assurance tests on a batch-by-batch basis. For tote or barrel delivery, it is unrealistic to expect what can amount to several hundred dollars of tests on each barrel. Instead, it is advisable to adopt a more pragmatic approach by randomly sampling (say every 10 or 20 barrels received) to, at a minimum, provide at least some accountability to the lubricant supplier.

Some tests can be performed fairly inexpensively, meaning that it may be justifiable to run these tests on each new barrel — particularly where the lubricant will be used in critical applications. Under these circumstances, tests such as viscosity, water content, particle count, and elemental additive content should

TABLE 7.4　Common New Oil Performance Properties Tests

Test	ASTM/ISO Procedure	Turbine Oils	Paper Machine Oils	Gear Oils	Engine Oils	Hydraulic Fluids
Kinematic viscosity	D445	•	•	•	•	•
Elemental analysis	D5185	•[a]	•	•	•	•
Water	D6304	•	•	—	—	•
Free water	Crackle	—	—	•	•	—
Particle count	ISO 4406:99	•	•	•	—	•
Acid number	D664	•	•	•	—	•
Base number	D4739 (or D896)	—	—	—	•	—
Demulsibility	D1401 (or D2711)	•	•	•	—	•
Foam stability/tendency	D892	•	•	•	—	•
RPVOT	D2272	•	—	—	—	—
Rust inhibition	D665	•	•	•	—	•
Copper strip corrosion	D130	•	•	•	—	•
Air release	D3427	•	•	—	—	•

[a] Turbine oils typically do not contain organometallic additives, though this test is still recommended to warn of cross contamination with other products.

TABLE 7.5   Common Oil Analysis Field Tests

Test For	Field Tests
Soot	Blotter spot
Particle contamination (patch)	Patch test kit, sediment
Fuel dilution	Odor, blotter spot, viscosity comparitor
Moisture contamination	Crackle, opacity, static sit, calcium hydride tester
Freon/refrigerant in oil	Crackle test and visgage
Glycol	Blotter spot, Schiff's reagent
Viscosity change	Viscosity comparitor
AN/BN (corrosion potential)	Field kits
Ferrous wear debris	Filter inspection, magnetic plug/probe, magnet patch test
Nonferrous wear debris	Filter inspection, patch test
Dispersancy	Blotter spot
Oxidation	Blotter spot, color, odor, AN kits, viscosity comparitor
Demulsibility	Blender comparitor
Antifoam	Blender comparitor
Additive depletion (overbase and antioxidant additives)	AN kits

all be conducted routinely, not just as a QC check, but also to provide a baseline for subsequent in-service oil analysis.

In many instances, it is desirable to be able to perform QC checks onsite, allowing immediate feedback and the ability to "accept" or "reject" batches or barrels without having to wait for data to be returned from the lab. While many more sophisticated users have stared using high quality onsite oil analysis instruments such as particle counters and viscometers, simple QC checks can be performed with little or no costs and just a small initial investment. Table 7.5 illustrates which test can be performed and how these can be achieved onsite.

Perhaps the most important property of any lubricating oil is viscosity. Insuring that any received lubricant is "in-grade" is easy to accomplish using a simple hand-held device known as a falling ball viscosity comparator (Figure 7.14). Though not as accurate as ASTM D445, this quick, simple field test

FIGURE 7.14   Falling ball viscometer — "visgage." (Courtesy of Louis C. Eitzen Co. Inc.)

is usually sufficient to determine whether new lubricant batches fall within their designated ISO or SAE viscosity grade.

Similarly, the presence of free water can be easily detected using a standard hot-plate crackle test. In this procedure, oil is dropped onto a hot plate around 325° F. Any significant quantities of free water cause water vapor bubbles to form, resulting in a scintillation or "crackle" effect much like placing wet food into a hot fry pan. Again, while not as accurate as ASTM procedures such as D6304 (Karl Fischer Moisture) or D95 (Dean and Stark distillation), it is a simple, easy-to-perform test that provides immediate feedback.

Determining the presence of particles in new oil can be a little more difficult. While the use of onsite electronic optical particle counters is becoming more and more prevalent, a simple field test involving a patch test kit can oftentimes be as effective for assessing the cleanliness of new oils. In this procedure, oil is passed through a filter patch (typically a 2 μm or smaller patch pore size is recommended for new oils). After the patch is solvent-rinsed to remove excess oil, the particle types and concentration can be examined with a small field microscope. Using a particle size and distribution comparator chart can also aid in providing a quantitative assessment of the degree of particle contamination, again allowing a "pass/fail" assessment to be made for any new lubricant batch received.

7.7 Conclusion

Of all the factors associated with lubrication, insuring that new lubricants meet minimum performance and cleanliness requirements is perhaps the most fundamental in insuring equipment reliability and longevity. While lubricant suppliers and manufacturers have a vital role to play in this process, deploying the practices outlined in this chapter should help insure that end-users of lubricants can achieve best practice in lubricant storage and handling and in turn insure their equipment achieves optimum levels of availability and reliability.

References

[1] ISO 4406:99 reference.
[2] API Standard 620 — Design and Construction of Large, Welded, Low-Pressure Storage Tanks.
[3] API 650 — Welded Steel Tanks for Oil Storage.
[4] EPA 40 CFR PART 280 — Technical Standards and Corrective Action Requirements for Owners and Operators of Underground Storage Tanks.
[5] "Best Practices in Bulk Lubricant Storage and Handling," Matthew Dinslage, Jim C. Fitch, and Sabrin Khaled Gebarin, *Lubrication Excellence 2004 Conference Proceedings*, Noria Corporation Publishing.

8

Conservation of Lubricants and Energy

Robert L. Johnson
and James C. Fitch
Noria Corporation

8.1 Introduction

Many of the materials essential to global industrial markets have been identified as being in potential short supply. A substantial number of those materials originate from regions of the world where interrupted supply is a real and present risk. Petroleum is a finite world resource with continuing supply and economic problems, with more special concerns for fuels than for petroleum-based lubricants.

It is the objective of this chapter to present specific conservation practices for lubricants and functional (hydraulics, coolants, etc.) fluids used in tribological components and for energy. The treatment will necessarily be brief, but references will point to more detail information.

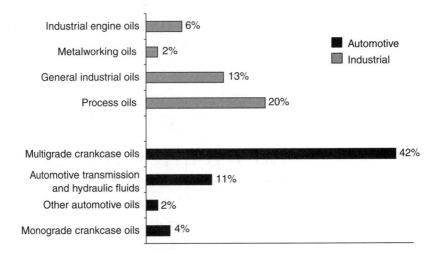

FIGURE 8.1   2002 sales of lubricants (total: 2.4 billion gal). (Taken from NPRA, Lubes'n'Greases Magazine. With permission.)

8.2   Conservation of Lubricants

8.2.1   Trends

Lubricants and hydraulic fluids are most commonly derived from petroleum sources. Petroleum serves as the base-stock in the majority of liquid lubricants and greases and very often is a raw material in the synthesis of unique lubricants (typically referred to as synthetics). Details of the distribution of the industrial and automotive lubricant markets in the United States are shown in Figure 8.1. A five-year trend of industrial and automotive lubricants sales is shown in Figure 8.2.

8.2.2   Improved Manufacturing and Formulation

The manufacturing methods used in refining lubricants are significant to material conservation and energy. Modern hydrogen processed bases oils (API Groups II and III) are popular because they

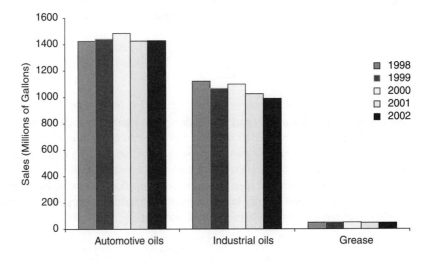

FIGURE 8.2   Five-year trend of total reported sales of lubricants. (Taken from NPRA, Lubes'n'Greases Magazine. With permission.)

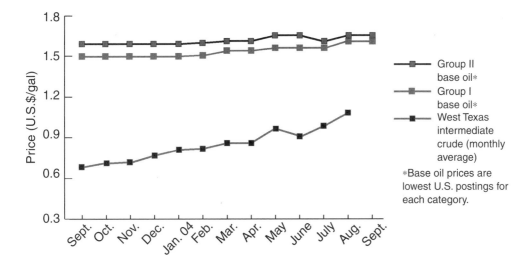

FIGURE 8.3 Base oil and crude prices. (Copyright 2004, Lubes'n'Greases Magazine. With permission.)

(1) minimize solvent, acid, and clay, (2) reduce by-product disposal problems, (3) increase yield, (4) lower costs, (5) permit use of wide range of crudes, (6) improve color, and (7) give higher viscosity index (VI). Figure 8.3 shows a 12-year base oil production trend for paraffinic and naphthenic stocks. Pricing trends in 2003–2004 for Group I mineral oils and Group II hydroprocessed oils are shown in Figure 8.4.

8.2.3 Packaging and Handling

Packaging and handling practices have a significant contribution to conservation of lubricants. Contamination commonly occurs when containers are left open in point-of-use and storage areas. The presence of moisture and particulates degrade the effectiveness of all types of lubricants, in many cases requiring that the lubricants be discarded. Recommended practice often requires packaging in sealed containers of the size, or fraction thereof, usually needed for the system. Where drum-sized or tote-bin

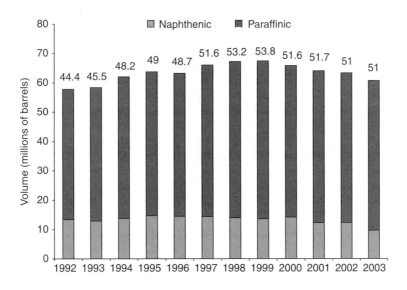

FIGURE 8.4 United States refinery production volumes. (Taken from NPRA, Lubes'n'Greases Magazine. With permission.)

containers are used, they should not be left open and storage position should be such that contamination cannot accumulate at the vent, drainage, or fill holes. Usually a horizontal drum position is indicated and such a simple consideration can conserve lubricants by minimizing discards and assuring expected performance.

8.3   Lubricant Utilization

8.3.1   Extended Life Lubricants

Extended life cycle use of lubricants is clearly conservation effective. The limiting life of many types of lubricants is determined by either intolerable contamination or additive depletion. Formulators, recognizing the importance of conservation and cost effectiveness, are providing additive packages for extended service. With sufficient life from improved additives and more robust base stocks, the primary importance is contamination and that may be controlled by vigilant contaminant exclusion and filtration. In considering the potential for extending the lubricant life cycle, the experience of airlines with turbine engine lubricants is of interest. Using synthetic lubricants, the airlines do not change lubricants between major engine maintenance events; for several thousand hours operation only minor makeup oil is added to compensate for leakage and consumption. That is the equivalent of millions of miles per oil change. Another illustration of extended life cycle lubrication is in hermetically sealed household refrigerators where continuous operation for over two decades is common. Such hermetically sealed systems are ideal for extended lubricant service life, emphasizing again the importance of contamination control.

8.3.2   Synthetic Lubricants and Functional Fluids

Conservation practices important to petroleum lubricants are also generally relevant to synthetic lubricants and functional fluids. Because synthetics can have greater oxidation stability than mineral-based lubricants, the possibilities for extended service life have even greater potential. The cost of synthetics is higher than refined petroleum products and the applications are more specialized and controlled than for petroleum lubricants. Airlines have used reprocessed phosphate ester hydraulic fluids; polyphenyl ether lubricants have also been recycled for some military use. In those cases, low initial cost as well as favorable collection and processing circumstances allowed recycling to be cost effective.

8.3.3   Greases

Most greases are compounded from mineral-based lubricants. Conservation and economics can be achieved by many of the same measures cited for petroleum oils. Extended operating life is gained through product improvement and contamination control. Improved sealing practices in applications offer major gains in conserving greases by minimizing contamination and leakage. More stable grease thickeners, as well as new and improved lubricant additives, have allowed significantly extended regreasing periods for mechanical components. Products with capabilities for high operating temperatures and others with resistance to displacement by water impingement are examples of lubricants that have extended relubrication intervals. For example, wear life of components have markedly improved while the incidence of mechanical failures have markedly decreased by modern thickeners, base oils, and additives.

8.3.4   Solid Lubricants

Solid lubricants are often used in greases, in slurries with liquid carriers, in bonded surface films, and in self-lubricating composite materials. Optimum concentrations have been determined in each application system. The extended use of solid lubricants in many applications has significant impacts on reliability and energy uses. Although the total volume of solid lubricants is less than one might consider in view of the many applications, even lesser quantities can often be used. The usual thickness of bonded films is in the range of 0.0002 to 0.0005 in., a thickness dictated by production control limitations of the coating processes.

Vacuum processes such as sputter-ion deposition allows improved film uniformity and, therefore, thinner films are used.

Similarly, with self-lubricating bulk solids, optimizing the required functions can conserve lubricating materials. In addition to friction and wear considerations in selection of self-lubricating solids, resistance to environment, mechanical stress, and thermal stress must be anticipated. Composite systems and specific function coatings can greatly extend the lives and hence conserve such materials.

8.4 Conservation of Energy

It seems counterintuitive that lubricants selected to optimize wear control may not be optimum when it comes to energy conservation. In fact, in view of today's growing pressure to reduce demand on nonrenewable energy resources and increase operating profits, we are increasingly facing a shift of emphases from past lubrication objectives. Energy-conserving lubrication offers motivation on several fronts. Consider the following:

1. When energy consumption is economized, equipment operating costs come down, translating to a boost in business profits, regardless of whether the energy source is renewable (hydro, solar, wind) or nonrenewable (coal or petroleum). For many industries, the cost of energy far exceeds the cost of maintenance, machine repair, and even downtime. A small percentage of reduction in energy consumption can translate into large returns.
2. Reduced demand on nonrenewable fossil fuels means cleaner air, reduced greenhouse gas emissions, and a healthier environment (of growing political and social importance in view of the Kyoto Protocol on global warming, ISO 14001, Clear Air Act, etc.). When fuels do not burn, there is no waste stream (smoke stack, tail pipe, etc.) and the risk of pollutants from emissions such as nitrogen oxides (the principle component of smog), sulfates, CO_2, and unburned hydrocarbons is reduced proportionally.
3. With few exceptions, lubricants and lubrication methods that reduce energy consumption will also reduce heat and wear debris generation; however, the reverse may not hold true. When heat and wear debris are reduced, less stress is imposed on additives and the base oil. The result will be longer thermal and oxidative stability, and in turn, longer oil drains, lower oil consumption, and the ancillary costs associated with oil changes (as much as 40 times the cost of the lubricant itself!).
4. When lubricant consumption is reduced, so too is the disposal of environmentally polluting waste oil and certain suspended contaminants, some of which may be hazardous and toxic; ethylene glycol (antifreeze), for example.
5. When there is better economy in the consumption of both petroleum fuels and mineral-based lube oil, there is reduced dependence on foreign sources of crude oil, including those from politically unstable countries.
6. In certain countries, including European Union nations, reductions in the consumption of nonrenewable fuels can avert energy tax penalties such as the Climate Change Levy in the United Kingdom.

In recent years, there has been growing interest in energy-conserving lubricants and energy-conserving lubrication. Note, energy-conserving lubricants relate to formulation (base stocks and additives) and their selection for machine application. In contrast, energy-conserving lubrication includes the use and application of lubricants (change intervals, delivery methods, lube volume, etc.). Both can have a marked impact on energy conservation.

Energy economy and wear control do not necessarily go hand-in-hand. In certain cases, they may be conflicting objectives. For many organizations, environmental factors and energy costs fall low on the list of priorities compared to productivity and machine reliability. In such cases, the principle objective of the practice of lubrication is to reduce wear and maximize reliability.

8.5   Energy-Conserving Fluid Properties

When formulating or selecting lubricants, the following properties are important in reducing friction and energy consumption.

8.5.1   Kinematic Viscosity

When it comes to energy economy, viscosity can be both an inhibitor and an enabler. Recalling the well-known Stribeck curve, the oil film produced by hydrodynamic lubrication is directly influenced by viscosity. However, too much viscosity causes churning losses (excessive internal oil friction) and heat production, especially in engines, gears, bearings, and hydraulics. In addition to energy losses, this increased heat can more rapidly break down the oil and its additives.

8.5.2   Viscosity Index

Kinematic viscosity by itself defines an oil's resistance only to flow and shear at a single temperature, typically 40 or $100°C$. However, in normal operation, lubricating oils transition through a wide range of temperatures. As such, it is the oil's VI combined with kinematic viscosity that defines what the viscosity will be at a specific operating temperature. Will it be too high when ambient start-up temperatures are low and too low when operating temperatures are high? Likewise, what will be the time-weighted average viscosity of the lubricating oil during the machine's service life? It is this average viscosity that defines energy consumption, not the occasional temperature-based viscosity excursions that may have a greater impact on wear (cold starts for instance). In general, the significance of VI on energy conservation and wear is often sharply underestimated.

8.5.3   Non-Newtonian Properties

Fluids that exhibit shear-dependent viscosity changes (known as the non-Newtonian fluids) are known to reduce energy consumption in many machines. Good examples are VI-improved motor oils (multigrades) and many all-season hydraulic fluids. As fluid movement increases (shearing) during service, the oil's effective viscosity self-regulates slightly downward, along with energy consumption. This, in part, explains why high-VI, multigrade motor oils are generally those that are designated energy-conserving by the API.

8.5.4   High-Temperature Shear Stability

Synthetics and other high-VI base oils perform best here, as do multigrade formulations with low-VI concentrations. Temperature and viscosity shear-back at high temperatures can lead to loss of critical lubricant film strength, leading to power losses and wear. However, temporary shear thinning can also reduce parasitic viscous drag in crankshaft bearings.

8.5.5   Pressure—Viscosity Coefficient

The role of pressure–viscosity (PV) coefficient on energy consumption is not well defined in the literature. However, it is widely understood that many base oils exhibit a sharp increase in viscosity as pressure rises; a necessary quality of lubricants in achieving effective elastohydrodynamic lubrication (EHD). Some oils, such as mineral oils and PAOs (polyalphaolefins), have higher PV coefficients than others, such as ester-based synthetics and water-based fluids. While high PV coefficients may be important at reducing contact fatigue wear, in some cases, this property may contribute to lower fuel economy. The high pressure-induced viscosity in sliding frictional zones and in hydraulic systems could result in exceedingly high viscous drag energy losses.

8.5.6 Bulk Modulus

A fluid that is sponge-like and easily compressed has low bulk modulus of elasticity. The more compressible a lubricant is, the more potential for lost energy and heat production. This is especially true in hydraulic and lube oil circulating systems.

8.5.7 Boundary Film-Strength Properties

Many lubricants and hydraulic fluids can gain considerable film strength under boundary and mixed-film lubrication from the base oil, without the need for additives. A phosphate ester synthetic is an example of a fluid with intrinsic lubricity. Most other lubricants rely on additives such as friction modifiers, antiwear agents, extreme pressure (antiscuff), solid lubricants, and fatty acids. The effectiveness of these additives at reducing wear, friction, and energy consumption can vary considerably between the different additive types employed. The performance of these additives also varies by machine and application (load, speed, metallurgy, temperature, and contact geometry).

8.5.8 Grease Consistency

The consistency of grease can have an impact on energy consumption in ways similar to viscosity. The energy needed to move grease in frictional zones and in adjacent cavities by moving machine elements is affected by its consistency and shear rate (grease is non-Newtonian). So, too, energy is required in some applications to pump grease to bearings and gears. Pumping energy losses is influenced, in part, by grease consistency and thickener type.

8.5.9 Grease-Channeling Properties

A grease that has good channeling characteristics helps keep the bulk lubricant away from moving elements, avoiding excessive churning and drag losses. Poor channeling characteristics may lead to increased energy consumption, heat production, and base oil oxidation.

8.6 Wear

Wear from boundary friction can have a near-term adverse effect on fuel economy and can generate heat. Wear is often the result of such things as lubricant starvation, low viscosity, poor or degraded antiwear additive performance, dirt and other contaminants, deposits (e.g., ring grooves), etc. One researcher identified an $8°C$ ($14.4°F$) increase in bearing oil temperature, which he attributed to solid contamination of the oil.

However, there is also a long-term effect especially in engines. Over a period of time, an engine loses so much metal (rings, cylinder walls, cam follower, cam lobe, etc.) that combustion efficiency is severely impaired. (This is discussed in greater detail in the next section.) Loss of combustion efficiency directly impacts fuel economy and tailpipe emissions. In this respect, the average fuel economy performance of a motor oil over a period of 100,000 miles or more is a better assessment of its life-cycle performance as opposed to snapshot energy consumption assessments of new engines and new oils.

8.7 The Surprising Role of Particle Contamination on Fuel Economy

When a lubricant degrades, it forms reaction products that become insoluble and corrosive. So, too, the original properties of lubricity and dispersancy can become impaired as the lubricant ages and additives deplete. Much has been published about the risks associated with overextended oil drains and the buildup of carbon insolubles from combustion blow-by, especially in diesel engines.

There have been surprisingly few studies published on the impact of fine abrasives in a motor oil as it relates to fuel economy over the engine's life. Yet, it is not hard to imagine numerous scenarios in which solid abrasives suspended in the oil could diminish optimum energy performance. Below is a list of several scenarios.

8.7.1 Antiwear Additive Depletion

High soot load of crankcase lubricants has been reported to induce abrasive wear and impair the performance of zinc dialkyldithiophosphate (ZDDP) antiwear additives. The problem is more pronounced in diesel engines. Some researchers believe that soot and dust particles exhibit polar absorbencies, and as such, can tie-up the antiwear additive and diminish its ability to control friction in boundary contacts (cam nose, ring/cylinder walls, etc.). However, there appears to be greater evidence that soot itself is highly abrasive in frictional zones where dynamic clearances are 1 μm. These include cam/follower and ring reversal areas on cylinder walls.

8.7.2 Combustion Efficiency Losses

Sooner or later, wear from abrasive particles and deposits from carbon and oxide insolubles will interfere with efficient combustion in an engine. Valve train wear (cams, valve guides, etc.) can impact timing and valve movement. Wear of rings, pistons, and cylinder walls influences volumetric compression efficiency and combustion blow-by resulting in power loss. Particle-induced wear is greatest when the particle sizes are in the same range as the oil film thickness (Figure 8.5). For diesel and gasoline engines, there are a surprising number of laboratories and field studies that report the need to control particles below 10 μm. One such study by General Motors concluded that, "controlling particles in the 3 μm to 10 μm range had the greatest impact on wear rates and that engine wear rates correlated directly to the dust concentration levels in the sump [14]."

8.7.3 Frictional Losses

When hard clearance-size particles disrupt oil films, including boundary chemical films, increased friction and wear will occur. One researcher reports that 40 to 50% of the friction losses of an engine are attributable to the ring/cylinder contacts, with two-thirds of the loss assigned to the upper compression ring. It has been documented that there is an extremely high level of sensitivity at the ring-to-cylinder zone of the engine to both oil- and air-borne contaminants. Hence, abrasive wear in an engine's ring/cylinder area translates directly to increased friction, blow-by, compression losses, and reduced fuel economy.

8.7.4 Viscosity Churning Losses

Wear particles accelerate the oxidative thickening of aged oil. High soot load and lack of soot dispersancy can also have a large impact on oil viscosity increases. Viscosity-related internal fluid friction not only increases fuel consumption, but also generates more heat, which can lead to premature degradation of additives and base oil oxidation.

8.7.5 Stiction Losses

Deposits in the combustion chamber and valve train can lead to restricted movements in rings and valve control. When hard particle contamination agglomerates with soot and sludge to form adherent deposits between valves and guides, a tenacious interference, called stiction, results. Stiction causes power loss and engine knock. It causes the timing of the port openings and closings to vary, leading to incomplete combustion and risk of backfiring. Advanced phases of this problem can lead to a burned valve seat.

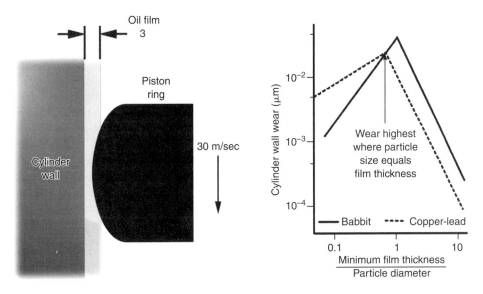

FIGURE 8.5   Common oil film thickness. (From Noria Corporation. With permission.)

8.8   Role of Lubrication Practices

While lubricant formulation and selection are important, energy conservation is also influenced by machine design and lubricant application factors. A superior lubricant cannot offer redemptive relief for poor lubrication practices and machine design. Even the very best lubricants cannot protect against destruction caused by dirt and water contamination. Overgreasing of bearings is known to increase frictional losses and raise bearing temperature. The same is true for bearings that are underlubricated. For bath lubricated bearings and splash lubricated gears, a change in oil level by as little as 1/2 in. (1.3 cm) can increase temperature by more than 10°C. This, of course, translates to greater energy consumption, shorter oil life, and increased wear.

Excessively aerated oils due to worn seals and wrong oil levels can have similar effects (loss of bulk modulus). There have also been studies showing the negative effects of overextended oil change interval on fuel economy in diesel engines. Additionally, overextended filter changes cause excessive flow resistance and fluid bypass. Both can often be corrected by the frequent and proper use of oil analysis in selecting the optimum oil and filter change interval, tailored to equipment type and its application.

8.9   Role of Machine Design

A machine's design and the quality of its manufacture can also impact energy economy. Together with operating load and speed, machine design influences the type of lubricant that must be employed for wear protection and energy efficiency. Already mentioned is the importance of viscosity films produced by hydrodynamic and elastohydrodynamic lubrication as well as boundary film strength from additives and polar base oil chemistry. These lubrication regimes relate to the contact dynamics associated with a machine's design and operating conditions. Additionally, specific film thickness, also known as lambda, brings into the picture the influence of surface roughness and shaft alignment.

Many users and suppliers have reported energy savings from total-loss lubricant delivery technologies such as oil mist and centralized lubrication systems. The amount of fluid that a machine uses to lubricate frictional surfaces at any moment is extremely small compared to the amount of fluid some machines must keep in continuous motion. The advantage of some total-loss lubrication systems is that there is minimal loss of energy from constant fluid churning and flow resistance of lubricants moving through lines. An example of internal fluid friction is observed when an oil is placed in a bottle and then shaken. The oil's temperature will rise.

In addition, bath, splash, and recalculating lubrication systems use the same oil over and over. As we all know, this reused oil over time can become impaired by loss of additives, base oil oxidation, and rising concentrations of contaminants. In contrast, when well engineered and in the right application, oil mist and other certain total-loss systems can provide a continuous supply of fresh, clean, and dry new oil. Energy consumption is also influenced by the size and type of fittings, oil lines, and filters.

8.10   Environmental Stewardship

In summary, lubricants, lubrication, and contamination play no small role in reducing energy consumption and the general wasteful use of petroleum products, including lubricants. Increasingly the selection and use of lubricants is going to stress greater importance on energy and environmental impact. At the same time, we will not lose sight of other vital objectives including machine reliability and safety.

References

[1] Anon, Technical Options for Conservation of Metals, Library of Congress Catalog Card Number 79-600172, *Congress of the United States*, Office of Technology Assessment, Washington, D.C., 1979.

[2] Barnett, R.S., *Molybdenum Disulfide as an Additive for Lubricating Greases — A Summary 1973*, Climax Molybdenum Company, Greenwich, CT, 1973.

[3] ASLE, *Proceedings of the 2nd International Conference on Solid Lubrication*, SP-6, 3978 Library of Congress Catalog Card Number 78-67090, American Society of Lubrication Engineers, Parkridge, IL, 1978.

[4] 2004–2005 *Lubricants Industry Sourcebook*, Lubes'n'Greases, Falls Church, VA, 2004, pp. 4–10.

[5] Addison, J. and Needelman, W., *Diesel Engine Lubricant Contamination and Wear*, Pall Corp, East Hills, NY, 1986.

[6] Andrews, G.E., Hall, J., Jones, M.H., Li, H., Rahman, A.A., and Saydali, S., The Influence of an Oil Recycler on Lubricating Oil Quality with Oil Age for a Bus Using In-Service Testing, Presented at *SAE 2000 World Congress* (SAE Paper 2000-01-0234), 2000.

[7] Ballentine, B., *Motor Oils — Fuel Economy vs. Wear*, Machinery Lubrication, 2003.

[8] Barris, M.A., Total Filtration: The Influence of Filter Selection on Engine Wear, Emissions and Performance, Presented at *SAE Fuels and Lubricants Meeting* (SAE paper 952557), 1995.

[9] Feldhaus, L.B. and Hudgens, R.D., Diesel Engine Lube Filter Life Related to Oil Chemistry, Presented at *SAE International Fuels and Lubricants Meeting* (SAE Paper 780974), 1978.

[10] Fitch, J.C., Troubleshooting Viscosity Excursions, *Practicing Oil Analysis*, May–June 2001.

[11] Foder, J. and Ling, F.F., Friction Reduction in an IC Engine Through Improved Filtration and a New Lubricant Additive, *Lubrication Engineering*, October 1985.

[12] Madhaven, P.V. and Needelman, W.M., Review of Lubricant Contamination and Diesel Engine Wear, Presented at *SAE Truck and Bus Meeting and Exposition* (SAE Paper 881827), 1988.

[13] McGeehan, J., Uncovering the Problems with Extended Oil Drains, *Machinery Lubrication*, September–October 2001.

[14] Staley, D.R., Correlating Lube Oil Filtration Efficiencies with Engine Wear, Presented at *SAE Truck and Bus Meeting and Exposition* (SAE Paper 881825).

[15] Troyer, D., Consider This (Sidebar to From under the Hood — Multigrade oil — To Use or Not to Use), *Machinery Lubrication*, July–August, 2001.

[16] Fitch, J.C., Energy-conserving lubrication — the endless debate, in *How to Select a Motor Oil and Filter for Your Car or Truck*, 2nd ed., Fitch, J.C. Ed., Noria Corp., Tulsa, OK, 2003, Chap. 7.

9

Centralized Lubrication Systems — Theory and Practice

Paul Conley and
Ayzik Grach
Lincoln Industrial

9.1   The Philosophy of Lubrication

The purpose of lubrication is to prevent metal to metal contact between two moving members, reduce friction, remove heat, and flush out contaminants. The philosophy of an automatic lubrication system is to deliver the right amount of lubricant at the right time. The method to do this is to deliver a small quantity of lubricant to a bearing often. The lubricant is delivered in small portions as the bearing consumes it. Applying more lubricant than the bearing can consume in a short period of time is often thrown off the moving components, causing housekeeping and environmental safety hazards. A properly designed automatic centralized system can deliver just the right amount of lubricant the bearing needs.

To illustrate this fact, Figure 9.1 shows the effect of manual lubrication in comparison to automatic lubrication. With manual lubrication, lubrication of the bearings is normally performed when the machine is not running. The even distribution of grease between the shaft and journal cannot be assured. With automatic lubrication, the distribution of grease between the shaft and journal is assured (see Figure 9.2.)

9.2   Using the Correct Grease

When designing an automatic grease lubrication system, the first priority is to use the correct lubricant. The lubricant must have the properties consistent for the application. For applications requiring grease, the base oil lubricant properties and the thickener must meet the application needs to include it in an automatic system.

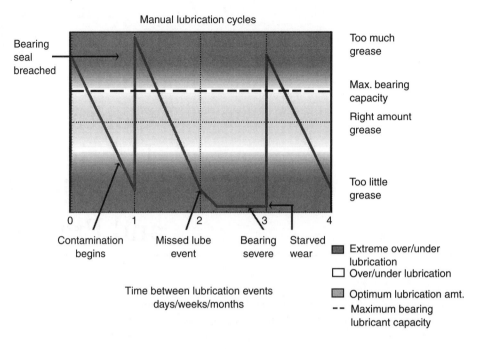

FIGURE 9.1   Manual method.

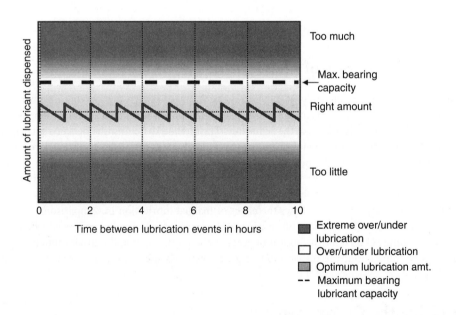

FIGURE 9.2   Automated method.

In a perfect world, the best grease to use is one that contains the proper base oil lubrication properties with the heaviest or stiffest thickener soap. The purpose of grease over oil is to make the lubricant stay put in the bearing. Grease is normally used in applications where there are heavy loads and slow relative motion between the shaft and bearing. Grease should also be used when strong shock loading is present, frequent starting and stopping, and when there is insufficient or no bearing seal.

Using oil in these applications is not advisable. The oil would run off or out of the bearing causing a drop-off in film thickness to prevent metal to metal contact. The use of grease prevents this from occurring. It is important to know that the soap thickener in the grease does not provide any lubrication. Oil is suspended in the thickener, providing the necessary lubrication. For high load situations, grease can suspend solid lubricant and extreme pressure (EP) additives in the base oil to provide additional protection against metal to metal contact.

Some of the desired characteristics of grease for use in automatic lubrication systems are:

- Good shear and mechanical stability (reference ASTM D217).
- Does not readily allow the base oil to separate out (reference ASTM D4425).
- Good reversibility defined as the characteristic of the grease to recapture the base oil should separation of oil from the thickener occur. Separation of oil may be induced under high load and pressure conditions in the bearing. A grease with good reversibility will return to its original consistency after the high load is removed.
- Possesses good washout resistance.
- Protects against corrosion.
- Good pumpability and ventability.

With the development of complex greases, the above characteristics can be achieved. With the use of synthetic base oils, the temperature ranges at which grease can operate have been expanded.

For grease used in automatic systems, one very important property to be considered is the measure of its pumpability or ventability. It would not be practical to produce a grease that is so stiff that it cannot be pumped or which produces such a resistance to flow that it cannot be used in an automatic system. The best combination is to use the thickest grease that can be pumped effectively. If the grease cannot be pumped, then it has no practical use; its resistance to flow is so great that the pressure produced in the supply lines would be excessive and thus prohibitive.

When centralized grease lubrication was introduced in the mid-1930s, the manufacturers of these centralized grease lubrication systems needed to understand the properties and flow characteristic of grease. Grease is classified as a non-Newtonian thixotropic pseudoplastic (viscosity decreased as the shear rate is increased) fluid. Most fluids such as water and oil are Newtonian fluids. That means the viscosity is constant as the shear rate changes. Figure 9.3 illustrates three types of viscosities.

Many grease manufacturers rate the grease by its National Lubricating Grease Institute (NLGI) rating. This is a method that only measures the general stiffness at one temperature. According to the ASTM test method, the grease stiffness is measured at 77°F.

The higher the NLGI rating of the grease, the stiffer the grease is and, in general, the higher apparent viscosity. The NLGI rating alone is not sufficient to determine its appropriate use in an automatic system. The user or designer of an automatic system must know the flow properties of the grease at the lowest temperature of the application. For example, a grease with an NLGI rating of 2 at 77°F will have an NLGI rating

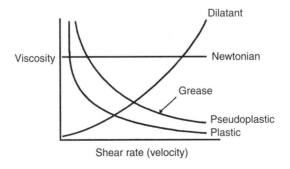

FIGURE 9.3 Flow behavior of different types of fluid.

FIGURE 9.4   Lincoln Ventmeter.

of 3 at 50°F. Another grease with an NLGI rating of 2 at 77°F may have an NLGI rating of 3 at 30°F. An NLGI number 1 grease will often behave like an NLGI number 2 grease at 30°F.

For this reason, the best indication of a grease's suitability for use in an automatic lubrication system is the Lincoln Ventmeter. Manufacturers and designers of automatic grease systems often use the Lincoln Ventmeter viscosity as the standard for selecting the correct grease for use in an automatic lubrication system.

To provide an understanding of the usefulness of the Lincoln Ventmeter, let us take a look at it (Figure 9.4). The Lincoln Ventmeter was developed as an instrument to measure the flow limits of grease. It is more precise than the classic NLGI number rating. It was developed in the early 1950s, and since 1965 has been used extensively in determining acceptable performance out of a single line injector type centralized lubrication system. Even today, samples of grease are tested and evaluated using the Lincoln Ventmeter by grease manufacturers and designers of centralized lubrication systems.

By measuring the flow ability of grease, an application engineer/technician or grease manufacturer can select the pump and line size to ensure good performance of the centralized grease lubrication system. The usefulness of the Lincoln Ventmeter is most noted in the following three ways:

1. What types of grease, according to consistency, can be used in a given grease supply line so that the pressure in the system will vent down sufficiently to successfully operate injectors?
2. What supply line length and diameter should be used for a specific type of grease in a centralized lubrication system?
3. When to utilize a lighter NLGI grade grease product so that the system will continue to operate correctly during colder temperatures?

The ability of the grease to vent is important for proper operation of the system. It is also good that the grease be vented during the off time of the system because the oil does tend to separate from the soap.

Figure 9.5 shows a schematic of the Lincoln Ventmeter and how the test is conducted. The yield pressure of the grease is obtained in the following manner: With valve 1 closed and valve 2 open, the sample of grease is filled with grease using a pump or grease gun. Valve 2 is then closed. Using a grease pump or a grease gun, the Ventmeter is charged to a pressure of 1800 psig, relief valve number 1 is opened and the grease is discharged. The discharge of the grease will reduce the pressure. The pressure left in the system is read 30 sec after the relief valve is opened. Thirty seconds is a sufficient time for the grease to stabilize. The gauge reading is the Ventmeter viscosity and is measured in pressure in pounds per square inch.

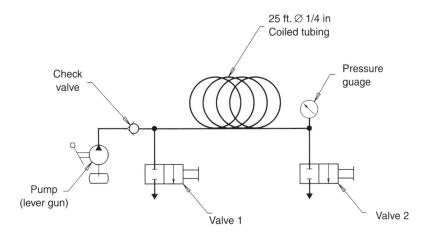

FIGURE 9.5 Lincoln Ventmeter schematic.

The test is conducted normally at ambient, 30 and 0°F. This way the yield pressure is obtained at various temperature conditions. Tests are often done at progressively lower temperatures to establish a value when the grease ceases to flow.

If the gauge reading goes to zero within the 30 sec, this is an indication that the grease has effectively no yield limit. Very light viscous grease behaves like oil and can be considered Newtonian. With lower temperatures or stiffer greases, the gauge reading will be some value other than zero.

Supply line charts have been developed according to the Lincoln Ventmeter for three types of injectors (see Tables 9.1, 9.2, and 9.3).

9.3 Centralized Grease Lubrication Systems

What do they do? The purpose of any automatic centralized lubrication system is to provide the correct amount of grease to the bearing at the right time. The method to do this is for a controller to turn on a pump that supplies grease to a positive displacement valve that will dispense a predetermined amount of grease to the bearing.

There are essentially three types of lubrication systems used in industry. The oldest of these are the progressive and dual line systems developed in the late 1800s during the industrial revolution. Single line systems were developed in the late 1930s as an enhancement to the progressive and dual line systems.

The main difference between one system and the other is the type of positive displacement valve. Each system is discussed to include principle of operations, features, design considerations, and strengths and weaknesses.

The importance of monitoring system performance cannot be understated. Each and every type of lubrication has some provisions for monitoring. The degrees in which the systems can be monitored vary from visual indicators to full electronic transducer feedback monitoring.

9.3.1 The Series Progressive System

The system gets its name from the serial and progressive nature in which the valves operate.

Basic components of the system:

- Pump
- Controller
- Progressive metering valve

TABLE 9.1   Supply Line Chart with Injectors that Require a Venting Pressure of 600 psi or Greater

	SL-1 and SL-11 supply line chart				SL-1 and SL-11 supply line chart		
NLGI Grease	Lincoln Ventmeter Reading (psi)	Nominal Pipe Size or ID of Tube or Hose (in.)	Max. Supply Line Length (ft)	NLGI Grease	Lincoln Ventmeter Reading (psi)	Nominal Pipe Size or ID of Tube or Hose (in.)	Max. Supply Line Length (ft)
#0	0–100	2.00	1100	#2	300–400	2.00	280
	0–100	1.50	875		300–400	1.50	200
	0–100	1.25	700		300–400	1.25	175
	0–100	1.00	575		300–400	1.00	140
	0–100	0.75	430		300–400	0.75	100
	0–100	0.50	270		300–400	0.50	70
	0–100	0.38	200		300–400	0.38	50
	0–100	0.25	130		300–400	0.25	30
#1	100–200	2.00	500	#3	400–500	2.00	230
	100–200	1.50	400		400–500	1.50	170
	100–200	1.25	350		400–500	1.25	140
	100–200	1.00	270		400–500	1.00	100
	100–200	0.75	210		400–500	0.75	80
	100–200	0.50	140		400–500	0.50	55
	100–200	0.38	110		400–500	0.38	40
	100–200	0.25	50		400–500	0.25	30
#2	200–300	2.00	350	#3	500–600	2.00	190
	200–300	1.50	280		500–600	1.50	140
	200–300	1.25	235		500–600	1.25	120
	200–300	1.00	180		500–600	1.00	90
	200–300	0.75	140		500–600	0.75	65
	200–300	0.50	90		500–600	0.50	45
	200–300	0.38	70		500–600	0.38	36
	200–300	0.25	40		500–600	0.25	15

The operation of a progressive system is straightforward. When the time for a lubrication event is necessary, a controller turns on a pump that sends lubricant to the progressive valves, which in turn meter the lubricant to the bearing. The output volume provided by the displacement of the metering piston determines the amount of lubrication to the bearing. The valves can be cycled once or multiple times to deliver an appropriate amount of grease to the bearing. The on-time selection and the pump output capacity determines the total amount of lubrication provided to the bearings.

Figure 9.6 is a schematic of a typical series progressive automatic lubrication system. Figure 9.7 is a schematic and the operation of progressive valves and how they work. More than a drilled manifold block, the valve incorporates a series of metering valves, which accurately dispense lubricant from each outlet, overcoming back pressure of up to 1000 psi. Visual monitoring is provided with an indicator pin, which confirms a valve has completed a full cycle. Progressive divider valves are available for grease or oil applications, and in carbon steel and 303 stainless steel for corrosive environments.

The inlet passageway is connected to all piston chambers at all times with only one piston free to move at any time.

1. With all pistons at the far right, lubricant from the inlet flows against the right end of the piston A (Illustration 1).
2. Lubricant flow shifts piston A from right to left, dispensing lubricant through connection passages to outlet 2 (Illustration 2).
3. Piston B shifts from right to left, dispensing lubricant through outlet 7. Lubricant flow is directed against the right side of piston C (Illustration 3).
4. Piston C shifts from right to left, dispensing lubricant through outlet 5. Lubricant flow is directed against the right side of piston D (Illustration 4).

TABLE 9.2   Supply Line Chart with Injectors that Require a Venting Pressure of 200 psi

	SL-32 and SL-33 supply line chart				SL-32 and SL-33 supply line chart		
NLGI Grease	Lincoln Ventmeter Reading (psi)	Nominal Pipe Size or ID of Tube or Hose (in.)	Max. Supply Line Length (ft)	NLGI Grease	Lincoln Ventmeter Reading (psi)	Nominal Pipe Size or ID of Tube or Hose (in.)	Max. Supply Line Length (ft)
#0	0–100	2.00	400	#2	300–400	2.00	100
	0–100	1.50	300		300–400	1.50	75
	0–100	1.25	250		300–400	1.25	63
	0–100	1.00	200		300–400	1.00	50
	0–100	0.75	150		300–400	0.75	37
	0–100	0.50	100		300–400	0.50	25
	0–100	0.38	75		300–400	0.38	18
	0–100	0.25	50		300–400	0.25	12
#1	100–200	2.00	200	#3	400–500	2.00	80
	100–200	1.50	150		400–500	1.50	60
	100–200	1.25	120		400–500	1.25	50
	100–200	1.00	100		400–500	1.00	40
	100–200	0.75	75		400–500	0.75	30
	100–200	0.50	50		400–500	0.50	20
	100–200	0.38	37		400–500	0.38	15
	100–200	0.25	25		400–500	0.25	10
#2	200–300	2.00	135	#3	500–600	2.00	65
	200–300	1.50	100		500–600	1.50	50
	200–300	1.25	80		500–600	1.25	40
	200–300	1.00	66		500–600	1.00	32
	200–300	0.75	50		500–600	0.75	24
	200–300	0.50	32		500–600	0.50	NR
	200–300	0.38	25		500–600	0.38	NR
	200–300	0.25	16		500–600	0.25	NR

5. Piston D shifts from right to left, dispensing lubricant through outlet 3. Piston D's shift directs lubricant through a connecting passage to the left side of piston A (Illustration 4).

Lubricant flow against the left side of piston A begins the second half-cycle, which shifts pistons from left to right, dispensing lubricant through outlets 1, 8, and 4 of the divider valve.

9.3.1.1   Crossporting a Divider Valve

Outputs from adjacent outlets may be combined by installing a closure plug in one or more outlets. Lubricant from a plugged outlet is redirected to the next adjacent outlet in descending numerical order. Outlets 1 and 2 must not be plugged since they have no crossport passage to the next adjacent outlet. In Figure 9.8, outlets 5 and 3 are crossported and directed through outlet 1. In this example, outlet 1 will dispense three times as much lubricant as outlet 7. The tube ferrules in outlets 1 and 7 block the crossport passage so that lubricant flow is only directed through outlets.

Typical outlets per cycle can range from as low as 0.0037 in.^{3} (0.06 cm^3) to 0.012 in.^{3} (0.20 cm^3) for nonmodular progressive valves. To increase the lube amount to the bearing, the valve will have to complete another cycle.

9.3.1.2   The Modular Lubrication System

The modular blocks are series progress valves that contain separate valve sections. With nonmodular valves, the measuring pistons are integral to the whole block. The separate valve sections provide more flexibility in selecting output volumes. This is different from a nonmodular valve. Modular valves are more complex and are more difficult to manufacture.

TABLE 9.3   Supply Line Chart with Quick Venting Injectors that Have a Venting Pressure of 1000 psi

| | SL-V supply line chart | | | | SL-V supply line chart | | |
NLGI Grease	Lincoln Ventmeter Reading (psi)	Nominal Pipe Size or ID of Tube or Hose (in.)	Max. Supply Line Length (ft)	NLGI Grease	Lincoln Ventmeter Reading (psi)	Nominal Pipe Size or ID of Tube or Hose (in.)	Max. Supply Line Length (ft)
#0	0–100	2.00	2000	#2	300–400	2.00	500
	0–100	1.50	1500		300–400	1.50	375
	0–100	1.25	1250		300–400	1.25	310
	0–100	1.00	1000		300–400	1.00	250
	0–100	0.75	750		300–400	0.75	185
	0–100	0.50	500		300–400	0.50	125
	0–100	0.38	375		300–400	0.38	90
	0–100	0.25	250		300–400	0.25	60
#1	100–200	2.00	1000	#3	400–500	2.00	400
	100–200	1.50	750		400–500	1.50	300
	100–200	1.25	625		400–500	1.25	250
	100–200	1.00	500		400–500	1.00	200
	100–200	0.75	375		400–500	0.75	150
	100–200	0.50	250		400–500	0.50	100
	100–200	0.38	185		400–500	0.38	75
	100–200	0.25	125		400–500	0.25	50
#2	200–300	2.00	660	#3	500–600	2.00	330
	200–300	1.50	500		500–600	1.50	250
	200–300	1.25	410		500–600	1.25	200
	200–300	1.00	330		500–600	1.00	160
	200–300	0.75	250		500–600	0.75	125
	200–300	0.50	160		500–600	0.50	80
	200–300	0.38	125		500–600	0.38	60
	200–300	0.25	80		500–600	0.25	40

For modular systems, individual outlets can be configured to vary in the amount of lubricant volume to the bearing. The modular blocks can be stacked to provide larger outputs per cycle of the valves to larger bearing and smaller outputs for bearing with smaller grease lubrication requirements.

To illustrate this point see Figure 9.9 and Figure 9.10 of a modular type progressive valve.

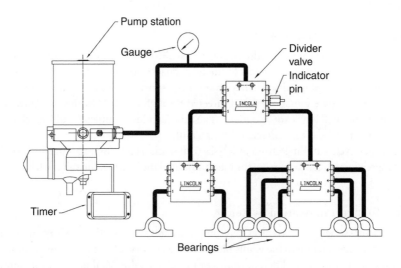

FIGURE 9.6   Typical progressive lubrication system.

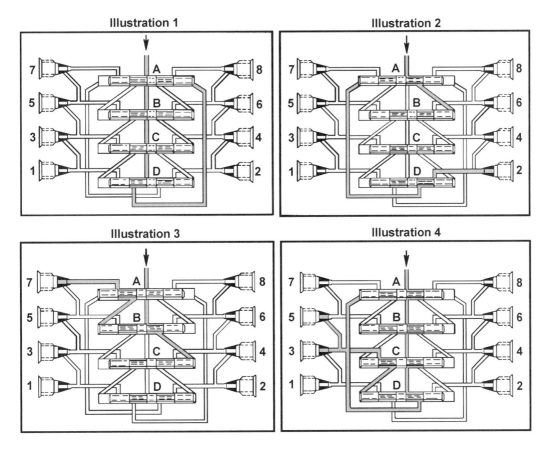

FIGURE 9.7   Valve sequence illustration.

The construction of modular type progressive valves in sections is shown in Figure 9.9. There is an inlet section, intermediate section, and end section all positioned to accept the divider valve.

Modular progressive valves can be rated for high pressure up to 7500 psi and for low pressure up to 3500 psi. High pressure valves contain very high manufacturing tolerances between the piston and cylinder to provide adequate sealing.

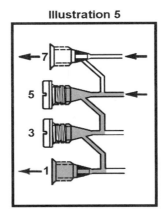

FIGURE 9.8   Crossport of divider valve.

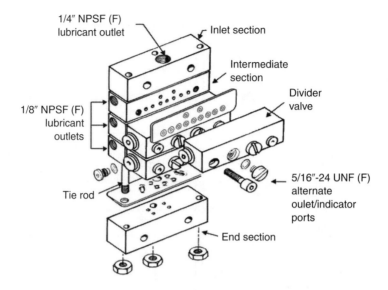

FIGURE 9.9   Modular valve construction.

Additionally, individual valves can be configured with a single or twin designation. The internal passages determine if the valve will function as either single or twin. For a single valve section, the displacement of the piston from both ends will deliver the lubricant to one outlet. In effect, the single outlet is capable of delivering twice as much lubricant as a valve configured as a twin; as the output from both sides of

FIGURE 9.10   Modular valve.

the piston is dispensed through one outlet. The operation of a twin valve configuration will deliver the lubricant to separate outlets. The face of the valve is usually marked with an "S" or "T" to designate if the valve section is single or twin.

Modular valves can also be configured to increase output delivered in a single cycle of the metering valve by crossporting one valve section to an adjacent valve section. Single valve sections can be crossported only on one side. Twin valve sections can be crossported on both sides. Manufacturers of these types of valves produce kits and design guides that show how and when the use of crossporting can be done.

9.3.1.3   Design Considerations for Progressive Systems

As with all systems, knowing the grease flow characteristics is very important at the lowest temperature the system is to work under. The grease flow characteristics can be obtained by the Lincoln Ventmeter. The amount of solid additives is important because with metal to metal fits, grease with high concentrations of solid lubricants such as low grade molybdenum disulfide, graphite, or copper antiseize additives can cause the progressive valves to lock up.

Applications where a cluster of bearings are grouped together and where there are relative short distances from pump to bearing point make this type of system most effective. The number of lubrication points can be extended by using a main supply block, called a master block or primary valve, to feed a secondary block. The lubricant is divided and distributed throughout the primary and secondary blocks to the bearing. With most progressive valves, the maximum number of outlets from one block is 18. With more than 18 lubrication points, secondary blocks will need to be added to cover the additional bearing points.

With progressive metering valves, the number of bearing points often does not match the number of outlets. For example, for a 7-point lubrication system, an 8-outlet valve minimally would be used. One outlet would have to be crossported to distribute the lubricant to 7 points. One bearing would then be getting twice as much lubricant as the other 6. All progressive valves are configured to have an even number of outlets with a minimum of 6. Progressive valves are usually available in 6, 8, 12, and 18 outlets.

Low output progressive blocks are not typically used for higher outputs of more than 0.10 in.^{3}/min for nonmodular valves and 5 in.^{3}/min of modular valves. The reason is that orifices in these valves are typically small and it is hard for grease to flow through. The orifices are usually kept small so that the amount of the piston shift can be controlled, thus keeping the accuracy of the grease volume within reasonable ranges.

Three valves at a minimum are necessary to complete the hydraulic logic and for the valves to operate. This is the reason why there can be a minimum of 6 outlets.

Never plug an output line in a series progressive valve. Plugging the outlet of a series progressive valve would be similar to causing a line blockage. With a progressive type valve, if one outlet is blocked, jammed, or prevented from moving, then all the pistons will be prevented from working and the metering block would cease to function. The dependency of one piston on another causes this to happen. If one goes, then they all go.

Crossporting can be done to match outlets with the number of bearings. This is done at the expense of some bearings getting more lubrication than the others.

9.3.1.4   System Monitoring

Because of the dependency of the proper operation of all pistons, the need for monitoring is more critical for a progressive system. Visual monitoring can be done by detecting movement of an indicator pin connected to the piston extended out through the valve body or through a sensor that can electronically detect the movement of the indicator pin. A failure of the progressive valve can be detected and an alarm can be signaled that the bearing is not getting lubricated.

As with all three types of systems, the pump can be driven pneumatically, hydraulically, or electrically. With grease systems, shovel type positive displacement reciprocating pumps are often used. (See Section 9.4 on pumping systems.) For progressive systems, smaller pumps that produce less output are normally used.

9.3.1.5   Strengths and Weaknesses

The weaknesses of a progressive system are:

1. Blockage of one outlet disables the whole system.
2. To use this system in large systems requires complex piping systems.
3. Progressive systems are not flexible to changes in the number of bearing lubrication points once the initial system is set up. When adding or removing lubrication points, the need to relay out the piping is necessary.
4. There is no easy and practical way to adjust the lubricant output to a bearing once the system is set up.
5. The output grease setting to a bearing is in multiples of the outlet volume of measuring piston. Discrete and individual bearing setting can be adjusted once the valve is installed.
6. The amount of grease flow through one valve is limited.
7. The use of close tolerance metal to metal fits makes the valves susceptible to malfunctioning when contaminants exist in the system. The use of grease with solid additives such as molybdenum disulfide or graphite is limited or not recommended.
8. The valves have to make complete cycles to distribute grease to any bearing that requires more grease than one outlet can provide in one cycle.

The strengths of the progressive system can be summarized as follows:

1. One valve instead of individual valves can be used to lubricant bearings. One valve can provide metered lubricant to a number of lubricant points.
2. System monitoring can detect a fault for every lube point should one valve or outlet be blocked.
3. There is no need for venting the system.
4. Elastomer seals are not used, which can sometimes fail prematurely.

9.3.2   Single Line Parallel Systems

The single line parallel gets its name because a single supply line is required and the measuring valves called injectors can operate independently. The heart of the system is the injector.

Basic components of the system are:

- Pump
- Controller
- Injectors
- Vent valve
- Pressure switch

The operation of a single line lubrication system is straightforward. When lubricant is needed, the controller opens an air solenoid to turn on the pump. The pump produces flow and builds up pressure in the line. When the pressure reaches 1800 psig, the injectors operate and meter a predetermined amount of lubricant to a bearing. A pressure switch usually located farthest away from the pump senses when the pressure has reached 1800 psig. Once reached, the pressure switch sends a signal to the controller indicating that the system pressure was achieved. The controller then turns off the air solenoid valve and thus the air supply to the pump. For electrically or hydraulically operated pumps, the controller will shut off electric or the flow of hydraulic fluid, respectively. For pneumatically or hydraulically operated pumps, when the air/hydraulic supply is turned off, a 3-way valve is activated, which directs any excess grease due to line expansion directly back to the reservoir. For electrical operation, the controller will shut off electric power to a 3-way vent valve. Thus, the pressure in the system can be bled off. This is known as venting the grease.

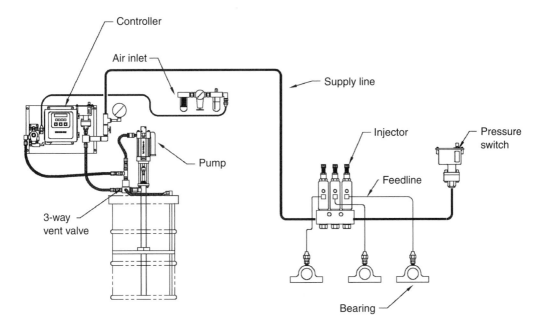

FIGURE 9.11   Single line lubrication system.

Figure 9.11 is a schematic of a traditional single line system using a pneumatically operated pump.

As with all three types of systems, the pump can be operated from all three power sources, pneumatically, hydraulically, or electrically. With grease systems, shovel type positive displacement reciprocating pumps are often used. (See Section 9.4 on pumping systems.) Because single line systems can range from small in the sense that the distance between the pump and the farthest injector is short (5 ft) to large in the sense that there are long distances between the pump and the farthest injector (500 ft), the size and power ratio of the pump can vary.

When the system is turned on, the 2-way valve is positioned to allow grease to flow to the injectors and thus to the bearings. After the injectors have metered the correct amount of grease to the bearing, the system is shut off by a controller turning off the pump. The 2-way valve is then shifted in a manner that bypasses the pump and redirects the grease back to the container, which is normally under atmospheric pressure only. This allows the line to bleed off the grease pressure or vent, thus allowing the injectors to reset and be ready for the next lube event.

The vent valve can be operated pneumatically, hydraulically, or electrically depending on the power source that the pump uses.

Figure 9.12 through Figure 9.15 are schematics of the operations of injectors with top adjustments for use in medium to large systems.

Each injector can be adjusted manually to discharge the precise amount of lubricant each bearing needs. A single injector can be mounted to lubricate one bearing, or grouped in a manifold with feedlines supplying lubricant to multiple bearings. In each case, injectors supplied with lubricant under pump, pressure pump lubricant through a single supply line. Two injector types are available: one with a top adjustment and one with a side adjustment. Both types can be used in the same system; their selection is made on the basis of bearing lubricant requirements and the general distances from pump to the last injector.

Stage 1 — The injector piston is in its normal or rest position. The discharge chamber is filled with lubricant from the previous cycle. Under the pressure of incoming lubricant, the slide valve is about to open the passage leading to the piston.

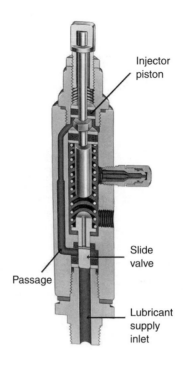

FIGURE 9.12   SL-1 or SL-11 type designation.

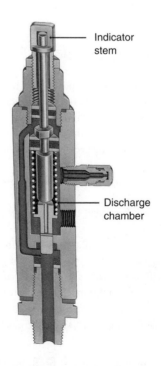

FIGURE 9.13   SL-1 or SL-11 type designation.

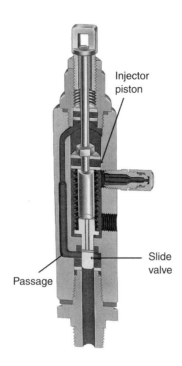

FIGURE 9.14   SL-1 or SL-11 type designation.

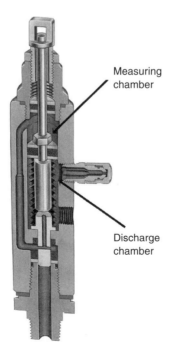

FIGURE 9.15   SL-1 or SL-11 type designation.

Stage 2 — When the slide valve uncovers the passage, lubricant is admitted to the top of the piston, forcing the piston down. The piston forces the lubricant from the discharge chamber through the outlet port to the bearing.

Stage 3 — As the piston completes its stroke, it pushes the slide valve past the passage, cutting off further admission of lubricant out the passage. Piston and slide valve remain in the position until lubricant pressure in the supply line is vented (relieved) at the pump.

Stage 4 — After pressure is relieved, the compressed spring moves the slide valve to the closed position. The piston opens the port from the measuring chamber and permits the lubricant to be transferred from the top of the piston to the discharge chamber.

Typical output for top adjustment injectors range from 0.008 in.^{3} (0.131 cm^3) to 0.080 in.^{3} (1.31 cm^3) for SL-1 type designation. Larger top adjustment injectors range from 0.050 in.^{3} (0.82 cm^3) to 0.500 in.^{3} (8.2 cm^3) for SL-11 type designation.

Figure 9.16 and Figure 9.17 are schematics of injectors with side adjustment used in small applications.

Stage 1 — Under pressure from the supply line, incoming lubricant moves the injector piston forward. The piston forces a precharge of lubricant from the discharge chamber through the outlet check valve to the feedline.

Stage 2 — When the system is vented (pressure relieved), the piston returns to the rest position, transferring lubricant from the measuring chamber to the discharge chamber.

The venting pressure of 200 psi is typical of the side adjustment injectors. Because of this lower venting pressure, less grease supply line distances can be realized.

Typical output of side adjustment injectors range from 0.001 in.^{3} (0.015 cm^3) to 0.008 in.^{3} (0.131 cm^3).

A typical injector incorporates seals. Only the bushing and plunger use metal to metal fits. The measuring chamber uses viton seals. The viton seals were chosen because their resistance to attachments by solvents or chemical additives in some greases and because of the wide tolerance in which these seals can operate.

9.3.2.1   The Quick Venting Single Line Injectors

What do they do? This is a new technology and is the state of the art in metering valves used in automatic lubrication. It is the most significant development in automatic lubrication since the invention of the traditional single line injector introduced in 1938. The quick venting injectors can reset at higher pressures, thus allowing a system to use smaller lines than traditional single line injector systems. With the quick venting injectors, the system can handle heavier grease, which is more desirable. The compromise to use a softer grease for most automatic lubrication systems is greatly reduced. Grease with NLGI rating of 2 can be used in much colder temperatures, often down to 20°F or less.

Figure 9.18 is a schematic of the basic operation of a quick venting injector and how it works.

Stage 1 — The injector is in its normal position. The discharge chamber is filled with lubricant from the previous cycle. Under pressure of incoming lubricant, lubricant is directed to both sides of the measuring piston through the slide valve. The port of the bearing is closed in this position, which prevents the measuring piston from moving. The indicator stem will be at its innermost position, having pulled away from the stop in the adjusting screw.

Stage 2 — Pressure has built up and has moved the slide valve in the position shown. This closes the flow to the upper side of the piston (larger diameter) while simultaneously opening the port to allow lubricant to flow out of the injector to the bearing. Pressure from the supply line continues to apply pressure to the low portion of the measuring piston, which causes a pressure difference across the measuring piston, thus allowing it to move upward.

Stage 3 — Movement of the measuring piston shown is caused by the pressure on the lower side of the measuring piston dispensing lubricant to the bearing. The indicator stem will move up against the stop in the adjusting screw when the lubricant has been delivered to the bearing.

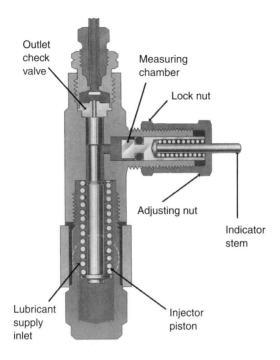

FIGURE 9.16   Side adjustment injector operation. SL-32 or SL-33 designation.

Stage 4 — As the pressure in the supply line is vented down to 1200 psi, the slide valve moves back to the piston shown. This closes the flow of lubricant to the bearing and simultaneously allows lubricant to flow to the upper (larger diameter) side of the piston. The remaining pressure in the line is directed to both sides of the piston and is equalized both on top and bottom of the piston. Because the piston diameter is larger

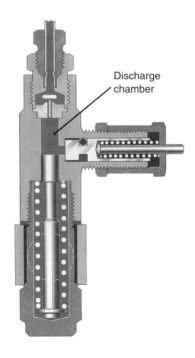

FIGURE 9.17   Side adjustment injector operation. SL-32 or SL-33 designation.

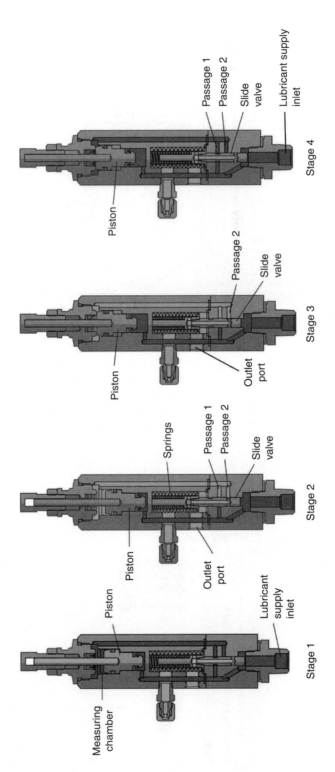

FIGURE 9.18   SL–V injector operation.

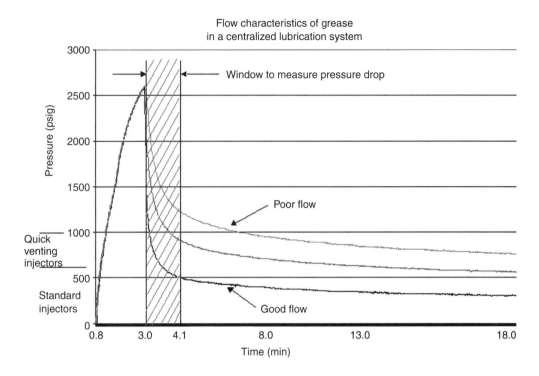

FIGURE 9.19   Flow characteristics of grease.

on the top, a net force results in the down direction causing the piston to move accordingly. The injector is recharged by the residual pressure in the supply line to the upper portion of the measuring chamber. The displacement of the fluid on the lower side of the measuring chamber is also allowed by the slide valve to flow to the upper side of the piston, thus completing the recharge of the injector. The resulting effect by absorbing lubricant from the supply line to recharge the injector is to reduce the pressure in the supply line close to zero.

The traditional limitation of a single line system is the necessity of the lubricant to vent back down under system pressure in order for the injector to reset. While this is still required for quick venting injectors, the venting now can be done at a much higher pressure. In other words, the pressure in the system only has to be vented to 1000 psi as opposed to 600 psi for traditional injectors. Because of this, systems requiring long runs from the pump to the furthest injectors are no longer a limitation.

To illustrate this point, consider the flow properties of grease in a centralized lubrication system. Figure 9.19 is a typical chart on how grease will flow when under pressure. The lubrication supply pump will build up pressure, normally above 2500 psi. At this pressure, the injectors will operate and lubricate the bearing. As the system is then vented, the pressure in the system drops as shown in the figure. At the point above 1000 psig, the grease will flow quickly. Below 1000 psig, the amount of grease that flows and vents back is slowed down considerably.

Tables 9.1, 9.2, and 9.3 show the venting performance of injectors and how long lengths that can be achieved with single line systems depend on the type of injector.

An important fact with all types of injectors is that they can be individually adjusted. The output of each injector can be adjusted by the position of the adjusting cap. The adjusting cap limits the travel of the piston, thus limiting the amount of the lubricant to be delivered to the bearing.

Figure 9.20 is a schematic of a top adjustment injector where (1) indicates the nut for limiting the travel of the injector pin (2). This can be screwed in and out, thus increasing or decreasing the output. A lock nut (3) is used to secure the setting after the adjustment is made.

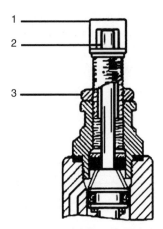

FIGURE 9.20   Top view of SL-1 or SL-11 type top adjustment injectors.

9.3.2.2   Design Considerations in a Single Line System

As with all systems, knowing the grease flow characteristics at the lowest temperature the system is to work under is very important. The grease flow characteristics can be obtained by the Lincoln Ventmeter. Knowing the Lincoln Ventmeter viscosity, the three supply line chart tables provided earlier can be used to determine the size of the supply lines. For single line systems, the work has all been done. Because of the use of seals, the ability to pass solid additives in the grease is greatly enhanced.

The sizing of the pump that is appropriate for the type of injectors used in the system is important. With long distances, the amount of lubricant needed to be pumped is often much greater that what will be dispensed out of the system. This is because of line expansion as grease is pumped into the system. For example, for a 28-point injector system using a total of 25 ft of 1/2 in. supply hose, the output of the injectors of the SL-1 type designation (if set at maximum output) would be 2.16 in.^{3} (0.08 in.^{3} × 27 = 2.16 in.^{3}). Using information from supply line expansion charts for SAE 100R2 hose, the amount of lubricant that would be absorbed by the hose expansion would be approximately 10 in.^{3}. For injector systems, the pump will need to supply initially twice the amount of lubricant for each injector for the purpose of dispensing and recharge, which would be 4.32 in.^{3} plus an additional 10 in.^{3} for supply line expansion, totaling 14.32 in.^{3}. Therefore, a pump would have to be selected that can produce 14.32 in.^{3} in the time needed before the next lubrication event. When the system is vented, the expansion grease is returned back to the reservoir.

It should be noted that for supply line sizes for steel pipe or tubing, the expansion would be much less. For a 1/2 in. schedule 40 pipe, the supply line expansion would be approximately 1 in.^{3}. A pump could be selected that would require 5 or 6 in.^{3}. Lincoln industrial design guide for single line systems contains all information for any possible system or system combinations for determining line sizing and pumping requirements.

Crossporting is another important feature that is available for single line injector type systems. Crossporting connects two or more injectors together allowing for additional outputs. Figure 9.21 is a schematic on how this is achieved. Output of the injectors can be doubled or tripled depending on how much crossporting can be achieved.

Crossporting is achieved by connecting the output of injector 1 to injector 2. The output of injector 2 is sent to the bearing. The output to the bearing in this configuration can be up to twice as much as with one injector.

9.3.2.3   System Monitoring

System monitoring can be done visually or through electronic detection of movement of the indicator pin. By visual inspection during the operation, the injector operation can be verified. For critical bearings,

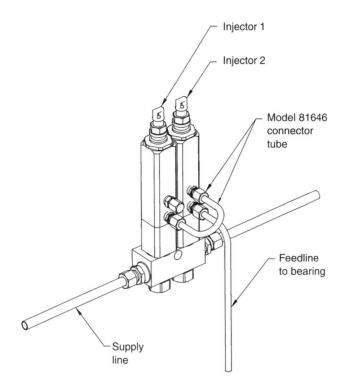

FIGURE 9.21   Quick venting injectors.

electronic detection can be incorporated by installing transducers to sense the movement of the pin. Should the sensor not detect any movement of the indicator pin, an alarm can be given indicting that the bearing is not getting grease.

Single line systems can also be used to detect if the bearing is getting grease. This is done by installing a grease flow sensor at the bearing. The grease flow sensor can detect small amounts of flow to the bearing or a lack of flow. If the sensor does not detect flow during a lube event, an alarm signal can be sent to a controller indicating a system problem.

9.3.2.4   Strengths and Weaknesses — Single Line Systems

The weaknesses of single line systems are summarized below:

1. For traditional systems, the venting of the injectors limits the distances that can be achieved from pump to the last injector in the system. Pipe and tubing diameter sizes required are larger than those of progressive or two line systems.
   For quick venting injectors, there is no practical limitation on distances and the piping sizes required. For long runs, quick venting injectors can outperform single line or dual line systems.
2. Monitoring a complete system is more difficult due to the fact that a separate transducer would have to be placed at each individual injector.
3. The elastomer seals used deteriorate faster with contamination.

The strengths of a single line system are summarized below:

1. Each injector services one lubrication point, making it easy to trace the metering valve to a single bearing.
2. Each injector output can easily be set individually to meet the lubrication requirements of a bearing.
3. Only one single line is needed, making the installation simple and straightforward.

4. The system is easiest to plan and understand.
5. The system is very flexible as lubrication points can be added or removed after the initial system is installed. To add lubrication points, just add injectors; to remove bearing lube points, just remove injectors.
6. The use of elastomer seals allows contaminants to pass through the system. Close metal to metal fits on the measuring chamber are not used.
7. The injectors can be used and are recommended for greases that contain solid lubricants such as molybdenum disulfide or graphite.
8. The grease in the systems does not remain under pressure. This is accomplished by venting of the grease. This reduces the changes of the base oil separating from the thickener.

9.3.3   Dual-Line Systems

The name implies that two main lubrication lines are used to set up, install, and operate the system. Correctly designed, the dual-line system can handle long lines, relatively high pressure, and more than 1000 lubrication points. A high pressure dual-line system is capable of lubricating bearings located at long distance from each other. Major components of the system are:

- Metering valves
- Reversing 4-way valve
- Pressure switch or transducer
- Lubricant pump
- Controller/timer

In the dual-line systems, a pump supplies the lubricant to the reversing 4-way valve. From the reversing valve, lubricant is supplied alternately into one of the two main lines. Dual-line systems can be combined with progressive single-line measuring valve as well. The system is suitable for either oil or viscous grease lubricants.

9.3.3.1   Basic Operating Principle

The dual line lubrication system works in two cycles. The central lubrication pump supplies the lubricant under pressure to main line "A" through the reversing 4-way valve. Main line "B" is connected to the reservoir. The metering valves are connected to the main supply lines "A" and "B." The lubricant is dispensed under pressure from one side of the metering valves to the point of application. As soon as the lubricant is dispensed from the last metering valve, the first half of the cycle is complete (see Figure 9.22). The lubrication pump will continue to operate, pressurizing the line "A" to the preset pressure. As soon as the preset pressure is reached, the reversing 4-way valve will switch the lubricant supply to the main line "B," connecting main line "A" to the reservoir.

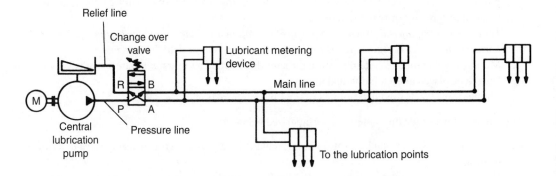

FIGURE 9.22   Dual-line lubrication system — first half cycle.

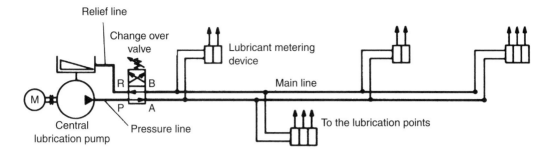

FIGURE 9.23   Dual-line lubrication system — second half cycle.

Now the pump supplies the lubricant under pressure to the main line "B." Line "A" is connected to the reservoir and pressure is relieved. The lubricant will be dispensed from the opposite side of the metering valves to the point of application (see Figure 9.23). The second half cycle is complete as soon as the lubricant is dispensed from the last metering valve. The pump will continue to operate until preset pressure has been reached. At this point, a signal from the end-of-line pressure switch or from the micro switch on the reversing 4-way valve will stop the pump, turning the system off.

9.3.3.2 Metering Valves

A dual-line metering valve is a positive displacement metering device with an adjustable stroke piston to dispense measured volumes of oil or grease. Figure 9.24 illustrates a schematic of a metering valve. The valve has two output ports. After adjusting the valve to the desired setting, it will dispense an equal volume of lubricant through each of two outlets. If application requires more lubricant, one outlet port can be closed and plugged and the remaining port receives twice the preset volume of lubricant. The valve is designed to withstand a harsh environment in steel, glass, and mining industries.

The valve body and internal parts construction are a carbon steel material. The seal material is a nitrile or fluorocarbon compound rubber. The valve consists of the pilot piston to direct the inlet lubricant flow, output metering piston with indicator pin, adjustment sleeve, two jam nuts, and crossporting rotary valve with lock nut. The sleeve has a transparent cover cap. The cover cap on the adjustment sleeve protects the seal of the indicator pin from dust and dirt in harsh environments. The movement of the indicator pin is used for visual confirmation.

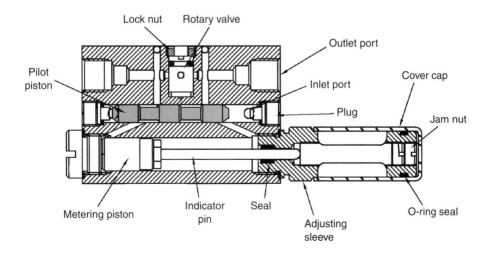

FIGURE 9.24   Metering valve.

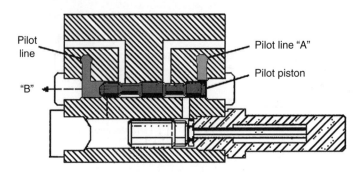

FIGURE 9.25   Stage 1.

There are two types of valves manufactured: high pressure valve for up to 5000 psi (340 bar) and medium pressure valve for up to 3500 psi (238 bar). Regardless of the pressure rating, the valve operation is identical.

9.3.3.2.1   *Operation of the Metering Valves*

Stage 1 — See Figure 9.25. Pressurized lubricant enters the valve through the pilot lines "A," forcing pilot piston to the left and opening the right pilot connecting port passage. A small amount of displaced lubricant is relieved or vented through pilot line "B," which is open to the reservoir.

Stage 2 — See Figure 9.26. When the pilot piston uncovers the left connecting port passage, lubricant enters the passage, pressurizing the top of the metering piston. The metering piston moves full stroke to the left, dispensing lubricant out through the outlet passage to the lubrication point. This completes the first half of the lubrication cycle.

Stage 3 — See Figure 9.27. With piston at the terminal position at the end of the stroke, lubricant pressure will rise to the preset point. A reversing 4-way valve switches the lubricant supply from main line "A" to main line "B." Now the pressurized lubricant enters the valve through pilot line "B," forcing the pilot piston to the opposite right position, opening the left pilot connecting passage. Again, a small amount of lubricant is displaced through line "A" now open to the reservoir.

Stage 4 — See Figure 9.28. Lubricant enters the left pilot connecting port passage, pressurizing the bottom of the metering piston. The metering piston moves full stroke to the right, dispensing lubricant out through the outlet passage to the lubrication point. This will complete the second half cycle.

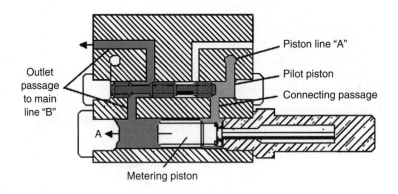

FIGURE 9.26   Stage 2.

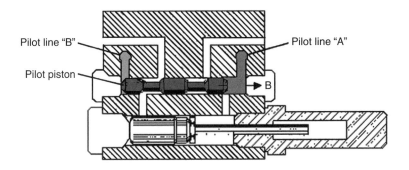

FIGURE 9.27   Stage 3.

9.3.3.2.2 *Valve Adjustment*

The valves are available in 2, 4, 6, or 8 outlet configurations. The output of the lubricant is infinitely adjustable from almost zero to the maximum specification volume of the valve. To adjust the output volume remove the sleeve cover, loosen and remove the upper jam nut, and turn the second nut in the desired position, limiting the travel of the indicator pin and metering piston. After the desired output has been set, replace the upper nut to lock the setting and install the sleeve cover back (see Figure 9.29).

9.3.3.3 **Reversing 4-Way Valve**

There are several reversing 4-way valve constructions. The most commonly used are:

- Hydraulic
- Electric

The hydraulic and electric valves are directional flow valves with pressure sensitive mechanisms to alternate the flow of lubricant from one line to the other at a predetermined pressure setting. Reversing action is automatic, actuated by pressure from the lubricant supply pump. The line that is not under pressure is vented back to the reservoir.

9.3.3.3.1 *Hydraulic Reversing 4-Way Valve*

The valve consists of the following basic components:

- Flow control piston with indicator pin
- Pressure sensing mechanism with adjustment
- Limit switch
- Pressure gauges (optional)

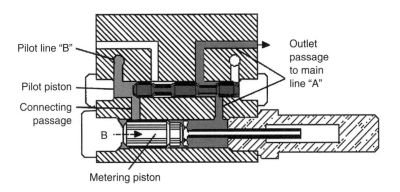

FIGURE 9.28   Stage 4.

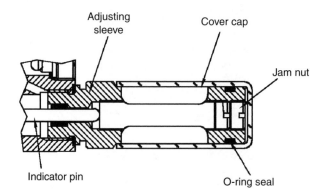

FIGURE 9.29   Output volume adjustment.

Turning the adjusting nut changes the preset of the pressure to reverse the flow of lubricant, from one main line to another. The adjusting nut will increase or decrease the spring force exerted on the pilot piston. The larger the spring force, the more pressure it takes to overcome it and to reverse the lubricant flow. The pressure sensing mechanism is acting as an overcentering device as well. As soon as the pilot piston crosses the middle position, it will complete the stroke using the full spring force. The optional pressure gauges are for visual monitoring of the switch over pressure only.

9.3.3.3.1.1   Operation of the Hydraulic Reversing 4-Way Valve — Nonreturned system

Stage 1 — The reversing piston directs the lubricant flow from the inlet port to supply line "A." Supply line "B" is connected to the reservoir (see Figure 9.30).

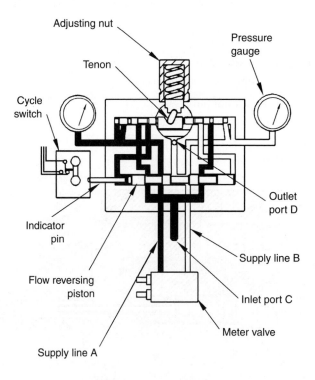

FIGURE 9.30   Nonloop stage 1.

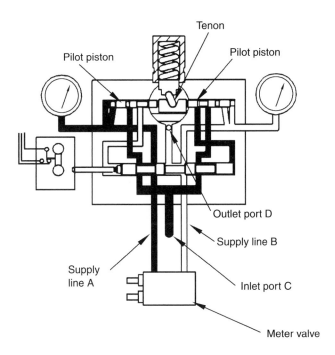

FIGURE 9.31 Nonloop stage 2.

Stage 2 — Rising pressure in supply line "A" forces the pilot piston to the right, overcoming the spring force of the tenon. The pilot piston shifts, pushing the excess lubricant to the reservoir through the outlet port "C" (see Figure 9.31).

Stage 3 — Pressure from the supply line moves the flow reversing piston to direct the lubricant flow from main line "A" to main line "B." When the reversing piston completes the full stroke, it trips the cycle switch to stop the pump. When the controller/timer starts the next lubrication cycle, main line "B" will be pressurized and line "A" will be relieved (see Figure 9.32).

9.3.3.3.2 Electric Reversing 4-Way Valve

The electrically controlled reversing valves use an electric signal from the pressure switch or the pressure transducer to reverse the lubricant flow from one to the other line. The pilot piston of the electric reversing valve is driven by electric motor with a camshaft or electromagnetic solenoid. When preset pressure has been reached, the signal from the pressure switch starts the electric motor or energizes the solenoid of the reversing valve to move the pilot piston and switches the lubricant flow from one main line "A" to main line "B." The principal of operation is similar to the hydraulic reversing valve.

9.3.4 Types of Two-Line Systems

There are three basic layouts of the two-line system that can be installed, depending on application:

- End-of-line system
- Dead-end system
- Loop system

Each of the layouts has advantages and disadvantages in cost, installation, and maintenance.

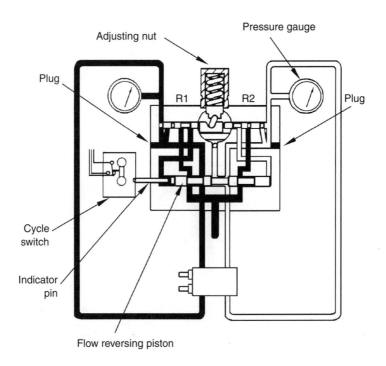

FIGURE 9.32   Flow reversing piston stage 3.

9.3.4.1   End-of-Line System

This system is preferred whenever the lubrication points are spread over long distances. Use of the electrical reversing 4-way valve is recommended. Figure 9.33 is a schematic of a typical end-of-line system. The pump station and electrical reversing valve are positioned in the middle to minimize the pressure drop and to equalize the pressure in both branches of the lubrication line. End-of-line pressure switches or pressure

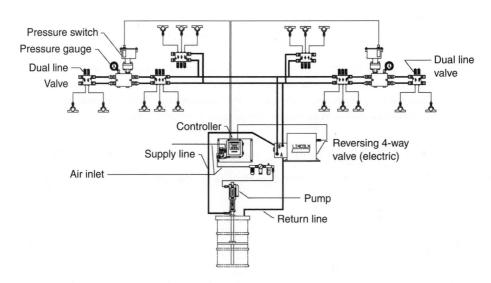

FIGURE 9.33   Typical end-of-line system.

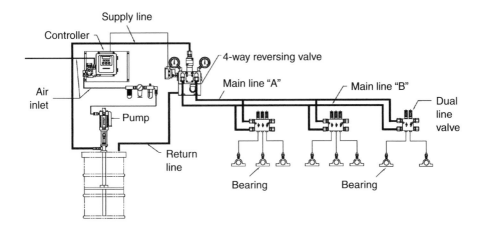

FIGURE 9.34 Typical dead-end (nonreturn) system.

transducers are installed to monitor the line pressure and give the signal to the electric valve to switch the flow at the preset pressure. The controller or timer starts the lubrication cycle and sets the lubrication frequency of the system. Upgraded systems are using flow and pressure monitoring accessories to alert the failures.

The cost of the system will depend on the selected pump and electrical reversing valve. In addition to the lubrication lines, electrical wires are needed to connect the pressure switches or transducers to the controller or timer, adding to the cost of material and installation.

The advantage of this type of system is that it does not depend on the temperature fluctuation and lubricant variations in viscosity.

9.3.4.2 Dead-End System

This system is preferred whenever lubrication points are located in a long line in close proximity to each other or in clusters/zones. The system can use either hydraulic or electrical reversing valve. Figure 9.34 is a schematic of a typical dead-end system. The pump station and the reversing valve are installed at one end of the system. The controller or timer starts the lubrication cycle and sets the on and off time of the system.

With a hydraulic reversing valve, this is the most economical system and commonly used in the steel and cement industries. The switch over pressure of the hydraulic reversing valve has to be preset considering the pressure fluctuations due to the temperature and lubricant viscosity variations.

The disadvantage is that it requires more maintenance and service. Seasonal adjustments of the hydraulic reversing valve switch over pressure are recommended. Adjustments are recommended if the lubricant brand or lubricant formulation changes as well. The system can be upgraded with additional accessories to monitor the lubricant flow and pressure.

9.3.4.3 Loop System

This type of system is preferred with the use of a hydraulic reversing valve and whenever better control of the line pressure is necessary. The main lubrication lines are connected to the reversing valve in a complete closed loop. Figure 9.35 is a schematic of a typical loop system. See the description of the valve for line connection in a loop system. The better control of the line is achieved by connecting each line to the opposite ends of the pilot piston and sensing the pressure at the end of the line. The system requires additional line connections, increasing the installation cost. The size and length of the feedlines to the metering valves have to be selected properly to maintain appropriate pressure drop for the valves to operate correctly. Properly designed, the advantage of the loop system is in more stable operation during temperature and lubricant property changes, without using costly monitoring accessories.

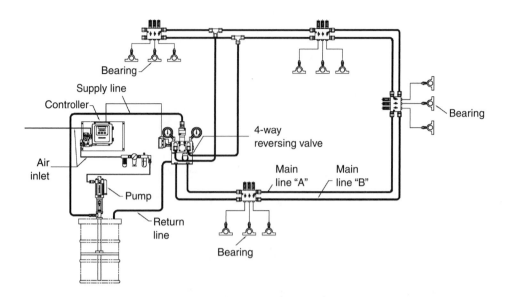

FIGURE 9.35   Typical loop system.

9.3.4.4   Design Considerations in a Two-Line System

Proper selection of the grease and components for the lubrication system is very important. Dual-line systems have certain limitations in selecting main supply lines. Total pressure drop between the lubrication point and hydraulic reversing valve should not exceed 2250 psig (153 bar). Pressure available at each metering valve should be 500 psig (34 bar) or greater. Pressure drop in the main supply line should not exceed 1500 psig (102 bar) for dead-end (nonreturn) systems and 2000 psig (136 bar) for loop systems.

Pressure drop of the branch lines should be less than half of the pressure drop in the main line. Selecting the pump consideration should be given to the pump output. Grease in long lines can compress up to 20% in volume. To better monitor the metering blocks, install the monitoring blocks parallel to the main line. Connect the block inlets and outlets to the same respective lines, so that after completion of one cycle or half-cycle the indicator pins will be in the same position.

9.3.4.5   Advantages and Disadvantages of the Dual-Line Systems

The advantage of the dual-line system is summarized below:

1. Long distances can be achieved.
2. Additional lubricant points can be added without changing the main lubrication lines.
3. Long history in industry.
4. Can be used in a system with a large number of lubrication points.
5. Uses metal to metal fits in the metering valves. This can be an advantage because there are no elastomer seals to wear.

The disadvantages of the dual-line system is summarized below:

1. Requires two lubrication lines and double the amount of fittings and mounting hardware.
2. Old technology.
3. Most two-line valves use metal to metal fits in the metering valve. Use of lubricants with solid additives may be prohibited. With metal to metal fits, the system is more susceptible to failure when contamination is present.

9.4 Pumping of Grease and Viscous Materials

In automatic grease lubrication systems, one common component is the pump. The pump must be able to remain primed (avoid cavitation) and deliver the correct amount of volume under back pressure. Because greases can have high apparent viscosity, back pressure is created to overcome the yield pressure of the grease and the friction loss that is produced when grease is flowing in the supply line. Typically, for NLGI number 2 grease, with a Lincoln Ventmeter viscosity of 400 psi, the apparent viscosity can be in the range of 70,000 to 100,000 cP or even higher.

The pump must have the capability to remain primed. That is to say, the amount of grease entering the pump must be sufficient to charge the pumping chamber. Almost all grease pumps operate with positive displacement reciprocating action. If the grease is too stiff or has too high of an apparent viscosity, grease may not flow into the pump. When this happens, the pump cavitates. Cavitation should be avoided as it can cause premature pump failure. Moreover, a pump that is cavitating cannot pump grease into the system, thus the bearing will not be getting lubrication. For pumps mounted in refinery drums such as a 55-gal (400 lb) or 120-lb drums, the pump must be able to pick up the grease under operating temperatures.

A positive displacement double acting shovel action pump is most used for pumping viscous material. The pumps are described as double acting because they output grease when the pump is in both the up- and downstroke. There are two pumping chambers, one for the upstroke and one for the downstroke. Positive displacement pumps create suction in their action. The pump operates in a piston/cylinder arrangement. The piston displacement creates a vacuum. This vacuum is used to create a pressure differential causing grease to flow. Because there is only 14.7 psi vacuum pressure possible, this may not be enough to produce flow from the grease reservoir to the pumping chamber. To overcome this situation, a mechanical shovel is used to mechanically push the grease into the pumping chamber. See Figure 9.36 and Figure 9.37 for a description on how they work.

The operation of a double acting shovel pump produces the same output when the piston is in the up- or downstroke. After the pump is inserted in a grease reservoir, the pump is first primed by turning on the pump and removing any air out of the pump. When the pump is turned on, the pump in the upstroke uses the mechanical shovel to force the grease into the pumping chamber. All volume of grease entering the pump occurs on the upstroke only. The pump does not accept grease on the downstroke.

There are two pumping chambers with double acting shovel pumps. The grease that is entering the pump tube first enters the first pumping chamber. The inlet check opens during the upstroke cycle. Simultaneously, the second pump chamber volume is compressed, forcing lubricant out of the pump.

During the downstroke, the outlet check closes and the pump piston fills the second pumping chamber while dispensing lubricant out of the pump. Because the displaced volume of the first pump chamber is twice that of the second, the grease fills the second pump chamber as it dispenses.

It is important that a pump produces the same amount of pressure and flow on the downstroke as the upstroke. If the pump ratio is, say 50:1, the pump should be able to generate the same pressure on either the up- or downstroke at that ratio.

9.4.1 Sealing Methods for Pumping Viscous Materials

The weakest point of any reciprocating pump is the gland area. This is the seal area at the top of the pump. There are two major factors affecting the gland of the reciprocation pump. The first factor is the environment of the exposed pump piston. The second is the fluctuation of internal pressure produced when the pump changes from an upstroke to a downstroke and vice versa. The exposed portion of the pump plunger can be contaminated and gradually destroys the gland seal. The internal pressure fluctuation, developed by pump during up- and downstrokes, deteriorates the gland seal contributing to possible premature failures.

To improve a pump's ability to maintain a good seal, there are several designs available. Some pumps use a staked seal design. The features of this seal arrangement allow for retightening when grease begins to leak out the top of the pump. This retightening reforms the seal and is often effective in stopping the leak.

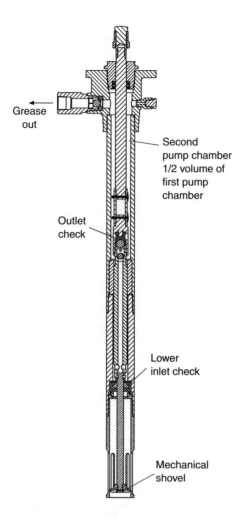

Grease
out

Second
pump chamber
1/2 volume of
first pump
chamber

Outlet
check

Lower
inlet check

Mechanical
shovel

FIGURE 9.36   Shovel pump in downstroke.

This is only good to a point and retightening cannot be performed infinitely. The most effective design is the special gland protection package incorporated into the pump.

The gland protection consists of a spring type scraper ring and a labyrinth bushing. The scraper sharp edge is in constant circumference contact with the plunger and cleans/scrapes the plunger surface from any particles that can damage the gland seal. The labyrinth bushing is protecting the gland seal from the pressure fluctuations due to the labyrinth path of the material before it reaches the gland seal (see Figure 9.38). This new technology is successfully proven and tested in many of the toughest environments.

Techniques are available to keep the pump primed when the grease being used is very stiff or has a high apparent viscosity. Often, the ambient temperature drops resulting in a drastic increase in the grease stiffness. The solution is to change to a lighter grade grease or to improve the priming capability of the pump. Those techniques are described below.

9.4.1.1 Create Positive Head Pressure by Using a Follower Plate

A follower plate can be used to create an additional pressure head, which will prevent a void from forming around the pump inlet. If grease is too stiff, and the pump draws in grease, a void could be created. This void will cause cavitation. The principal behind this technique is using differential pressure produced by the pump's ability to produce a vacuum. When the pump displacement causes a vacuum, some grease will flow into the pump chamber. Simultaneously, this will create a pressure differential across the follower

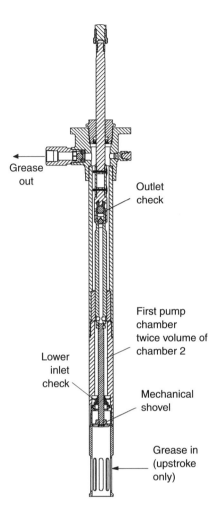

FIGURE 9.37 Shovel pump in upstroke.

plate. The pressure differential across the follower plate may be small, but the net force produced becomes large, creating positive head that will prevent voids or pockets. The simple relationship can be illustrated as follows. A typical follower plate for a 55-gal refinery drum may be 24 in. in diameter. This results in an area of 452 in.^{2}. If the pump can produce just a small vacuum of 2 psi, the net force acting on the follower plate would be F = Pressure × Area or 2 psi × 452 = 904 lb. This net downward force will cause any void in the grease to collapse. Figure 9.39 is an example of a pumping arrangement using a follower plate.

9.4.1.2 Create Positive Pressure by Using a Pressurized Reservoir

The grease reservoir can be pressurized within the structural limits of the reservoir. A contained grease reservoir can be pressurized to, say, 10 psig. This 10 psig plus ambient pressure can be added to force the grease into the pumping chamber. Good positive displacement pumps should be able to produce 12 psi of vacuum pressure. This would result in a differential pressure between the grease and the pumping chamber of 12 psi vacuum plus 10 psig to develop 22 psig of differential pressure.

9.4.1.3 Use Pressure Primer to Force the Grease into the Pumping Chamber

For more severe conditions, a pressure primer can be used. A pressure primer uses mechanical actuators pressurized by air to apply a downward force onto the follower plate. This action forces the grease into the pumping chamber. Pressure primers are often used on applications where the grease is stiff due to

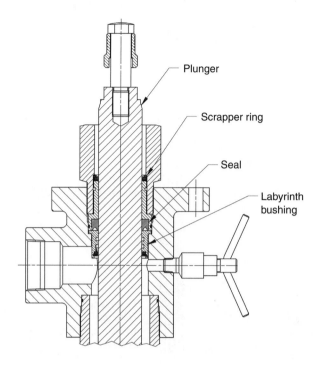

FIGURE 9.38   High pressure pump seal configuration.

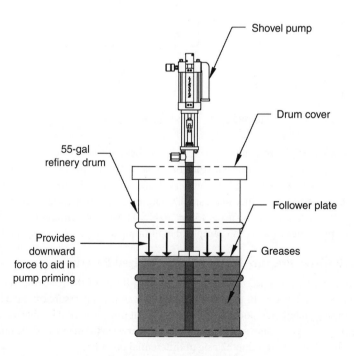

FIGURE 9.39   Grease drum follower plate.

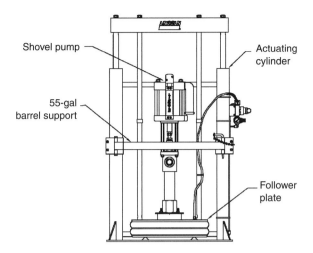

FIGURE 9.40   High pressure primer configuration.

the requirement of the application combined with cold temperature environments. Figure 9.40 is an illustration of a pump set up in a pressure primer system.

The illustration shows a pressure primer with the actuating cylinder in the retracted position. When the actuation cylinder is extended, a 55-gal refinery drum containing the viscous material is placed under the follower plate. The air control valve is switched to the retraction position, thereby producing a continuous downward force on the follower plate, thus providing positive head pressure to keep the pump primed. When the 55-gal refinery drum is empty, the actuating cylinder is extended pushing the follower plate out of the drum. The barrel supports keep the drum in place while the follower plate is being removed.

Grease pumps are usually specified by the amount of pressure that can be developed. For pneumatically and hydraulically operated pumps, the pump ratio is the amount of grease pressure the pump can develop over the amount of air/hydraulic pressure supplied to the motor. For example, a 50:1 pressure ratio means that if 80 psi of air pressure is supplied to the motor, the pump will be able to produce 4000 psi. The pump would stall out at this pressure.

When specifying pumps, manufacturers produce flow and output curves based on the pressure needed to be developed. Figure 9.41 is an example of typical pump curves for positive displacement pneumatically driven pumps.

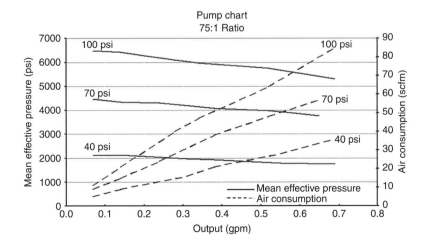

FIGURE 9.41   Mean effective pressure vs. output chart.

For electric driven grease pumps, the ratio is not as straightforward. For DC and AC operated grease pumps, the manufacturer will specify the maximum pressure the pump will achieve, depending on the motor horsepower and gear transmission design of the pump.

Because grease pumps can produce high pressure, safety must be a concern when designing a system. The designer of the system must select the hardware and piping to handle the amount of pressure the system will develop. Relief valves must be installed in the system at the correct locations to vent out any excessive pressure before damage or harm could occur.

10

Used Oil Recycling and Environmental Considerations

Dennis W. Brinkman
Indiana Wesleyan University

Barbara J. Parry
Newalta Corporation

10.1 Introduction

Lubricants are unique among all petroleum products in that often they are not consumed during use. Solid lubricants, such as greases, are not considered here as they are seldom in a form to be easily recovered at the end of their useful life. Most fluid lubricants, such as crankcase oil and transmission fluid, are collected and replaced as part of normal equipment service. These "used oils" are now available to become either a useful resource if handled properly or a potentially hazardous waste if ignored or discarded.

Used oil reclamation is not a new idea. As early as World War I, there was recognition that used oils represented a resource that could be accumulated and utilized instead of thrown away. Starting with World War II, there was a concerted effort to recycle used oils, such that by 1960 there were 150 separate businesses throughout the United States producing several hundred million gallons of crankcase oils and other petroleum products each year. Similar infrastructure was in place throughout the world [1]. "Today, there are around 400 known re-refining facilities [worldwide] with an overall capacity of 1800 kt/yr" [2].

Tremendous changes in technology and environmental concerns have significantly impacted the used oil recycling industry in the past 50 years. Lubricants have become much more complex in composition and function. This makes the challenge of processing these fluids after use increasingly difficult. Additionally, the increased use of bio-based oils and fuels is a challenge to the re-refining industry, since these are incompatible with most current technologies.

Used oils have been shown to have the potential for serious environmental damage as well as health and safety risks if not handled properly. This has led to a significantly higher intensity of government regulations, which is usually more than a small business can handle. Thus, the industry has gone through several decades of consolidation and enhancements that have resulted in collection of used oil in larger volumes to the betterment of the environment.

Many studies have been done on used oil characteristics, recycling technologies, and related topics [3,4]. The most recent wave of research into used oil recycling technology peaked in the 1970s and 1980s, during the period of oil embargoes. Once the OPEC crude oil production rates had stabilized at a rate acceptable to the developed countries, government funding for this research was eliminated and the economic driving force for private investment faded. More recently, used oil studies have focused on environmental stewardship and life cycle analysis. Interest in finding ways to increase collected used oil volumes and improve the quality of recycled products is again on the rise. This renewed interest is primarily due to pollution control and prevention advocacy initiatives such as the Kyoto Protocol, risk averse management practices, and recognition that used oil is, arguably, the largest single source of hazardous recyclable material.

When one does an Internet search using the term "used oil recycling," thousands of entries are found. With a few exceptions, each entry represents a government agency or program that is intended to inform its constituents about the potential harm caused by uncontrolled dumping of used oil and filters, new and proposed legislation regarding recycling and stewardship programs, and suggestions for recycling. The websites often contain links to other information sources as well as listings of locations that accept used oil. The number of sites underscores the importance that environmental agencies have placed on maximizing the proportion of used oil generated that is collected for recycling.

10.2   Terminology

After collection, used oil recycling consists of a wide variety of activities yielding a range of products. Thus, it became necessary to appropriately define terms used to describe various recycling approaches. While regulatory agencies have often crafted independent definitions, consensus terminology was created and published by American Society for Testing Materials (ASTM) International in the 1980s. These are found in ASTM Standard D4175 and are reproduced in Table 10.1 [5].

From a practical standpoint, the following are typical descriptive terms in common use:

1. Used oil — oil that has, through use or contamination, become unsuitable for continued use in its current application, but which is likely to have some use in another application or as a feedstock for a process that generates a useful product.
2. Waste oil — oil that has become so degraded or contaminated that it is impractical to recover anything useful from it other than its heat content (if that can be done in an environmentally sound manner). With the increased understanding of lubricant chemistry, a focus on pollution

TABLE 10.1   Consensus Terminology

Reclaiming	The use of cleaning methods during recycling primarily to remove insoluble contaminants, thus making the oil suitable for further use. The methods may include settling, heating, dehydration, filtration, and centrifuging.
Recycling	The processing of oil that has become unsuitable for its intended use, in order to regain useful material.
Re-refining	The use of refining processes during recycling to produce base stock for lubricants or other petroleum products from used oil.
Used oil	In petroleum product recycling, oil whose characteristics have changed since being originally manufactured, and which is suitable for recycling (see also waste oil).
Waste oil	In petroleum technology, oil having characteristics making it unsuitable either for further use or for economic recycling.

Source: ASTM Annual Book of Standards, Vol. 5.02, D4175, Standard Terminology Relating to Petroleum, Petroleum Products, and Lubricants.

prevention, and the enhanced sophistication of recycling facilities, very little lubricant needs to end up here.

3. Recycling—This is usually the umbrella term that covers all aspects of used oil collection, processing, and reuse. Thus, reprocessing, reclaiming, and re-refining are all subsets of recycling.

4. Reprocessing — used oil recycling where the primary objective is producing fuels, whether for burning in small space heaters or large industrial boilers. This may include simple settling and filtration techniques to remove bulk water and solids.

5. Reclaiming—used oil recycling where the primary objective is rejuvenating a lubricant so that it can be reintroduced into the original application. This is especially useful for hydraulic fluids and other industrial lubricants that have relatively simple compositions and less demanding applications. This involves simple dehydration, settling, and filtration techniques followed by additive replenishment.

6. Re-refining — used oil recycling when the primary objective is a clean base oil equivalent to virgin base oil from which any and all petroleum-based lubricants can be blended. This involves sophisticated processing and testing.

10.3 Quantifying the Resource

Worldwide production of lubricants in 2002 was estimated to be 37 million tonnes (roughly 11.2 billion gal). Table 10.2 shows the approximate breakdown by region [6].

Regional volumes have changed over time but the total volume has not changed significantly in the last 10 years. While exact numbers are difficult to break out, one can assume these volumes are split roughly equally between automotive and industrial lubricants. In round numbers, the production of petroleum-based lubricants in the United States is about 2.5 billion gal/yr [7].

However, not all used oil is recoverable. Many oils are either consumed during processing (e.g., quench oils) or are lost during use (e.g., chain saw and 2-cycle oils), but of the estimated 65 to 70% that is recoverable [8], most information agencies agree that approximately 40% of the original product is collected; roughly equivalent to 15 million tonnes (4.5 billion gal). The balance disappears into the environment potentially creating nonpoint source pollution.

In addition to used lubricants themselves, used oil filters and containers usually hold residual, recoverable hydrocarbon fluids and can also be recycled. Most jurisdictions that have legislation pertaining to used oil also have provisions for recycling used oil filters provided they have been thoroughly drained. Recycling plastic bottles with lube oil residues is more problematic. Capacity to recycle plastic containers into such items as new plastic containers, flower pots, pipe, fencing, and patio furniture is slowly growing in most jurisdictions. Draining the bottles prior to disposal may be the optimum scenario for the present.

TABLE 10.2   Estimated Regional Lubricant Production

Region	Estimated Lubricant Production (2003) billion gal/million tonnes
North America	2.7/8.9
Central and South America	1.0/3.2
Western Europe	1.5/5.1
Central/Eastern Europe	1.4/4.9
Near/Middle East	0.6/2.0
Africa	0.6/1.8
Asia Pacific	3.4/11.2

Source: Adapted from Europalub publication extracts "Consumption by Countries 1996–1999" July 2004, Singh, H., "Lubricant Technology Today" Science in Africa online Science Magazine Nov 2002, and Cheuveux Germany Annual Report for FuchsPetrolube Q2, 2004.

10.4   Common Contaminants

Used oils most often come to the end of their usefulness not because the hydrocarbons in the base oil have broken down, but because of a combination of additive depletion and contaminant accumulation. The additives provide many (if not most) of the desired lubrication properties. The combination of additive depletion and introduction of contaminants creates the potential for damaging the hardware being lubricated.

Contaminants include solids (sludge, varnish, rust, and wear debris), additive degradation by-products (oxidized additive molecules and sheared viscosity improver polymers), water (free, emulsified, and dissolved), fuels, and process chemicals. Once the used lubricant is isolated from a crankcase, hydraulic cylinder, or other application — and is awaiting collection and recycling — further contamination often occurs as other wastes are added to the container. This secondary adulteration is often more problematic than the original, since the sources are less predictable and often are unrelated to the petroleum-based lubricant.

While no two truckloads of used oil are exactly alike, typical parameters for used oil arriving at a collection tank farm and processing plant might include:

- Water content   10 to 30%
- Flash point   > 100°C
- Acid number   2 to 3
- Chlorine   500 to 3000 ppm
- Sulfur   1000 to 4000 ppm
- Nondistillables   5 to 15%

A major breakthrough for used oil recycling came during the 1970s and 1980s when researchers at the U.S. Department of Energy at its Bartlesville (OK) Energy Technology Center demonstrated that hydrocarbons in the base oils of lubricants were not being broken down or oxidized during use. Evidence was clear that the lubricants were just becoming dirty and, therefore, could be cleaned up to their original condition [9]. This was significant because it implied that high-quality base oil could be recovered with standard refinery techniques. This research was foundational to the development of the modern rerefineries and acceptance of re-refined base oils.

10.4.1   Contaminants of Special Interest

Water and sludge are contaminants that affect product quality, but do not contribute to any health concerns. There are, however, other contaminants present at much lower levels, which provide significant motivation for the controlled collection and disposition of used oils [10,11].

Polychlorinated biphenyl compounds (PCBs) were used extensively as the fill fluids in large electrical transformers, as heat transfer fluids, and as hydraulic fluids. While PCBs are attractive due to their resistance to oxidation, the entire family of compounds (all 209 congeners) are listed as probable human carcinogens [12].

Because PCBs tend to look and behave like oils and are used in the same locations as oils, past practice has been to combine waste PCBs with used oils. This custom was halted in the 1970s as the health concerns of PCBs became more widely publicized and regulations were put in place that led to the banning of their production in 1977. Because PCBs are specifically regulated in the United States under the Toxic Substances Control Act (TSCA), the used oil recycling industry has been forced to incorporate specific testing and control measures to make sure the used oil collected and processed contains no more than 50 ppm of total PCBs. Similar legislation exists in other regions of the world.

Other halogenated compounds are also of interest. The most prominent are halogenated solvents that are often used as cleaning agents, such as perchloroethylene (tetrachloroethane), methylene chloride, trichloroethylene, and 1,1,1-trichloroethane. Since these compounds are never components of lubricants, they are indicators of adulteration of the used oil. Further, they have toxicity issues of their own, although concentrations found are almost always below 100 ppm, and often below detection limits for modern analytical instruments [6].

Ethylene glycol is widely used as antifreeze in automotive cooling systems. As such, it is another waste stream generated in the automotive servicing industry and is often found blended into used oils for disposal. Some modern used oil re-refineries isolate and market the ethylene glycol contained in their feed stream. While antifreeze is typically not a regulated waste, ethylene glycol is toxic and contributes to the overall toxicity of used oil.

Lead and benzene contamination in used oil originally came from gasoline that found its way into the oil as blowby in engines (mist pushed past the piston rings). In the mid-1980s, environmental legislation in North America forced the elimination of lead in gasoline. A direct result of this decision was the decreased lead content of used oils. The primary source now is aftermarket additives and wear metal. Benzene, a known carcinogen, is present at levels of 0.5 to 3% in gasoline. Thus, any gasoline contamination incorporated into used oils contributes a volatile carcinogen, albeit one that is handled safely every day by millions of people.

Benzene is the simplest example of a family of hydrocarbons known as aromatics. These compounds of hydrogen and carbon exist as rings with alternating single and double bonds connecting the carbons. If more than one of these rings are connected together, polycyclic aromatic hydrocarbons (PAHs) are formed. These are found in high-boiling fractions of crude oil (such as asphalt) and are often mutagenic and carcinogenic. These are closely related to polynuclear aromatic hydrocarbons (PNAs), which tend to be mutagenic and carcinogenic [13] and also contain oxygen, sulfur, nitrogen, and other elements.

Toxic metals can be introduced during use or from contamination after use. The four of primary interest in used oil are arsenic, cadmium, chromium, and lead. Of these, only chromium and lead are found at above 1 ppm with any frequency, and both tend to be below 50 ppm.

While specific contaminants found in used oils are of some interest, the primary concern is the overall health effect on those exposed to used oils. Further, the primary pathway for transmission into humans can be assumed to be through dermal exposure. Relatively simple tests are now available for testing used oils to determine their relative carcinogenicity [14]. Testing has clearly shown that more advanced used oil processing techniques can remove the carcinogenic properties of used oils as well as the heavy metals.

10.5 Typical Uses

After collection, used lubricating oil is used in or sold as feed for a variety of uses. Prior to 1988, the most prevalent use for used oil was fuel for energy recovery and dust control, such as road oiling. Since that time, environmental legislation has ensured that road oiling with used oils has almost completely disappeared in developed countries.

Today, the primary use worldwide is as a fuel. It is estimated that at least two-thirds of all collected used oil in North America is eventually burned for energy recovery either directly in space heaters and industrial boilers, as blended fuels used in commercial applications or from reprocessing facilities, which produce fuels for various applications. Worldwide this number may be closer to 90%.

Space heaters are small furnaces that provide heat in automotive garages and manufacturing locations. Often the user also provides the fuel in the form of used oil generated from on-site operations. There are two basic types: aspirating (fuel aspirated with air into a flame) and vaporizing (fuel converted into a vapor, which is then burned). The former has caused some concerns due to air emissions of metals contained in the used oil.

Industrial burners and commercial boilers operate in the same manner but on a much larger scale. In North America, these units are usually required to comply with local air emission standards and in some regions must register the fuels they are burning with regulatory agencies. In order to encourage a positive experience with the use of fuels blended with used oils in industrial and commercial boilers, ASTM International issued two separate fuel specifications [15,16]. These specifications provide guidance on what properties should be monitored when purchasing fuels made with used oils to ensure good performance.

Road oiling with used oils has almost completely disappeared in developed countries. The most famous event involving used oil applied to roads occurred in 1982 in Times Beach, MO, when the Environmental Protection Agency (EPA) purchased all properties, closed the town of 2000 people, and demolished all the homes due to extensive road oiling with used oil contaminated with dioxin and PCBs [17]. As part of the cleanup effort, 265,000 t of soils were incinerated. In the years that followed, road oiling with used oil has been banned in almost every state in the United States and most developed countries throughout the world.

Another widespread use for used oil is in hot-mix asphalt. While incorporating used oil into asphalt might seem related to road oiling, it is not. Some asphalt plants buy used oil for use as a substitute fuel with the residues incorporated into the asphalt. Asphalt manufacturers also purchase the vacuum distillation bottoms from re-refiners for direct blending into their product as an "asphalt extender" or "asphalt flux." It has been found that these bottoms contribute positively to the properties of the resulting asphalt. Further, the asphalt binds trace metal contaminants to prevent leaching into the environment as they would with road oiling and volatile organics are consumed in the manufacturing process.

Used oil is mixed with crude oil as refinery catalytic cracking feedstock. It also provides feedstock for reprocessing facilities, in Europe particularly, that produce distillate fuels for such applications as industrial fuel, home heating oil, and diesel fuel additives.

Of all the potential uses, re-refining (recycling used oil to produce lubricant base oil) is often viewed as the optimum pathway for used oil. By making a product that can be used over and over, re-refining saves a valuable natural resource while diverting a potentially hazardous waste from loss into the environment.

The modern re-refinery now uses vacuum distillation with a finishing stage, such as catalytic hydrotreating, just like a modern crude oil refinery. It is therefore not surprising that the quality of the products made from re-refined oils is equivalent to that made from virgin oils [18]. This has been demonstrated multiple times with the passage of engine sequence tests and the certification of specific formulations by such organizations as the American Petroleum Institute (API) and the European Automobile Manufacturers Association (Association des Constructeurs Européens d' Automobiles; ACEA).

10.6 Technologies

In-plant recycling of used oil is a specific form of reclaiming in which the life of the oil is extended, often through the simple steps of filtration to remove solids, heating to remove water, and refortification of the depleted additives. If the performance requirements for the original application are too stringent for efficient reclaiming (too much processing and testing required), alternative uses can often be found for the oil, for example, as a metalworking fluid. The fact that no outside party is involved and no transportation of the used oil is required can often mean elimination of any regulatory paperwork (such as manifests) and reduction in process costs.

Reclaiming using outside resources can still result in the lubricant returning to the original producer. More likely, the used oil generator has spent material removed and recycled product delivered continuously such that there is no identification of the source of the material in use at the time. The processes used by the commercial reclaimer are similar to those that would be used in an in-house operation, but the cost benefits of larger-scale operations can make a third-party, central operation attractive.

Reprocessing normally involves chemical or physical treatment of the used oil to produce a fuel oil that meets customer requirements. While this also means removal of solids and water as in reclaiming, the standards may well be less stringent. The large volumes of fuel oil purchased by industrial users also lends itself to the blending of used oils into the virgin fuel oil stream at low proportions to minimize any handling difficulties or change in emissions characteristics.

Alternatively, many fuel producers are using variations of re-refining techniques without the final finishing step used in base oil production. These methods include catalytic cracking, distillation, thin-film evaporation, thermal de-asphalting, and propane or direct hydrogen extraction. These processes are more costly to build and operate, but the fuels produced are of consistently higher quality and generally provide

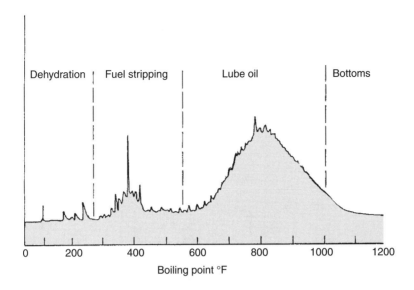

FIGURE 10.1 Simulated distillation curve showing typical cut points used by re-refiners.

a higher rate of return. An emerging trend in fuel production is processing used bio-oils (lubricant fluids made from renewable plant materials) that have been segregated from used mineral oils into bio-fuels for transportation and energy recovery applications. These fuels are generally blended into virgin diesel at 5 to 20%.

In the used oil recycling hierarchy, re-refining to make high quality base oil is the highest tier and requires the use of more sophisticated technology. When looking at the distillation characteristics of used oil, as shown in Figure 10.1, it becomes obvious that many of the contaminants can be removed by distillation. The base oil is the fraction boiling between roughly 315°C (600°F) and 540°C (1000°F). Many approaches to isolating commercially attractive lubricant products have been attempted over the years. These include [19]:

- Blending used oil into crude oil refinery streams
- Sulfuric acid/clay treatment
- Vacuum distillation/solvent treatment/extraction
- Vacuum distillation/clay treatment
- Vacuum distillation/hydrotreatment

While there are a number of specific proprietary processes currently in place around the world, they all tend to have a few similar basic steps [18–20]. A sequential distillation process, as shown in Figure 10.2, allows the rerefiner to first remove the water and solvents, then to remove the distillate fuels, and finally to volatilize the base oil away from the distillation bottoms, a very high viscosity material will be blended into asphalt. The base oil is then passed over a precious metal catalyst along with high-pressure hydrogen to produce a clean, stable hydrocarbon base oil. Additional fractionation after hydrotreating allows the re-refinery more control over the viscosity and volatility of the final products.

A benefit of this modern approach is that re-refining no longer produces hazardous by-products, such as the sludges resulting from treatment with sulfuric acid and clay that was so prevalent in the middle of the last century and still is used in less developed countries where lubricant volumes or collection networks are too small to support a high technology plant. Light hydrocarbon by-products are used as fuels, water is treated and released, and even the catalyst can be recycled for reuse in the process.

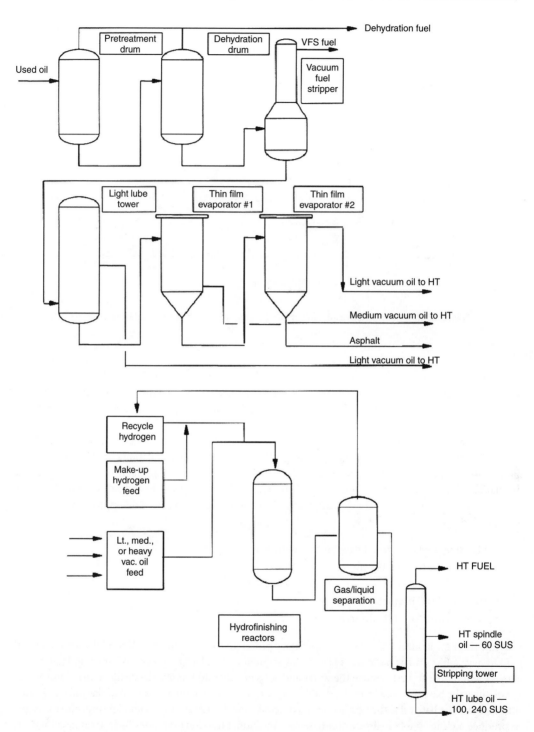

FIGURE 10.2   Used oil re-refinery schematic showing stepwise distillation followed by catalytic hydrofinishing.

10.6.1   Re-Refining Process

The re-refining process illustrated in Figure 10.2 may not reflect any particular facility, but it does represent the common steps found at any modern rerefinery. In sequence, these are:

1. *Pretreatment* — Incoming used oils often require some chemical treatment to protect the plant against corrosive attack and the hydrocarbons in the oil against chemical degradation during the process. Most often, this pretreatment involves a mild caustic wash, but it could include demulsifier, antioxidant, or other chemical agents. (*Note:* one advantage the much larger rerefineries of today have over their predecessors of 50 years ago is sheer size. The large volume throughput requires massive storage tanks, which are continuously being filled and drained. This provides a natural averaging of the composition of the feed, such that the process operates in relative equilibrium.)

2. *Dehydration* — Water is typically a major constituent of the used oil stream. It comes from metalworking fluids, antifreeze, and contamination of the used oil as it resides in drums and tanks waiting for collection. An atmospheric distillation is usually sufficient to remove most water and light hydrocarbons (e.g., gasoline, solvents, and glycols). The water is treated for plant use or discharge while the light hydrocarbons are often used as fuel within the plant due to their high odor level. The by-product produced may represent 10 to 30% of the incoming volume depending on the level of contamination.

3. *Stripper Tower* — The next stage involves distillation at a moderate vacuum to remove distillate fuels (e.g., fuel oils, jet fuel, and diesel fuel). These fuels can be sold on the industrial market to blend into boiler fuel, but again are often used within the re-refining process. This stream is usually less than 5% of the original input.

4. *Vacuum Distillation* — The distinction of most rerefineries is the manner in which they perform the distillation step(s) that volatilize the base oil present in the used oil. A maximum vacuum (typically 10 to 50 mm Hg) is desired to keep the oil temperatures below that resulting in thermal cracking (225°C). Many installations use some variation of a thin-film evaporator that combines short residence time with efficient heat transfer. The by-product of this stage is the nondistillable bottoms, which are sold to the asphalt industry as an asphalt extender or flux. The bottoms can be as much as 10% of the input to the plant.

5. *Finishing Step* — The vacuum distillate contains too many unsaturated and polar molecules to be stable in typical lubricant applications. Thus, some type of final treatment is required to convert or remove these reactive compounds. In the past, this has been accomplished with activated clay (attapulgus clay with bonding sites converted to the acid form). However, this generates a large quantity of solid waste and has some limits on the quality of the final product. Thus, most large rerefineries now utilize catalytic hydrotreating (passing the oil distillates over a fixed bed of precious metal catalyst at temperatures of 350°C or more and pressures of 600 to 1000 psi). The final base stocks usually add up to be 60 to 70% of the volume of the original used oil feed.

Fractional distillation or multiple fractionation towers are used to generate the series of distillate cuts required to produce a variety of base oil viscosities. Some oil components that are required for final lubricant blends, such as the nondistillable bright stock, are purchased from crude oil refineries and the additive packages are purchased from special chemical suppliers. The base oils from these modern re-refineries are sufficiently rejuvenated and stabilized so that they respond in an identical manner to the additives as with virgin base oils.

As with any manufacturing process, the economics of this process depend on efficient operation and minimal downtime. However, somewhat unique to used oil re-refining, the operating component that often causes financial troubles is the feedstock. Reliably obtaining enough good quality used oil to keep the feed tanks reasonably full (to benefit from blend averaging) and the plant operating at full capacity can be more difficult than new entrepreneurs might suspect. If the scale of the plant and the volume of available used oil are well-matched, the economics of re-refining can be attractive [28].

10.7   Environmental Regulations

With the advent of enhanced environmental consciousness in the 1960s and 1970s, disposal and recycling practices for waste streams such as used oil came under a new level of scrutiny. Given the volumes involved and levels of contamination described above, it is not surprising that used oil was specifically targeted for pollution control. As already mentioned, an excellent summary of the toxicology of used oils was released in 1997 by the U.S. Department of Health and Human Resources [11]. A broader discussion of the backdrop to environmental regulations in Europe is provided in a report from CONCAWE [4].

Most developed countries have enacted federal, regional, and local environmental protection legislation. This legislation varies considerably in the treatment of used oil from generic pollution prevention and spill response strategies to detailed collection and treatment requirements to mandated use of recycled products.

In the United States, the direct regulation of the disposition of used oils is usually considered to have begun with language in the Energy Policy and Conservation Act (EPCA) of 1975, followed immediately by the Resource Conservation and Recovery Act (RCRA) in 1976. These established guidelines determine which wastes would be considered hazardous and require special handling. Used oil was then addressed specifically in the Used Oil Recycling Act of 1980. All of these tried to make sure that used oil was considered a resource for further use and not disposed of in landfills.

These three legislative actions resulted in a sizeable body of regulation issued by the U.S. EPA and contained within Part 279 of Volume 40 of the U.S. Code of Federal Regulations (40CFR279). The fact that an entire section of regulations was devoted exclusively to used lubricants illustrates the significance given to this commodity.

A primary focus of the regulations is used oil burned for energy recovery. The used oil specifications currently in place for fuels are provided in Table 10.3. If there are more than 1000 ppm total halogens, but less than 4000 ppm, an analysis showing that the contaminants are not halogenated solvents listed by EPA as hazardous will allow the material to be used as a fuel.

On a positive note, Executive Orders 12873 (1993) and 13101 (1998) issued by President Clinton encourages the use of recycled products, including re-refined oil, which has demonstrated compliance with purchase specifications [21,22]. The direct pressure on government agencies to procure re-refined lubricants has provided a market for these products, as well as many others, the most successful of which has been paper.

In Europe, governments have taken a more activist role. Countries such as Germany and Italy subsidize the collection and re-refining of used oils [23]. This direct encouragement has led to an output of approximately 100,000 t of base oil in an Italian market of around 600,000 t. In Germany about 35,000 t of base oils are produced by re-refiners in a market of 1,100,000 t. Those numbers would suggest that even government subsidies are not enough to divert used oils from the less difficult pathway of use as substitute

TABLE 10.3   Used Oil Fuel Regulatory Limits — U.S. EPA

Constituent	Allowable Limit
Flash point	60°C (min.)
Lead	100 ppm (max.)
Chromium	10 ppm (max.)
Arsenic	5 ppm (max.)
Cadmium	2 ppm (max.)
PCBs	50 ppm (max.)
Total halogens	1000 ppm (max.)[a]

[a] 4000 ppm if nontoxic source (see 40CFR279 for details).

Source: United States Code of Federal Regulations, Vol. 40, Part 279, Section 11 (40CFR279.11).

fuels. A report released in 2001 considers the fairly radical step of prohibiting the burning of used oils in the United Kingdom. As stated in the report, this would have to be implemented very carefully, since the re-refining capacity to absorb that volume does not exist [24].

Another example of comprehensive environmental legislation can be found in Australia, where the federal and state governments have published and enforced extensive Codes of Practice for the management of used oil, filters, and oil containers. Federally funded public education programs and tiered tax incentives for processors have created a positive environment for the used oil recycling industry.

Canada regulates used oil at the provincial level. Most provinces have environmental legislation, which includes sections on used oil, and currently four of the provinces have incentive-based stewardship programs for the collection of used oil and filters managed by nonprofit organizations. Environmental handling fees are charged on oil sales. The fees are then used to pay registered collectors as a return incentive. These programs are supported by provincial recycling and used oil management legislation and are being considered for implementation across all provinces.

Japan, having no onshore petroleum resources, currently is studying the most effective legislative method specifically to encourage used oil re-refining.

10.8   Pollution Prevention/Lifecycle Assessment

Environmental protection and pollution prevention legislation exits in various forms in most developed countries. Recognition that used oil management is crucial to pollution prevention and resource conservation while also having a positive economic impact has been a major turning point in both government regulation and private sector interest.

As has been previously shown, the environmental hazards, health, and safety aspects of used oil have been well defined. Current studies are now focusing on two main areas: (1) defining and quantifying the economics and marketplace drivers for used oil recovery and (2) the life cycle and environmental fate of used oil [25].

Life cycle analysis is a complex project, which must take into consideration such things as original production and sale, detailed use analysis (including loss mechanisms), recoverable volumes after use, available recycling options, transportation impacts (including emissions, from transport vehicles), energy use and emissions, and residues from the recycling process whether from burning or re-refining. Figure 10.3 is a very simple schematic to illustrate the life cycle and environmental fate analysis process for lubricant fluids.

10.9   Conclusion

Regardless of the motivation mechanism, government and industry now share to some extent the twin goals of reducing the amount of waste going to landfill while preserving nonrenewable resources through recycling and removal of hazardous materials from air, land, and water. The evidence is found in regulatory documents and in environmental policy statements in corporate literature across the world.

The plethora of suggested methods to achieve these goals indicates that a universal solution has yet to be identified. Indeed, based on waste diversity and quantity, technological advancement and local economies, there may be many routes chosen to achieve the best possible recycling solutions. Ultimately, business interest in maximizing profits and environmental interest in minimizing the generation of wastes should converge to optimize the use and recycling of lubricants. Looking at more than just the initial price of lubricants is a first step. Starting with high quality re-refined lubricants, maximizing the lifetime of these lubricants through good system maintenance, and collecting the used oils in such a way as to preserve them for optimum recycling all lead to decreased costs and increased natural resource conservation.

The steadily increasing lubricant prices, feedstock shortages, and the cost of legal disposal make conservation in use and recycling of used oil a necessity. By assessing the entire life cycle of a lubricant, one

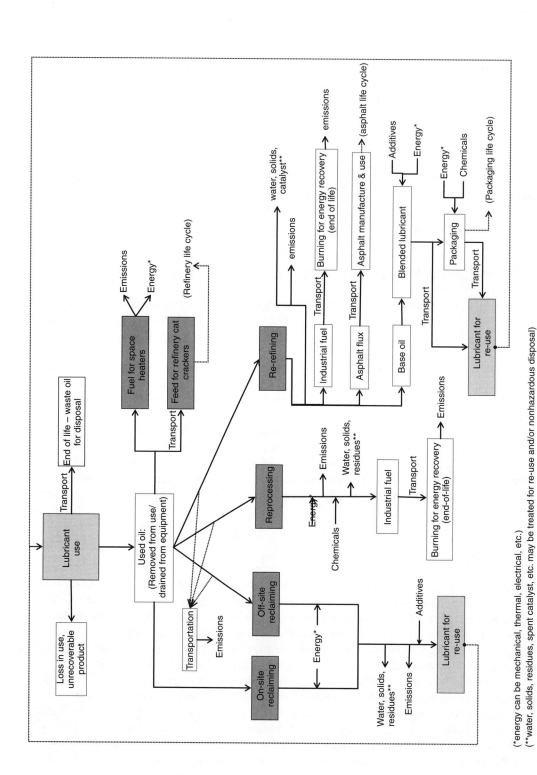

FIGURE 10.3   Life cycle and environmental fate analysis for lubricant fluids.

(*energy can be mechanical, thermal, electrical, etc.)
(**water, solids, residues, spent catalyst, etc. may be treated for re-use and/or nonhazardous disposal)

can minimize the overall costs incurred. By contracting with a responsible recycler, one can be assured of not contributing to pollution once the lubricant is no longer useful.

References

[1] Brinkman D. W., Used oil: Resource or pollutant?, *Technology Review*, 88(5), 1985, p. 47.

[2] European Commission, Integrated Pollution Prevention and Control, Draft Reference Document on the Best Available Techniques for the Waste Treatment Industries, Draft — February 2003.

[3] Cotton, F.O., Waste Lubricating Oil: An Annotated Review, U.S. Department of Energy, DOE/BETC/IC-82/4, 1982, 84 pages.

[4] Pedenaud, M., A. Bruce, M. Cayla, B. Descotis, G. Fisicaro, M.R.S. Manton, R. Prince, F.J. Sheppard, and J. Smith, Collection and Disposal of Used Lubricating Oil, CONCAWE Report No. 5/96, November 1996, 107 pages.

[5] ASTM Annual Book of Standards, Vol. 5.02, D4175, Standard Terminology Relating to Petroleum, Petroleum Products, and Lubricants.

[6] Fitzsimons, D., P. Lee, and N.J. Moreley, Oakdene Hollins, Ltd., WASTE OILS — Policy Options in the Light of German and Italian Experience, Final Report to UK Dept. for Environment, Food, and Rural Affairs, March, 2003, p. 20.

[7] Sullivan T., U.S. Lubes Dipped in 2002, Lube Report, LNG Publishing Co., November 2003.

[8] American Petroleum Institute, National Used Oil Collection Study, May 1996.

[9] Cotton, F.O., M.L. Whisman, J.W. Goetzinger, and J.W. Reynolds, Analysis of 30 used motor oils, *Hydrocarbon Processing*, 56(9), 1977, p. 131.

[10] Brinkman, D.W. and J.R. Dickson, Contaminants in used lubricating oils and their fate during distillation/hydrotreatment re-refining, *Environmental Science and Technology*, 29, 1995, p. 81.

[11] U.S. Department of Health and Human Services, Toxicological Profile for Used Mineral-Based Crankcase Oil, September 1997, 175 pages (available at www.atsdr.cdc.gov/toxprofiles/tp102.pdf).

[12] Sax, N.I., *Dangerous Properties of Industrial Materials*, 7th ed., Van Nostrand Reinhold, New York, 1989, p. 2815.

[13] International Agency for Research on Cancer (IARC), IARC Monographs on the Evaluation of Carcinogenic Risk to Humans; Polynuclear Aromatic Compounds; Part 1: Chemical, Environmental and Experimental Data, Vol. 32, 1983, pp. 57–62.

[14] Dickson, J.R., D.W. Brinkman, and G.R. Blackburn, Evaluation of the dermal carcinogenic potential of re-refined base stocks using the modified Ames assay, PAC analysis, and the ^{32}P-postlabeling assay for DNA adduct induction, *Journal of Applied Toxicology*, 17(2), 1997, p. 113.

[15] ASTM D6823, Specification for Commercial Boiler Fuels with Used Lubricating Oils.

[16] ASTM D6448, Specification for Industrial Boiler Fuels from Used Lubricating Oils.

[17] U.S. EPA, http://www.epa.gov/superfund/programs/recycle/success/1-pagers/timesbch.htm.

[18] Brinkman, D.W., Large grassroots lube rerefinery in operation, *Oil and Gas Journal*, 89(33), 1991, p. 60.

[19] Pyziak, T. and D.W. Brinkman, Recycling and re-refining used lubricating oils, *Lubrication Engineering*, 49(5), 1993, p. 339.

[20] Brinkman, D.W., Technologies for re-refining used oil, *Lubrication Engineering*, 43(5), 1987, p. 324.

[21] Executive Order 12873, 10/20/1993, Federal Acquisition, Recycling, and Waste Prevention, Sec. 506.

[22] Executive Order 13101, 9/14/1998, Greening the Government Through Waste Prevention, Recycling, and Federal Acquisition, Sec. 507.

[23] Fitzsimons, D., P. Lee, and N.J. Morley, Oakdene Hollins, Ltd., http://www.oakdenehollins.co.uk/pdf/wasteoils2.doc.

[24] Fitzsimons, D., N. Morley, and P. Lee, UK Waste Oils Market 2001, Oakdene Hollins, Ltd., http://www.oakdenehollins.co.uk/pdf/wasteoilsreport.doc.

[25] Boughton, B. and A. Horvath, Environmental assessment of used oil management methods, *Environmental Science and Technology*, 38(2), 2004, p. 353.

[26] ICIS LOR Base Oils Conference 2003 (based on estimated 2001 regional differences) [see also www.europalube.org].

[27] United States Code of Federal Regulations, Vol. 40, Part 279, Section 11 (40CFR279.11).

[28] McKeagan, D.J., Economics of rerefining used lubricants, *Lubrication Engineering*, 48(5), 1992, p. 418.

Index